HIV/AIDS Prevention Formulation Design and Optimization and Its Pharmacokinetic-Pharmacodynamic Evaluation

HIV/AIDS Prevention Formulation Design and Optimization and Its Pharmacokinetic-Pharmacodynamic Evaluation

Guest Editors

José Gerardo García-Lerma
S. Rahima Benhabbour

MDPI

Basel • Beijing • Wuhan • Barcelona • Belgrade • Novi Sad • Cluj • Manchester

Guest Editors
José Gerardo García-Lerma
Division of HIV Prevention
Centers for Disease Control
and Prevention
Atlanta
USA

S. Rahima Benhabbour
Lampe Joint Department of
Biomedical Engineering
The University of
North Carolina
Chapel Hill
USA

Editorial Office
MDPI AG
Grosspeteranlage 5
4052 Basel, Switzerland

This is a reprint of the Special Issue, published open access by the journal *Pharmaceutics* (ISSN 1999-4923), freely accessible at: https://www.mdpi.com/journal/pharmaceutics/special_issues/1STR546UJF.

For citation purposes, cite each article independently as indicated on the article page online and as indicated below:

Lastname, A.A.; Lastname, B.B. Article Title. *Journal Name* **Year**, *Volume Number*, Page Range.

ISBN 978-3-7258-7973-1 (Hbk)
ISBN 978-3-7258-7974-8 (PDF)
https://doi.org/10.3390/books978-3-7258-7974-8

Contents

About the Editors

José Gerardo García-Lerma

José Gerardo García-Lerma leads the Antiretroviral Prophylaxis team at the Laboratory Branch, Division of HIV Prevention, Centers for Disease Control and Prevention (CDC). His team investigates pre-exposure prophylaxis (PrEP) strategies to prevent HIV infection in humans. His work provided proof-of-concept evidence supporting four of the six currently approved PrEP regimens, including daily and "on-demand" emtricitabine and tenofovir disoproxil fumarate, daily emtricitabine and tenofovir alafenamide, and long-acting cabotegravir. He is named as an inventor on 12 patents related to the use of antiretroviral drugs for PrEP and post-exposure prophylaxis. When taken as prescribed, PrEP can reduce the risk of acquiring HIV from sex by up to about 99%. He has published more than 125 papers in peer-reviewed journals and has been invited to deliver talks at leading scientific conferences. His Team received the CDC/ATSDR award for Excellence in Laboratory Research in 2022, one of the CDC's highest honors. An internationally recognized expert on PrEP, drug pharmacology, and antiretroviral drug resistance, García-Lerma has advised multiple national and international organizations—including NIH, USAID, the Bill & Melinda Gates Foundation, and the WHO—and has served on numerous NIH study sections and external scientific advisory committees.

S. Rahima Benhabbour

S. Rahima Benhabbour is an Associate Professor in the UNC-NCSU Lampe Joint Department of Biomedical Engineering. Her lab focuses on harnessing innovative engineering and chemistry tools to develop next-generation drug delivery technologies for disease treatment and prevention. S. Rahima Benhabbour is listed as the inventor for 5 patents and 5 patent applications, and has received multiple awards and honors, including her induction into the CRS College of Fellows, the Yang Family Biomedical Scholars Award, CRS Women in Science Award, Alexander R. Matzuk Award, CRS Member of the Year Award, and many more. She is also the Founder and Director of Anelleo, Inc., a startup company pioneering innovative 3D-printed intravaginal ring platform technology for unmet women's health applications. Her development of highly translational drug delivery technologies has been highlighted in her multiple publications in high-impact scientific journals (*Nature Communications*, *Nature Reviews Bioengineering*, *Biomaterials*, *JCR*), with her latest paper discussing an innovative technology, which was featured among the Editors' Highlights pages in *Nature Communications* and was nominated for the CDC's 2024 Charles C. Shepard Science Award.

Preface

Increasing the number of options for HIV pre-exposure prophylaxis (PrEP) is essential because no single product can meet the diverse needs and preferences of all individuals. Challenges such as adherence to daily oral pills, acceptance of long-acting injectables or vaginal rings, and the complexity of on-demand dosing schedules highlight the importance of offering a broader range of PrEP choices. This Special Issue of *Pharmaceutics* presents research covering an expanding landscape of innovative PrEP options and is intended for experts in the field of drug formulation, delivery devices, and HIV prevention. The Reprint offers a compilation of cutting-edge research on novel long-acting and ultra-long-acting PrEP strategies, including monoclonal antibodies with extended durability; new drug formulations, such as semi-solid prodrug nanoparticles; and biodegradable implants delivering antiretrovirals alone or combined with contraceptives. The Reprint also highlights user-friendly delivery platforms suitable for self-administration, such as microarray patches and topical inserts that can be used on demand for rectal and vaginal protection against HIV. Expanding the toolbox of PrEP options will allow for more personalized prevention strategies that can better accommodate different lifestyles, preferences, and circumstances, ultimately improving adherence and overall PrEP uptake.

José Gerardo García-Lerma and S. Rahima Benhabbour
Guest Editors

pharmaceutics

MDPI

Article

Preclinical Evaluation of Long-Acting Emtricitabine Semi-Solid Prodrug Nanoparticle Formulations

Paul Curley [1], James J. Hobson [1], Neill J. Liptrott [1], Edward Makarov [2], Amer Al-khouja [3], Lee Tatham [1], Christopher A. W. David [1], Helen Box [1], Megan Neary [1], Joanne Sharp [1], Henry Pertinez [1], David Meyers [3], Charles Flexner [3], Caren L. Freel Meyers [3], Larisa Poluektova [2], Steve Rannard [1] and Andrew Owen [1,*]

[1] Centre of Excellence in Long-Acting Therapeutics (CELT), University of Liverpool, Liverpool L7 8TX, UK; pcurley@liverpool.ac.uk (P.C.); jayhob@liverpool.ac.uk (J.J.H.); liptrott@liverpool.ac.uk (N.J.L.); tatham@liverpool.ac.uk (L.T.); cdavid@liverpool.ac.uk (C.A.W.D.); mneary@liverpool.ac.uk (M.N.); joanne.livermore@liverpool.ac.uk (J.S.); pertinez@liverpool.ac.uk (H.P.); srannard@liverpool.ac.uk (S.R.)

[2] Department of Pharmacology and Experimental Neuroscience, University of Nebraska Medical Centre, Omaha, NE 68198, USA; makarove@unmc.edu (E.M.); lpoluekt@unmc.edu (L.P.)

[3] Department of Pharmacology and Molecular Sciences, Johns Hopkins School of Medicine, Baltimore, MD 21205, USA; amkhouja@gmail.com (A.A.-k.); dmeyers7@jhmi.edu (D.M.); flex@jhmi.edu (C.F.); cmeyers@jhmi.edu (C.L.F.M.)

* Correspondence: aowen@liverpool.ac.uk; Tel.: +44-151-794-8211; Fax: +44-151-794-5656

Abstract: Long-acting injectable (LAI) formulations promise to deliver patient benefits by overcoming issues associated with non-adherence. A preclinical assessment of semi-solid prodrug nanoparticle (SSPN) LAI formulations of emtricitabine (FTC) is reported here. Pharmacokinetics over 28 days were assessed in Wistar rats, New Zealand white rabbits, and Balb/C mice following intramuscular injection. Two lead formulations were assessed for the prevention of an HIV infection in NSG-cmah$^{-/-}$ humanised mice to ensure antiviral activities were as anticipated according to the pharmacokinetics. Cmax was reached by 12, 48, and 24 h in rats, rabbits, and mice, respectively. Plasma concentrations were below the limit of detection (2 ng/mL) by 21 days in rats and rabbits, and 28 days in mice. Mice treated with SSPN formulations demonstrated undetectable viral loads (700 copies/mL detection limit), and HIV RNA remained undetectable 28 days post-infection in plasma, spleen, lung, and liver. The in vivo data presented here demonstrate that the combined prodrug/SSPN approach can provide a dramatically extended pharmacokinetic half-life across multiple preclinical species. Species differences in renal clearance of FTC mean that longer exposures are likely to be achievable in humans than in preclinical models.

Keywords: HIV; long-acting; pre-exposure prophylaxis; nano; emtricitabine

1. Introduction

HIV (human immunodeficiency virus) infection is now commonly considered as a chronic disease rather than a terminal condition. In the absence of a cure, treatment requires a lifelong commitment to antiretroviral medication [1]. While there have been significant steps to reduce pill burden, and current treatment regimens can now be a single daily tablet, patient adherence to medication is an ongoing challenge over a life-time commitment to therapy [2].

Sub-optimal patient adherence has the potential to reduce efficacy and lead to treatment failure [3], and sub-therapeutic plasma exposure can allow continued viral replication and facilitate the development of mutations conferring drug resistance [4,5]. In some cases, a single mutation can confer class-wide resistance and compromises multiple agents in one event [6]. In addition to therapy, patient non-adherence has the potential to negatively impact pre-exposure prophylaxis (PrEP). Recently, mathematical simulations were used to explore the impact of not only reduced patient adherence but also patterns of adherence (simulations to predict efficacy following daily doses that were skipped periodically, randomly, or in large blocks). This study, along with other studies in HIV and schizophrenia, strongly indicates that pill-taking patterns exacerbate the reduction in efficacy [7–10].

Amongst the treatment strategies under development, long-acting injectable (LAI) formulations have received a great deal of interest. By increasing the dosing interval, the reliance on patient adherence is reduced. In addition to a clear need for alternative treatments, it has been demonstrated that patients would prefer the option of LAI. In a survey of 400 HIV-infected patients, 73% indicated they would definitely or probably try injectable nanoformulated antiretroviral therapy. Patient interest was demonstrated to increase with increasing the dosing interval; with 61%, 72%, and 84% interest in weekly, bi-weekly, and monthly dosing, respectively [11]. Patient preferences evaluations in clinical trials of existing LAI agents have also confirmed the high level of acceptability [12].

LAI formulations have been developed to address adherence issues in a number of other therapeutic areas [13,14]. Patient adherence has been a key driver to the development of LAI formulations, with ongoing research aiming to further increase the duration of therapy [14,15]. Given the proven utility in HIV and other treatment areas, interest in the development and deployment of LAI for HIV therapy and prevention is set to increase.

The recent approval of rilpivirine (RPV) and cabotegravir (CAB) for HIV treatment and/or prevention has further accelerated interest in LAI approaches for HIV [16]. Both RPV and CAB formulations provide plasma concentrations > 20 weeks in patients [17,18], although clinically they are used for shorter durations. Preclinical data describing tenofovir, maraviroc, and dolutegravir long-acting interventions have also emerged [19–21]. Nucleoside/nucleotide reverse transcriptase inhibitors (NRTIs) have conventionally been a core component of successful therapy and prevention [22].

Given the clear utility of NRTIs, an LAI formulation has the potential for application for the treatment and prevention of HIV. Accordingly, previous work described a unique prodrug/nanoparticle formation strategy to facilitate development of injectable aqueous semi-solid prodrug nanoparticle (SSPN) formulations of emtricitabine (FTC) [23]. The data presented herein describe the in vivo PK analysis in multiple preclinical species and antiviral activity of these SSPN LAI formulations of FTC in humanised mice.

2. Materials and Methods

2.1. Materials

Wistar rats, BALB/c mice, and New Zealand white rabbits were purchased from Charles River (Oxford, UK). NSG-cmah$^{-/-}$ mice were bred at the University of Nebraska Medical Centre. Chloroform was purchased from Fisher Scientific (Loughborough, UK). FTC (>98% purity) was purchased from WISChem (Shanghai, China). All other consumables were purchased from Merck Life Science UK Ltd. (Dorset, UK).

2.2. Prodrug Synthesis

Structures for the three FTC prodrugs investigated in this study are shown in Figure 1. FTC prodrugs were prepared using standard coupling conditions as previously described [23]. Briefly, in a flame-dried 25 mL round-bottom flask, cooled under argon, FTC (1.0 eq., 0.5 M) was suspended in DCM. Pyridine (3.0 eq., 1.5 M) was then added to the flask, and the resulting mixture was cooled to 0 °C in an ice-water bath. The reaction was initiated by the dropwise addition of the alkyl chloroformate (2.1 eq., 1.05 M). The reaction mixture was allowed to warm to room temperature with stirring. The reaction was deemed complete after 3 h, as monitored by TLC. Volatiles were removed from the reaction mixture under reduced pressure. The resulting residue was purified via silica flash chromatography (30% EtOAc in Hexane for 5 min, then 30–100% EtOAc over 6 min, then 100% EtOAc for 3 min).

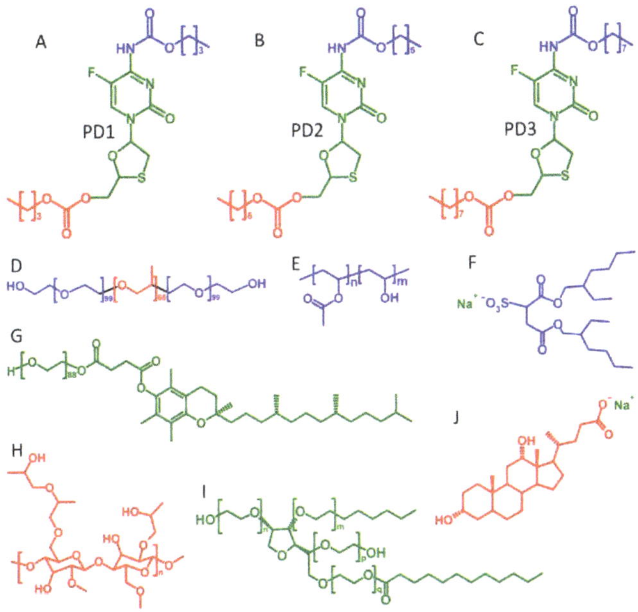

Figure 1. Structures of the varying chain length carbamate/carbonate FTC prodrugs, polymers, and surfactants used in SSPN manufacture: (**A**) prodrug 1, C_4 derivatisation; (**B**) prodrug 2, C_6 derivatisation; (**C**) prodrug 3, C_8 derivatisation; (**D**) pluronic F127; (**E**) poly(vinyl alcohol) (PVA); (**F**) sodium dioctylsulfosuccinate (AOT); (**G**) d-alpha-tocopheryl polyethylene glycol succinate (TPGS); (**H**) hydroxypropyl methylcellulose (HPMC); (**I**) polysorbate-20 (Tween-20); and (**J**) sodium dexoxycholate (NDC).

2.3. SSPN Formulation

Polymers were weighed into glass vials and dissolved to a final concentration of 13.3 mg/mL (for 50 wt% loaded SSPNs) or 10 mg/mL (for 70 wt% loaded SSPNs), whilst surfactants were dissolved to 10 mg/mL (for 50 wt% loaded SSPNs) or 5 mg/mL (for 70 wt% loaded SSPNs) both in distilled water. Solutions were rolled overnight to ensure thorough dissolution. Each prodrug was stored at −20 °C and removed only immediately prior to SSPN synthesis, at a final concentration of 50 mg/mL or 70 mg/mL (for 50 wt% and 70 wt% loading, respectively) in chloroform. For SSPNs loaded with 50 wt% prodrug, 100 μL of surfactant, 300 μL of polymer, and 100 μL of prodrug was added to a 4 mL glass sample vial, whilst for SSPNs loaded with 70 wt% prodrug, 200 μL of surfactant, 200 μL of polymer, and 100 μL of prodrug was added to a 4 mL glass vial. This emulsion consisting of chloroform-in-water was sonicated for 15s using a Covaris S220X, and then immediately transferred to liquid nitrogen to ensure rapid freezing. The frozen samples were placed into a VirTis BenchTop K freeze dryer (SP Scientific, Ipswich, UK) set to −100 °C and vacuum of <40 μbar for 48 h, after which they were kept in an ambient temperature desiccator. The compositions of the formulations used in the current study are shown in Table 1.

Table 1. This table shows the formulation number, excipients (polymer and surfactant), prodrug designation percentage drug loading, and physical characterisation parameters.

Formulation	Polymer	Surfactant	Prodrug	Prodrug Loading (wt%)	Z-Average nm (±SD)	PDI (±SD)
SSPN-1	F127	TPGS	PD3	70	267 (26)	0.314 (0.022)
SSPN-2	PVA	Tween 20	PD1	70	412 (15.5)	0.347 (0.045)
SSPN-3	F127	TPGS	PD1	50	702 (111)	0.305 (0.090)
SSPN-4	F127	TPGS	PD2	50	782 (199)	0.450 (0.152)
SSPN-5	F127	TPGS	PD3	50	239 (50)	0.323 (0.042)
SSPN-6	PVA	Tween 20	PD1	50	298 (21)	0.339 (0.038)
SSPN-7	PVA	Tween 20	PD2	50	261 (44)	0.295 (0.019)
SSPN-8	PVA	Tween 20	PD3	50	293 (9)	0.294 (0.028)
SSPN-9	HPMC	NDC	PD2	50	196 (6)	0.345 (0.027)
SSPN-10	HPMC	NDC	PD3	50	261 (17)	0.376 (0.038)
SSPN-11	HPMC	AOT	PD2	50	217 (10)	0.306 (0.022)
SSPN-12	HPMC	AOT	PD3	50	237 (5)	0.296 (0.021)

2.4. Pharmacokinetics of SSPN Formulations in Wistar Rats

Wistar rats, New Zealand white rabbits, and BALB/c mice (Charles River UK) were used for PK analysis of FTC. PK experiments were conducted in accordance with the Animals (Scientific Procedures) Act 1986 (ASPA), implemented by the United Kingdom Home Office. NSG-cmah$^{-/-}$ were obtained from the established breeding colony and housed under pathogen-free conditions in accordance with ethical guidelines for care of laboratory animals at the National Institutes of Health and the University of Nebraska Medical Centre.

An initial screen of 12 candidate formulations (Table 1) was conducted in Wistar rats. Adult male Wistar rats (~300 g) were randomised into 12 groups, (N = 1 per group). Groups were dosed with SSPN formulations at 10 mg/kg (FTC equivalent) via a single intramuscular (IM) injection in the musculus biceps femoris (0.1 mL volume). Blood samples were collected (500 µL) post-dosing from the tail vein every 24 h over 7 days and stored at −80 °C until further analysis. At the point of termination, the rats were sacrificed using a rising gradient of CO_2 followed by cervical dislocation. Plasma concentrations of FTC were quantified using LC-MS/MS.

Three lead formulations (Table 1; SSPN-9, -10 and -12) were selected on the basis of PK performance in screening. These formulations were progressed to a longer-term study in adult male Wistar rats (~300 g). Animals were randomised into 3 groups (n = 3 per group), and dosed with SSPN formulations at 40 mg/Kg (FTC equivalent) via 2 IM injections in both the left and right musculus biceps femoris (0.1 mL volume per site). Blood samples (100 µL) were collected at 1.5, 3, 6, 24, 48, 96, 168, 240, 336, 504, and 672 h from the tail vein and stored at −80 °C until further analysis. At the point of termination, the rats were sacrificed using a rising gradient of CO_2 followed by cervical dislocation. Plasma concentrations of FTC were quantified using LC-MS/MS as described below.

2.5. Pharmacokinetics of Lead SSPN Formulations in New Zealand White Rabbits

The three lead formulations (Table 1 SSPN-9, -10 and -12) were further characterised in adult male New Zealand white rabbits (~3 kg), randomised into 3 groups, (n = 3 per group). Groups were dosed with SSPN formulations at 40 mg/Kg (FTC equivalent) via 2 IM injections in both the left and right musculus biceps femoris (1 mL volume per site). Blood samples (500 µL) were collected at 1.5, 3, 6, 24, 48, 96, 168, 240, 336, 408, 504, 576, and 672 h from the marginal ear vein and stored at −80 °C until further analysis. At the point of termination, the rabbits were sacrificed via intra-venous injection of an anaesthetic agent. Plasma concentrations of FTC were quantified using LC-MS/MS as described below.

2.6. Pharmacokinetics of Lead SSPN Formulations in BALB/c Mice

SSPN-9 and -10 were selected for pharmacokinetic and efficacy studies in mice because they enabled the evaluation of near identical formulations for two different prodrugs. Balb/C mice (aged 11–12 weeks, ~25 g), were randomised into 3 groups, (n = 3 per group). Groups were dosed with SSPN formulations at 140 mg/Kg (FTC equivalent) via 2 IM injections in both the left and right musculus biceps femoris (50 μL volume per site). A serial sacrifice experimental design was utilised. Mice were sacrificed at each time point via rising CO_2 gradient. Blood was collected post-mortem via cardiac puncture and transferred to heparinised sample tubes. Samples were collected at 1.5, 3, 6, 24, 48, 96, 168, 240, 336, 504, and 672 h and stored at −80 °C until further analysis. Plasma concentrations of FTC were quantified using LC-MS/MS as described below.

2.7. Quantification of FTC in Plasma

FTC was detected using a fully validated method adapted from [24]. Quantification was achieved via LC-MS/MS (TSQ Endura, Thermo Scientific, Waltham, MA, USA) operating in positive mode. The following ions were monitored for quantification in selected reaction monitoring scan: FTC (m/z 248 > 113 and 130). A stock solution of 1 mg/mL FTC was prepared in H_2O and stored at 4 °C until use. A standard curve was prepared in plasma by serial dilution from 500 ng/mL to 1.9 ng/mL and an additional blank solution. Chromatographic separation was achieved using a multi-step gradient with a Synergi polar C18 column (4 mm:150 mm:2.0 mm; Phenomenex, Cheshire, UK) using mobile phases A (100% H_2O, 0.1% formic acid) and B (100% ACN, 0.1% formic acid). Chromatography was conducted over 6 min at a flow rate of 400 μL/min. At the start of each run, mobile phase A was 95% until 0.2 min when mobile phase B was increased to 30% at 0.5 min. Mobile phase B was then maintained at 30% until 3 min. Mobile phase B was then reduced to 5% at 3.2 min which was held until 6 min. Inter- and intra-assay variance was assessed by 3 levels of independent standards. The coefficient of variation of accuracy and precision were <15% in all assays (Supplementary Materials Table S1).

2.8. Pharmacokinetic Modelling in Rats, Rabbits, and Mice

Plasma concentration–time profiles from all three species were each fitted with a one-compartment disposition PK model with first order absorption input from depot release using the R data analysis environment (v 4.1.1) [25], treating the composite PK profile from all rats, rabbits, or mice for a given treatment as a naïve pool. Fittings made use of the Pracma library [26] and the lsqnonlin function for nonlinear regression. Apparent clearance (CL/F), apparent volume of distribution (V/F), and absorption rate constant (KA) parameters were estimated in the absence of IV data to quantify absolute bioavailability in each species in this study.

2.9. Efficacy of SSPN Formulations in NSG-Cmah$^{-/-}$ Mice

A human immune system was reconstituted in NSG-cmah$^{-/-}$ mice, as previously described [27,28]. In brief, mouse pups (aged 0–4 days) first received full body irradiation (1Gy). CD34+ human stem cells (5×10^4 cells) were then delivered via intra-hepatic injection 6 h post-irradiation. Reconstitution of the human immune cells was monitored via flow cytometry (anti-human CD45-FITC BD Pharmigen, #555482, CD3-Alexa Fluor 700 BD Biosciences 557943, CD4-APC BD Pharmingen #555349, CD8-BV421 BD Horizons #562428). Humanised NSG-cmah$^{-/-}$ mice were then used to investigate the efficacy of SSPN formulations when applied to PrEP. Humanised mice were randomised into 2 groups: a 7-day challenge and a 14-day challenge. Each group was subdivided into 3 treatment conditions (8 mice per treatment): untreated control, SSPN-9, and SSPN-10 (Table 1). Mice were treated at day 0 with SSPN-9 or SSPN-10 (140 mg/Kg FTC equivalent) via 2 IM injections in the left and right musculus biceps femoris (50 μL volume per site). HIV-1_{ADA} challenge was administered 7- or 14-days post-drug administration via intraperitoneal injection (10^4 50% tissue culture infectious dose; $TCID_{50}$). Blood samples were taken at 14 days post-infection to determine viral load via PCR (COBAS® AmpliPrep TaqMan® HIV-1 Test, v2.0, Roche Molecular Systems Inc., Pleasanton, CA, USA). All animals were terminated at 28 days post-infection. Plasma samples were collected for determination of viral load via PCR (COBAS® AmpliPrep TaqMan® HIV-1 Test, v2.0, Roche Molecular Systems Inc., Pleasanton, CA, USA) and characterisation of human immune cells by flow cytometry. The acquisition was conducted on BD LSRII flow cytometer cells and analysed on gated leukocytes using FLOJO analysis software (v 10.2 Tree Star, Ashland, OR, USA).

Following termination, samples from spleen, liver, and lung tissues were collected and stored at −80 °C until analysed. Tissue processing was carried out as previously described [19]. In brief, tissue samples were homogenised using a bead mill homogeniser (Qiagen TissueLyzer II Valencia, CA, USA) and Qiagen AllPrep DNA/RNA mini kit (Hilden, Germany). cDNA was produced from RNA using ThermoScientific Verso cDNA synthesis kit (Vilnius, Lithuania) as per manufacturer's instructions. RT-PCR HIV-1 Gag p24 was conducted using primers and probe assays purchased from Life Technologies (forward primer ACATCAAGCAGCCATGCAAAT, reverse primer ATCTGGCCTGGTGCAATAGG, probe sequence 6FAMCATCAATGAGGAAGCTGCAGAATGGGATAGATAMRA). Thermal cycling was conducted using a QuantStudio™ 3 system (Applied Biosystems®, Waltham, MA, USA) programmed for standard cycling conditions (95 °C for 10 min, followed by 40 cycles of 95 °C for 15 s, and 50 °C for 1 min).

Following termination, samples from spleen, liver, and lung tissue were collected and fixed in paraformaldehyde (4%). The 5 μm sections were collected and stained for HIV-1 p24 (1:10, DAKO, Carpinteria, CA, USA) or HLA-DQ/DP/DR (clone CR3/43, 1:100, DAKO, Carpinteria, CA, USA) and incubated overnight at 4 °C as previously described [29]. Tissue samples were washed 3× in TBS-Tween and then incubated with secondary reagent the polymer-based horse radish peroxidase (HRP) and 3,3′-Diaminobenzidine (DAB) DAKO EnVision system for 15 min. Tissue samples were counterstained in filtered hematoxylin for 1 min. Cells were visualised at 20× and 40× on a Nuance EX camera fixed to a Nikon Eclipse E800 microscope using Nuance software (Cambridge Research & Instrumentation, Woburn, MA, USA).

Differences in viral load in plasma and tissues at timepoints 14 and 28 days following challenge 7 or 14 days after dosing of SDN9, SDN10, or control/no treatment were tested for significance using a one-sided, pairwise Mann–Whitney–Wilcoxon test, with a Bonferroni correction factor of 3 (for 3 pairwise comparison tests per challenge/tissue/timepoint group).

3. Results

3.1. Characterisation of Formulations Used in the Studies

All of the SSPN formulations displayed narrow, monomodal particle size distributions (Figure 2) with z-average (Dz) diameters ranging from 196 to 782 nm and polydispersity index (PDI) values varying from 0.295 to 0.450 across the three different prodrugs (Table 1). These criteria were selected as uniformity was expected to confer reproducible drug release.

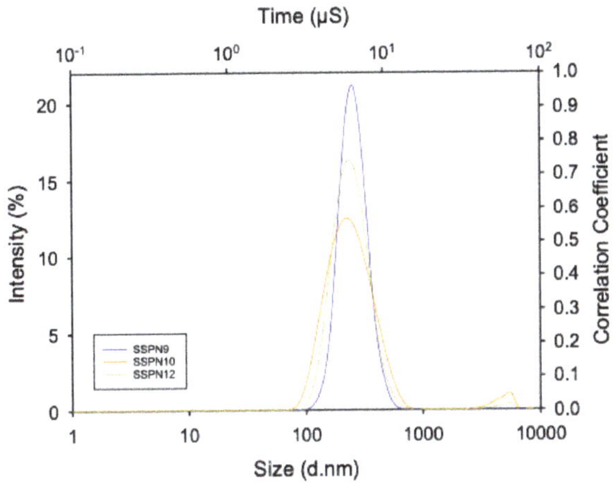

Figure 2. Representative z-average hydrodynamic diameter particle size distribution of SSPN−9 (blue), SSPN-10 (orange), and SSPN-12 (yellow). Data are shown as the average of 3 measurements from dynamic light scattering analysis.

3.2. Initial Pharmacokinetic Screening of SSPN Formulations in Wistar Rats

An initial screen of 12 candidate formulations (Table 1) comprising 3 FTC prodrugs (PDs) and dual combinations of three polymers and four surfactants were tested for in vivo PK in Wistar rats. Each candidate was administered via a single IM injection (10 mg/kg based on FTC content, $n = 1$). Plasma samples were collected daily for 7 days (Table 2). SSPN 1–8 demonstrated poor plasma exposure, Cmax ranged between 19.8 ng/mL and <2 ng/mL. Plasma concentrations in all eight formulations fell below the limit of detection (2 ng/mL) within 4 days. SSPN-11 demonstrated high burst release (Cmax 88.5 ng/mL) before FTC concentrations rapidly decreased to 2.45 ng/mL by day 7. The remaining 3 formulations (SSPN-9, -10, and -12) all demonstrated favourable PK for LAI, Cmax (SSPN-9 42.8 ng/mL, SSPN-10 33.9 ng/mL, and SSPN-12 47.1 ng/mL), and Cmin (SSPN-9 22.2 ng/mL, SSPN-10 24.1 ng/mL, and SSPN-12 16.9 ng/mL) over 7 days. SSPN formulations SSPN-9, SSPN-10, and SSPN-12 were therefore progressed to a longer-term study.

Table 2. This table shows the 12 SSPN candidates. The table shows PK parameters (Cmax, Tmax, and AUC) over 7 days following a single IM injection (10 mg/kg based on FTC content).

Formulation	C_{max} (ng/mL)	C_{min} (ng/mL)	T_{max} (h)	AUC (ng/h/mL)
SSPN-1	2.3	<2.0	72	170.5
SSPN-2	<2.0	<2.0	24	89.0
SSPN-3	<2.0	<2.0	24	86.0
SSPN-4	19.8	<2.0	24	802.3
SSPN-5	6.0	<2.0	48	305.9
SSPN-6	3.5	<2.0	24	236.1
SSPN-7	12.1	<2.0	24	443.2
SSPN-8	3.3	<2.0	72	169.1
SSPN-9	42.8	22.2	24	3449.7
SSPN-10	33.9	24.1	24	3731.0
SSPN-11	88.5	2.5	24	3960.9
SSPN-12	47.1	16.9	24	3695.1

3.3. Longer-Term Pharmacokinetic Evaluation of Lead SSPN Formulations in Wistar Rats

Three lead formulations (SSPN-9, -10, and -12 derived from C6 and C8 FTC-PDs) were progressed to a longer-term study. Each candidate was administered via 2 IM injections (20 mg/kg each = 40 mg/kg total dose based on FTC content, n = 3) and plasma samples were collected over 28 days (Figure 3A). Cmax was reached by 6–12 h (SSPN-9 127 ± 52.9 ng/mL, SSPN-10 95 ± 40.6 ng/mL, and SSPN-12 119 ± 21.0 ng/mL) and plasma concentrations were detectable for all three formulations until 14 days (SSPN-9 21 ± 16.2 ng/mL, SSPN-10 27 ± 0.05 ng/mL, and SSPN-12 29 ± 22.7 ng/mL). Plasma concentrations were below the validated limit of detection (2 ng/mL) by 21 days for all three formulations (Figure 3A, Table 3). The rat data were adequately described by a one-compartment disposition PK model with first order depot release input, with acceptable precision of parameter estimates. As expected for this formulation, "flip-flop" PK was demonstrated [30] where the observed, long terminal phase half-life of the profiles (T1/2, SSPN-9 89 h, SSPN-10 113 h, and SSPN-12 100 h) reflects the kinetics of slow release from the depot rather than the elimination rate constant which is intrinsic to the disposition of FTC as shown following IV dose [31]. This is also reflected in the parameter estimates where the absorption rate constant KA is smaller (and therefore slower) than elimination rate constant as calculated from CL divided by V.

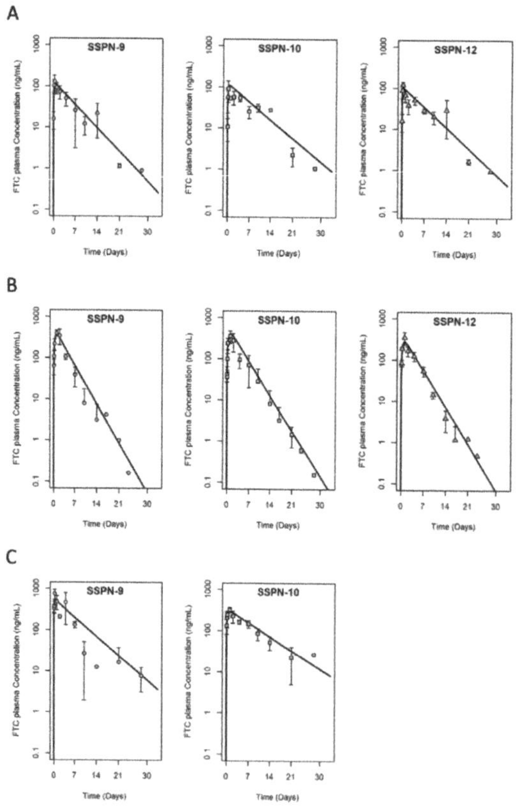

Figure 3. PK profiles of the three lead SSPN candidates over 28 days in (**A**) Wistar rats (**B**) New Zealand white rabbits, and (**C**) Balb/C mice. Animals were administered 2 IM injections (1 injection in each leg, given simultaneously) of 40 mg/kg for Wistar rats, 40 mg/kg for New Zealand white rabbits, and 140 mg/kg for Balb/C mice based on FTC content. Error bars represent standard deviation. Solid line is the PK model fitting to the naïve pooled data (1-compartment disposition with 1st order depot release input).

Table 3. PK parameters for the three lead SSPN candidates in Wistar rats, New Zealand white rabbits, and Balb/C mice (Cmax [maximum plasma concentration], C 14 [plasma concentration at day 14], Tmax [time maximum plasma concentration was reached], AUC [area under the curve], CL/F [clearance], V/F [volume of distribution], KA [absorption rate constant], and $T_{1/2}$ [terminal half-life] from the PK model fitting) over 28 days following 2 IM injections (1 injection in each leg, given simultaneously, 40 mg/kg total dose based on FTC content). SD = standard deviation; % RSE = "% relative standard error of estimate".

Formulation		C_{max}	C_{14}	T_{max}	AUC	CL/F	V/F	KA	Terminal $T_{1/2}$
	Species	(ng/mL) (SD)	(ng/mL) (SD)	(h) (SD)	(ng·h/mL) (SD)	(L/h/kg) (% RSE)	(L/kg) (% RSE)	(h^{-1}) (% RSE)	(h)
SSPN-9	Wistar rat	127 (52.9)	21 (16.2)	6 (0.0)	15,136 (6306.1)	2.44 (11)	5.02 (38)	0.0078 (7)	89
	NZW rabbit	416 (93.6)	3.1 (4.3)	32 (13.9)	33,680 (8817.7)	1.05 (4)	10 (1)	0.0129 (6)	54
	Balb/C mice	747 (206.5)	7.7 (4.6)	6 (0.0)	60,387 (21,111.2)	1.56 (12)	2.38 (55)	0.0064 (9)	109
SSPN-10	Wistar rat	95 (40.6)	27 (0.1)	12 (10.4)	16,003 (3220.3)	2.06 (12)	8.72 (30)	0.0061 (8)	113
	NZW rabbit	356 (116.9)	8 (9.0)	24 (0.0)	34,444 (5587.3)	0.87 (10)	10.07 (12)	0.0117 (4)	59
	Balb/C mice	330 (41.5)	26.6 (0.63)	17 (12.1)	57,870 (11,698.8)	1.97 (10)	9.99 (27)	0.0047 (11)	148
SSPN-12	Wistar rat	119 (21.0)	29 (22.7)	6 (0.0)	15,096 (3208.3)	2.58 (9)	5.30 (35)	0.0069 (6)	100
	NZW rabbit	348 (117.3)	3.8 (2.0)	24 (0.0)	28,414 (4674.4)	1.33 (4)	12.85 (11)	0.0122 (4)	57
	Balb/C mice	-	-	-	-	-	-	-	-

3.4. Long-Acting PK Study in New Zealand White Rabbits

Multiple species are employed in preclinical screening due to species differences dictating differences in drug exposure. The confirmation of LAI plasma exposure was, therefore, also evaluated in rabbits for lead formulations. Each candidate was administered via 2 IM injections (20 mg/kg each = 40 mg/kg total dose based on FTC content, $n = 3$) and plasma samples were collected over 28 days (Figure 3B). Cmax was reached at 24–48 h (SSPN-9 416 ± 93.6 ng/mL, SSPN-10 356 ± 116.9 ng/mL, and SSPN-12 348 ± 117.3 ng/mL) and plasma concentrations were detectable for all three formulations until 14 days (SSPN-9 3.1 ± 4.3 ng/mL, SSPN-10 8 ± 9.0 ng/mL, and SSPN-12 3.8 ± 2.0 ng/mL). Plasma concentrations were below the validated limit of detection (2 ng/mL) by 21 days for all three formulations (Figure 3B, Table 3). As with the rat data, the rabbit plasma PK data were adequately described by a one-compartment disposition model with first order depot release input, with acceptable precision of parameter estimates. Flip-flop PK was again demonstrated with a slow KA estimate and slow terminal half-life; however, for all three formulations, this terminal phase was ~2× faster in rabbits than in rats, implying a faster release from the depot in rabbits than rats (T1/2, SSPN-9 54 h, SSPN-10 59 h, and SSPN-12 57 h).

3.5. Long-Acting PK Study in BALB/c Mice

Two candidates were selected for efficacy evaluation in humanised mice and supportive pharmacokinetic evaluation was conducted ahead of efficacy studies in Balb/c mice. For this, SSPN-9 and -10 were selected because they enabled evaluation of near identical formulations for two different prodrugs. Each candidate was administered via 2 IM injections (70 mg/kg each = 140 mg/kg total dose based on FTC content, $n = 3$) and plasma samples were collected over 28 days (Figure 4C). Cmax was reached within 24 h (SSPN-9 747 ± 206.5 ng/mL and SSPN-10 330 ± 41.5 ng/mL) and plasma concentrations were detectable for all three formulations until 28 days (SSPN-9 7.7 ± 4.6 ng/mL and SSPN-10 26.6 ± 0.63 ng/mL) (Figure 4C, Table 3). As with the rat and rabbit, the mouse plasma PK data were adequately described by a one-compartment disposition model with first order depot release input, with acceptable precision of parameter estimates. Again, flip-flop PK is demonstrated with a slow KA estimate, which is further reflected in the slow terminal half-life, which is more similar in magnitude to the rat than the rabbit for the two lead formulations examined (T1/2, SSPN-9 109 h, SSPN-10 148 h).

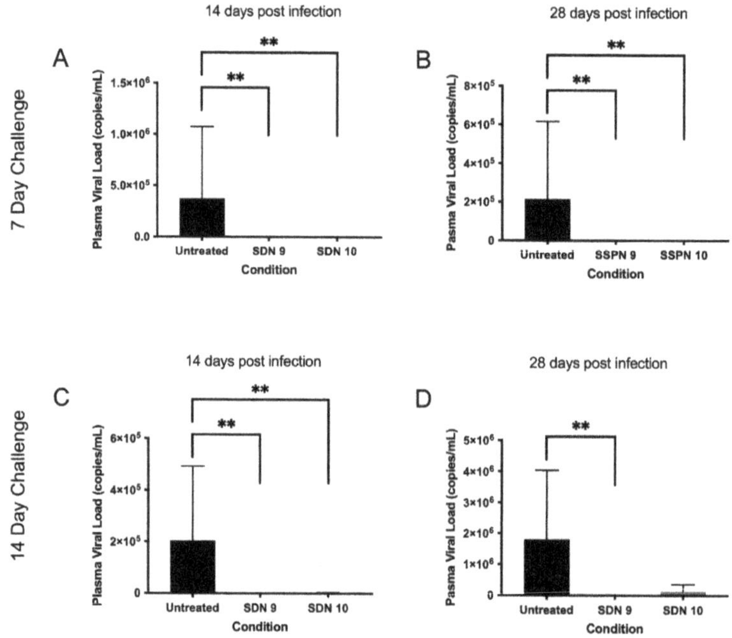

Figure 4. Plasma viral load following HIV challenge 7 (**A**,**B**) and 14 days (**C**,**D**) post-injection. Subfigures show viral load (copies/mL) at 14- and 28-days post-infection IM injection (700 copies/mL detection limit). ** denotes p value > 0.01 determined using significance using a one-sided, pairwise Mann–Whitney–Wilcoxon test, with Bonferroni correction factor.

3.6. In Vivo Protection against HIV-1 Infection by Lead SSPN Formulations

In order to assess the efficacy of the SSPN formulations in preventing an HIV infection, a humanised mouse model of HIV infection was employed. Humanised mice were divided into 2 groups: a 7-day challenge and a 14-day challenge. Humanised mice were treated at day 0 with SSPN-9 or SSPN-10 (140 mg/Kg FTC equivalent) via 2 IM injections. Virologic challenge was administered 7- or 14-days post-drug administration (10^4 $TCID_{50}$). Plasma viral load was undetectable in mice treated with SSPN-9 and SSPN-10 VL (700 copies/mL detection limit) 14 days and 28 days post-infection (Figure 4A,B). HIV RNA was also undetectable 28 days post-infection in spleen, lung, and liver tissue in all animals for the 7-day challenge (Figure 5A–C). Further tissue samples were stained for HLA- DR (human leukocyte surface antigen) and Gag-p24 (HIV capsid protein). The histopathology of these samples confirmed the presence of human leucocytes while confirming undetectable viral protein in the SSPN treatment groups (Figure 5G,H).

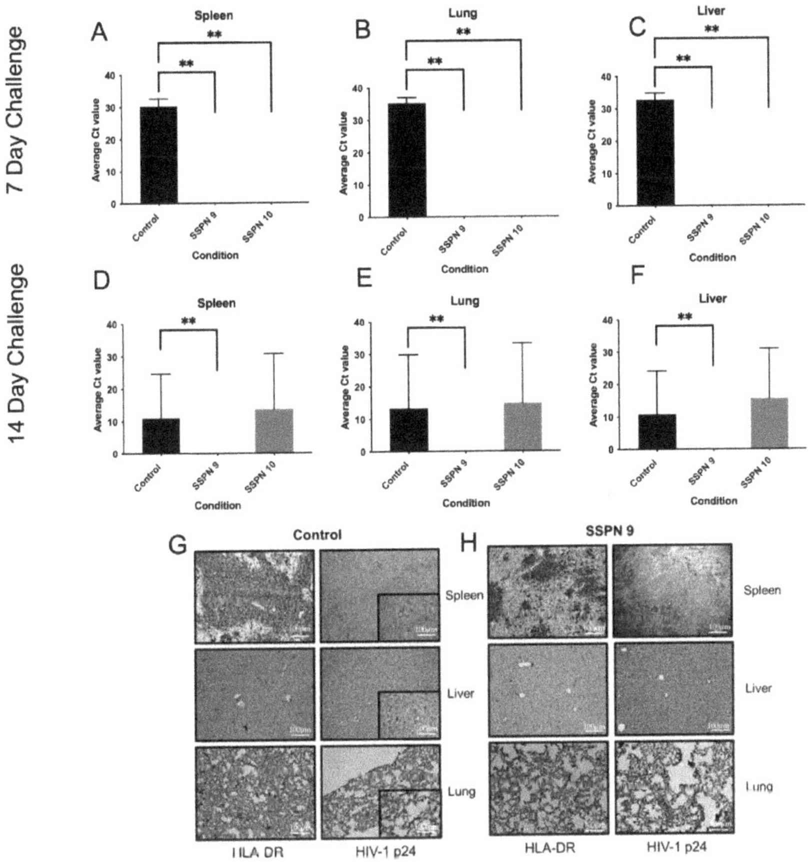

Figure 5. This table shows detection of HIV-1 Gag-p24 RNA via PCR in humanised mice following HIV infection 7 (**A–C**) and 14 (**D–F**) days post-IM injection. Tissue samples from spleen, lung, and liver were taken 28 days post-HIV infection. Also shown are representative histopathology samples taken from spleen, liver, and lung tissues 7 days post-IM injection from the (**G**) control group and (**H**) SSPN-9-treated group. Tissue samples were collected 28 days post-HIV infection. Samples were stained for HLA-DR and Gag-p24. Gag-p24-positive staining in the control group (brown). No Gag-p24 was detected in the SSPN-9-treated group. ** denotes p value > 0.01 determined using significance using a one-sided, pairwise Mann–Whitney–Wilcoxon test, with Bonferroni correction factor.

Following the 14-day challenge, mice treated with SSPN-9 demonstrated undetectable HIV in plasma 14 days and 28 days post-infection (Figure 4C,D). Viral RNA was also undetectable 28 days post-infection in spleen, lung, and liver tissue (Figure 5D–F). Mice treated with SSPN-10 demonstrated that 2 mice had detectable plasma VL (4.77×10^3 copies/mL) and 3 mice showed presence of HIV RNA in plasma and proteins in spleen, lung, and liver on day 28 (Figure 5D–F). HIV was detectable in all untreated animals. p-values summarising the significance of the differences in VL between SDN9 and SDN10 treatment vs. untreated following 7- and 14-day viral challenges are broadly consistent with visual presentation of the data in Figures 4 and 5 (Supplementary Materials Table S2). For a day 7 challenge post-dose, both SDN9 and SDN10 treatment give significantly lower plasma VL than untreated (and are not significantly different from each other) both 14 and 28 days post-challenge infection. SDN9 treatment also has significantly lower VL in plasma and tissues vs. untreated at both 14 and 28 days post-infection after a day 14 challenge post-dose. Due to greater inter-animal variability in VL following a 14-day challenge after SDN10 dosing, *p*-values indicate greater ambiguity in distinguishing SDN10 treatment from both untreated and the SDN9-treated groups.

4. Discussion

Existing antiretroviral LAI utilise milling to produce particle dispersions where controlled release is achieved by slow dissolution as a result of low drug solubility. As with many NRTI's, FTC is highly water-soluble, making it incompatible with technologies such as milling. In order to circumvent this limitation, three varying chain-length carbamate/carbonate prodrugs of FTC were produced to mask key hydrophilic groups (Figure 1) to decrease aqueous solubility. These highly lipophilic prodrugs are semi-solid at ambient temperatures and are thus also not compatible with bottom-down processing technologies such as nanomilling. However, compatibility with emulsion-templated freeze drying (ETFD) has been explicitly demonstrated to form particle dispersions from this PD class via this approach [23]. The in vivo data presented here demonstrate that the SSPN approach can provide a half-life extension for water-soluble FTC in multiple pre-clinical species. This approach represents an interesting avenue of research to allow the exploitation of antiretrovirals that would otherwise be incompatible for processing or formulation with LAI.

In order for FTC to exert efficacy, it must be taken up and intracellularly undergo phosphorylation to the active FTC-TP. In the current study, intracellular concentrations of FTC-TP were not examined. Guidelines for the treatment of laboratory animals limit the sample volumes that may be taken over the experimental period, precluding the option to assay for both FTC and FTC-TP in mice and rats [32,33]. However, it can be concluded that the SSPN formulations successfully deliver FTC without abrogation of phosphorylation from the demonstration of the effectiveness in the mouse model of HIV infection. The plasma half-life in humans of FTC following oral administration is 7.4 h, whereas the intracellular half-life of FTC-TP is 39 h [34,35]. The active FTC-TP is present in vivo over 5 times as long as FTC, indicating plasma concentrations alone may underestimate the efficacy of the LAI formulations presented here. The lead formulation (SSPN-9) presented here demonstrated an 18.5-fold (4.79 h vs. 89 h) and 19.5-fold (5.6 h vs. 109 h) plasma half-life extension with SSPN compared to oral administration in mice and rats, respectively [31,36]. No published pharmacokinetics data for orally administered FTC are available in rabbits, but consistency in the half-life extension between mouse and rat is reassuring. The oral plasma half-life of FTC in humans is approximately 10 h, and the presented data therefore support a half-life longer than 185 h for SSPN-9 in humans if the relationship holds.

The data presented here show some clear patterns in the SSPN composition and PK exposure. Amongst the excipients tested, the formulations that performed optimally contained HPMC (with either AOT or NDC). The importance of including HPMC is not entirely unexpected as HPMC has been used frequently to modulate drug release from oral formulations [37]. In addition to the selection of polymer and surfactant, the inclusion of different prodrugs appears to affect the release profile. Formulations SSPN-11 and SSPN-12 both consist of HPMC and AOT but differ in prodrug composition. SSPN-11 contained C6 carbamate/carbonate while SSPN-12 contained the C8 carbamate/carbonate prodrug. Despite the similar formulation, the resulting PK behaviour differed. SSPN-12 demonstrated a modulated steady release, providing plasma concentrations up to 14 days; whereas, SSPN-11 showed a high burst release followed by rapid decrease in plasma exposures. Interestingly, this relationship was not observed when PK for SSPN-9 and SSPN-10 were compared. Both formulations comprised HPMC with NDC and either C6 or C8 carbamate/carbonate and show very similar PK profiles. These data demonstrate the careful consideration required when selecting polymer, surfactant, and active pharmaceutical agent, as the performance may change in an unpredictable manner.

While the current data do not enable robust conclusions to be drawn about how prodrug structures and formulation compositions might impact PK performance, both formulations studied in humanised mice were able to protect from the acquisition of an HIV infection. Importantly, there were differences in PK exposure between SSPN-9 and SSPN-10, with a higher C_{max} and AUC observed for SSPN-9 in Balb/c mice. In agreement with these PK observations, the duration of protection for SSPN-9 was longer than that for SSPN-10, validating the pharmacological strategy for selection of lead candidates. SSPN-9 and SSPN-10 differed only regarding the prodrug contained therein, and both formulations were composed of HPMC and NDC, with similar Z-average diameter and PDI. It is, therefore, reasonable to conclude that the chain length of the carbamate/carbonate prodrug moiety is an important contributor to overall performance of the SSPN formulation.

The development of LAI is hindered by the paucity of knowledge of the mechanisms involved in the release from the depot. Many of the physiological, anatomical, and formulation-specific characteristics involved are poorly characterised, making targeting successful LAI problematic. Recent data indicate several mechanisms within the depot site that may contribute to LAI PK performance. IM administration of paliperidone palmitate has been shown to illicit a granulomatous response with macrophage penetration. Furthermore, co-administration of sunitinib completely suppressed the granulomatous reaction and effectively inhibited the neovascularization of the paliperidone palmitate LAI depot. This resulted in a slower systemic exposure with delayed and lower C_{max} [38]. As a better understanding of depot site mechanisms emerges, it may be possible to tune prodrug and formulation composition towards critical mechanisms that optimize LAI behaviour.

5. Conclusions

The data presented here show the successful in vivo application of SSPN-based LAI to provide half-life extension for lipophilic bio-reversible PDs of water-soluble drugs. Species differences in clearance would indicate that the formulations in humans may provide even greater exposure. In general, smaller species have a higher rate of clearance [39]. FTC specifically is renally cleared at a rate of 22.3 mL/kg/min in rats and 3.5 mL/kg/min in humans [31,40]. The difference in clearance would indicate the potential for plasma concentrations of FTC to remain in circulation longer in humans than those observed in preclinical species.

This formulation strategy shows the potential to add a new class of antiretrovirals to LAI treatment or prevention options. In addition to potential applications towards co-formulation with other antiretrovirals, LAI FTC may have applications in PrEP. Future work will aim to assess the applicability of this approach to other water-soluble drugs for HIV and other indications.

Supplementary Materials: The following supporting information can be downloaded at: https://www.mdpi.com/article/10.3390/pharmaceutics15071835/s1, Table S1 Accuracy and precision of the validated LCMS assay. Table S2 Statistical assessment for significance of Protection against HIV-1 Infection by Lead SSPN Formulation.

Author Contributions: A.O., S.R., C.L.F.M. and C.F. conceived the work secured the funding, and supervised the research. L.P. supervised the humanized mouse model. P.C., L.T., H.B., M.N., J.S., J.J.H., N.J.L., C.A.W.D., A.A.-k., D.M. and E.M. conducted the experimental research. H.P. conducted the pharmacokinetic modelling. P.C. and A.O. wrote the manuscript with contributions from all authors. All authors have read and agreed to the published version of the manuscript.

Funding: This work was funded by the US National Institutes of Health (NIH) under grant number R01AI114405-01.

Institutional Review Board Statement: In vivo experminets were conducted under the following ethical approval Ethic Committee Name: Institutional Animal Care and Use Committee, Approval Code: IUCAC # 17-107-10-FC, Approval Date: 1 May 2018. The remaining Animal studies were conducted at the University of Liverpool and the following ethics apply: Ethic Committee Name:

University of Liverpool Animal Welfare and Ethical Review Board (AWERB) followed by approval from the UK Home Office. Approval Code: Project licence PP9284915, Approval Date: 27 May 2020.

Informed Consent Statement: Not applicable.

Data Availability Statement: Data available on request due to restrictions privacy and intellectual property. The data presented in this study are available on request from the corresponding author. The data are not publicly available due to privacy restrictions.

Acknowledgments: The authors thank Yan Cheng, Amanda Branch Woods and Weimin Wang for technical assistance in the generation and experimental use of humanised mice.

Conflicts of Interest: A.O., S.R., P.C., J.H., C.L.F.M, A.A., D.J.M. and C.F. are co-inventors of WO2018178722A1. A.O. and S.R. are co-directors of Tandem Nano Ltd. A.O. has been co-investigator on funding received by the University of Liverpool from Bicycle Therapeutics, ViiV Healthcare and Gilead Sciences in the past 3 years. A.O. has received personal fees from Gilead and Assembly Biosciences in the past 3 years. S.R. has received support/funding from Gilead, Merck, Janssen, ViiV Healthcare, AstraZeneca, and Pfizer. S.R. has received personal fees from Gilead.

References

1. Bhatti, A.B.; Usman, M.; Kandi, V. Current Scenario of HIV/AIDS, Treatment Options, and Major Challenges with Compliance to Antiretroviral Therapy. *Cureus* **2016**, *8*, e515. [CrossRef]
2. Landovitz, R.J.; Grinsztejn, B. Long-Acting Injectable Preexposure Prophylaxis for HIV Prevention in South Africa: Is There a Will and a Way? *J. Infect. Dis.* **2016**, *213*, 1519–1520. [CrossRef]
3. Rosenbloom, D.I.; Hill, A.L.; Rabi, S.A.; Siliciano, R.F.; Nowak, M.A. Antiretroviral dynamics determines HIV evolution and predicts therapy outcome. *Nat. Med.* **2012**, *18*, 1378–1385. [CrossRef]
4. Ssemwanga, D.; Lihana, R.W.; Ugoji, C.; Abimiku, A.; Nkengasong, J.; Dakum, P.; Ndembi, N. Update on HIV-1 Acquired and Transmitted Drug Resistance in Africa. *AIDS Rev.* **2014**, *17*, 3–20.
5. Abram, M.E.; Ferris, A.L.; Das, K.; Quinones, O.; Shao, W.; Tuske, S.; Alvord, W.G.; Arnold, E.; Hughes, S.H. Mutations in HIV-1 reverse transcriptase affect the errors made in a single cycle of viral replication. *J. Virol.* **2014**, *88*, 7589–7601. [CrossRef]
6. Luber, A.D. Genetic barriers to resistance and impact on clinical response. *J. Int. AIDS Soc. Vol.* **2005**, *7*, 69. [CrossRef]
7. Dimitrov, D.T.; Masse, B.R.; Donnell, D. PrEP Adherence Patterns Strongly Affect Individual HIV Risk and Observed Efficacy in Randomized Clinical Trials. *J. Acquir. Immune Defic. Syndr.* **2016**, *72*, 444–451. [CrossRef]
8. Scarsi, K.K.; Swindells, S. The Promise of Improved Adherence with Long-Acting Antiretroviral Therapy: What Are the Data? *J. Int. Assoc. Provid. AIDS Care* **2021**, *20*, 23259582211009011. [CrossRef]
9. D'Amico, R.; Cenoz Gomis, S.; Moodley, R.; Van Solingen-Ristea, R.; Baugh, B.; Van Landuyt, E.; Van Eygen, V.; Min, S.; Cutrell, A.; Foster, C.; et al. Compassionate use of long-acting cabotegravir plus rilpivirine for people living with HIV-1 in need of parenteral antiretroviral therapy. *HIV Med.* **2023**, *24*, 202–211. [CrossRef]
10. Romagnoli, A.; Santoleri, F.; Costantini, A. Long-acting Injectable vs. Oral Antipsychotics: Adherence, Persistence and Switching over three Years of Real-life Analysis. *Curr. Rev. Clin. Exp. Pharm.* **2021**, *16*, 109–116. [CrossRef]
11. Williams, J.; Sayles, H.R.; Meza, J.L.; Sayre, P.; Sandkovsky, U.; Gendelman, H.E.; Flexner, C.; Swindells, S. Long-acting parenteral nanoformulated antiretroviral therapy: Interest and attitudes of HIV-infected patients. *Nanomedicine* **2013**, *8*, 1807–1813. [CrossRef]
12. Titus-Lay, E.N.; Ansara, E.D.; Isaacs, A.N.; Ott, C.A. Evaluation of adherence and persistence with oral versus long-acting injectable antipsychotics in patients with early psychosis. *Ment. Health Clin.* **2018**, *8*, 56–62. [CrossRef]
13. Brissos, S.; Veguilla, M.R.; Taylor, D.; Balanza-Martinez, V. The role of long-acting injectable antipsychotics in schizophrenia: A critical appraisal. *Adv. Psychopharmacol.* **2014**, *4*, 198–219. [CrossRef]
14. Tolley, E.E.; McKenna, K.; Mackenzie, C.; Ngabo, F.; Munyambanza, E.; Arcara, J.; Rademacher, K.H.; Lendvay, A. Preferences for a potential longer-acting injectable contraceptive: Perspectives from women, providers, and policy makers in Kenya and Rwanda. *Glob. Health Sci. Pract.* **2014**, *2*, 182–194. [CrossRef]
15. Hard, M.L.; Mills, R.J.; Sadler, B.M.; Wehr, A.Y.; Weiden, P.J.; von Moltke, L. Pharmacokinetic Profile of a 2-Month Dose Regimen of Aripiprazole Lauroxil: A Phase I Study and a Population Pharmacokinetic Model. *CNS Drugs* **2017**, *31*, 617–624. [CrossRef]
16. Curley, P.; Liptrott, N.J.; Owen, A. Advances in nanomedicine drug delivery applications for HIV therapy. *Future Sci. OA* **2018**, *4*, FSO230. [CrossRef]
17. Verloes, R.; Klooster, G.v.; Baert, L.; Velsen, F.v.; Bouche, M.-P.; Spittaels, K.; Leempoels, J.; Williams, P.; Kraus, G.; Wigerinck, P.; et al. TMC278 long acting—A parenteral nanosuspension formulation that provides sustained clinically relevant plasma concentrations in HIV-negative volunteers. In Proceedings of the 17th International AIDS Conference, Mexico City, Mexico, 3–8 August 2008.
18. Spreen, W.R.; Margolis, D.A.; Pottage, J.C., Jr. Long-acting injectable antiretrovirals for HIV treatment and prevention. *Curr. Opin. HIV AIDS* **2013**, *8*, 565–571. [CrossRef]
19. Sillman, B.; Bade, A.N.; Dash, P.K.; Bhargavan, B.; Kocher, T.; Mathews, S.; Su, H.; Kanmogne, G.D.; Poluektova, L.Y.; Gorantla, S.; et al. Creation of a long-acting nanoformulated dolutegravir. *Nat. Commun.* **2018**, *9*, 443. [CrossRef]

20. Kraft, J.C.; McConnachie, L.A.; Koehn, J.; Kinman, L.; Collins, C.; Shen, D.D.; Collier, A.C.; Ho, R.J. Long-acting combination anti-HIV drug suspension enhances and sustains higher drug levels in lymph node cells than in blood cells and plasma. *AIDS* **2017**, *31*, 765–770. [CrossRef]
21. Tatham, L.M.; Savage, A.C.; Dwyer, A.; Siccardi, M.; Scott, T.; Vourvahis, M.; Clark, A.; Rannard, S.P.; Owen, A. Towards a Maraviroc Long-Acting Injectable Nanoformulation. *Eur. J. Pharm. Biopharm.* **2018**, *138*, 92–98. [CrossRef]
22. Cory, T.J.; Midde, N.M.; Rao, P.; Kumar, S. Investigational reverse transcriptase inhibitors for the treatment of HIV. *Expert Opin. Investig. Drugs* **2015**, *24*, 1219–1228. [CrossRef]
23. Hobson, J.J.; Al-Khouja, A.; Curley, P.; Meyers, D.; Flexner, C.; Siccardi, M.; Owen, A.; Meyers, C.F.; Rannard, S.P. Semi-solid prodrug nanoparticles for long-acting delivery of water-soluble antiretroviral drugs within combination HIV therapies. *Nat. Commun.* **2019**, *10*, 1413. [CrossRef] [PubMed]
24. Jackson, A.; Moyle, G.; Watson, V.; Tjia, J.; Ammara, A.; Back, D.; Mohabeer, M.; Gazzard, B.; Boffito, M. Tenofovir, emtricitabine intracellular and plasma, and efavirenz plasma concentration decay following drug intake cessation: Implications for HIV treatment and prevention. *J. Acquir. Immune Defic. Syndr.* **2013**, *62*, 275–281. [CrossRef] [PubMed]
25. R Core Team. *R: A Language and Environment for Statistical Computing*; R Foundation for Statistical Computing: Vienna, Austria, 2020; Available online: https://www.r-project.org/ (accessed on 13 June 2023).
26. Borchers, H.W. Pracma: Practical Numerical Math Functions. R Package Version 2.2.9. 2019. Available online: https://CRAN.R-project.org/package=pracma (accessed on 13 June 2023).
27. Dash, P.K.; Gendelman, H.E.; Roy, U.; Balkundi, S.; Alnouti, Y.; Mosley, R.L.; Gelbard, H.A.; McMillan, J.; Gorantla, S.; Poluektova, L.Y. Long-acting nanoformulated antiretroviral therapy elicits potent antiretroviral and neuroprotective responses in HIV-1-infected humanized mice. *AIDS* **2012**, *26*, 2135–2144. [CrossRef] [PubMed]
28. Dagur, R.S.; Branch-Woods, A.; Mathews, S.; Joshi, P.S.; Quadros, R.M.; Harms, D.W.; Cheng, Y.; Miles, S.M.; Pirruccello, S.J.; Gurumurthy, C.B.; et al. Human-like NSG mouse glycoproteins sialylation pattern changes the phenotype of human lymphocytes and sensitivity to HIV-1 infection. *BMC Immunol.* **2019**, *20*, 2. [CrossRef]
29. Su, H.; Sravanam, S.; Gorantla, S.; Kaminski, R.; Khalili, K.; Poluektova, L.; Gendelman, H.E.; Dash, P.K. Amplification of Replication Competent HIV-1 by Adoptive Transfer of Human Cells From Infected Humanized Mice. *Front. Cell. Infect. Microbiol.* **2020**, *10*, 38. [CrossRef] [PubMed]
30. Yanez, J.A.; Remsberg, C.M.; Sayre, C.L.; Forrest, M.L.; Davies, N.M. Flip-flop pharmacokinetics—Delivering a reversal of disposition: Challenges and opportunities during drug development. *Ther. Deliv.* **2011**, *2*, 643–672. [CrossRef]
31. Nirogi, R.; Bhyrapuneni, G.; Kandikere, V.; Muddana, N.; Saralaya, R.; Komarneni, P.; Mudigonda, K.; Mukkanti, K. Pharmacokinetic profiling of efavirenz-emtricitabine-tenofovir fixed dose combination in pregnant and non-pregnant rats. *Biopharm. Drug Dispos.* **2012**, *33*, 265–277. [CrossRef]
32. Wolfensohn, S.; Lloyd, M. *Handbook of Laboratory Animal Management and Welfare*; John Wiley & Sons: Hoboken, NJ, USA, 2008.
33. Diehl, K.H.; Hull, R.; Morton, D.; Pfister, R.; Rabemampianina, Y.; Smith, D.; Vidal, J.M.; van de Vorstenbosch, C.; European Federation of Pharmaceutical Industries Associations; European Centre for the Validation of Alternative Methods. A good practice guide to the administration of substances and removal of blood, including routes and volumes. *J. Appl. Toxicol.* **2001**, *21*, 15–23. [CrossRef]
34. Molina, J.M.; Peytavin, G.; Perusat, S.; Lascoux-Combes, C.; Sereni, D.; Rozenbaum, W.; Chene, G. Pharmacokinetics of emtricitabine, didanosine and efavirenz administered once-daily for the treatment of HIV-infected adults (pharmacokinetic substudy of the ANRS 091 trial). *HIV Med.* **2004**, *5*, 99–104. [CrossRef]
35. Saag, M.S. Emtricitabine, a new antiretroviral agent with activity against HIV and hepatitis B virus. *Clin. Infect. Dis.* **2006**, *42*, 126–131. [CrossRef]
36. Labarthe, L.; Gele, T.; Gouget, H.; Benzemrane, M.S.; Le Calvez, P.; Legrand, N.; Lambotte, O.; Le Grand, R.; Bourgeois, C.; Barrail-Tran, A. Pharmacokinetics and tissue distribution of tenofovir, emtricitabine and dolutegravir in mice. *J. Antimicrob. Chemother.* **2022**, *77*, 1094–1101. [CrossRef] [PubMed]
37. Al-Tabakha, M.M. HPMC capsules: Current status and future prospects. *J. Pharm. Pharm. Sci.* **2010**, *13*, 428–442. [CrossRef] [PubMed]
38. Darville, N.; van Heerden, M.; Marien, D.; De Meulder, M.; Rossenu, S.; Vermeulen, A.; Vynckier, A.; De Jonghe, S.; Sterkens, P.; Annaert, P.; et al. The effect of macrophage and angiogenesis inhibition on the drug release and absorption from an intramuscular sustained-release paliperidone palmitate suspension. *J. Control. Release* **2016**, *230*, 95–108. [CrossRef] [PubMed]
39. Owen, A.; Curley, P. Species similarities and differences in pharmacokinetics and distribution of antiretroviral drugs. In *Humanized Mice for HIV Research*; Springer: New York, NY, USA, 2014; pp. 339–360.
40. Rajoli, R.K.; Back, D.J.; Rannard, S.; Freel Meyers, C.L.; Flexner, C.; Owen, A.; Siccardi, M. Physiologically Based Pharmacokinetic Modelling to Inform Development of Intramuscular Long-Acting Nanoformulations for HIV. *Clin. Pharm.* **2015**, *54*, 639–650. [CrossRef] [PubMed]

pharmaceutics

Perspective

Preclinical and Early Clinical Development of Tenofovir Alafenamide/Elvitegravir Topical Inserts for Effective On-Demand Vaginal and Rectal HIV Prevention

M. Melissa Peet, Vivek Agrahari, Meredith R. Clark and Gustavo F. Doncel *

CONRAD, Eastern Virginia Medical School, Norfolk, VA 23507, USA; mpeet@conrad.org (M.M.P.); vagrahari@conrad.org (V.A.); mclark@conrad.org (M.R.C.)
* Correspondence: doncelgf@evms.edu; Tel.: +1-757-446-5026

Abstract: HIV/AIDS remains a global public health issue, and products available for the prevention of HIV infections are limited, especially those for short-acting, on-demand, user-controlled applications. Topical inserts are products that can be applied vaginally or rectally and have been explored as drug delivery systems. To fill the gap in the HIV prevention product pipeline, CONRAD has developed a topical insert containing tenofovir alafenamide fumarate (TAF) and elvitegravir (EVG), two potent and synergistic antiretrovirals, as a simple, low-cost, and discreet option that can be self-administered vaginally and/or rectally, before and after coitus. In this review, we have described the development path of the TAF/EVG insert up to its current point in clinical testing, highlighting findings from acceptability, preclinical safety, pharmacokinetics, and efficacy evaluations and early clinical studies. In summary, the TAF/EVG inserts are stable, easy to manufacture, low-cost, acceptable, and show highly promising preclinical and clinical results for on-demand topical pre- or post-exposure HIV prevention.

Keywords: HIV pre-exposure and post-exposure prophylaxis; topical PrEP and PEP; antiretrovirals; acceptability; safety

1. Introduction

The goal of the HIV Prevention 2025 Roadmap is to end AIDS as a public health threat by 2030 and reduce new HIV infections to fewer than 370,000 by 2025 [1]. According to UNAIDs (2023), 1.3 million (between 1 and 1.7 million) new infections were reported in 2022 (https://thepath.unaids.org/, accessed on 7 December 2023). With the need to reduce the yearly incidence by over 1 million, in addition to continued provision of antiretrovirals (ARVs) for treatment, expanded development of HIV prevention products (pre-exposure prophylaxis; PrEP) that are not only safe and effective but are also affordable, desirable, and meet the needs of end-users is critically needed. Barrier methods (mostly male condoms) are widely available due to their affordability [2] but do not fit every user's preferences. Oral PrEP in the form of a daily tablet containing tenofovir disoproxil fumarate (TDF) and emtricitabine (FTC) is highly effective but, particularly in women, only with high adherence [3]. This regimen is burdensome and carries with it a stigma as pills are associated with the treatment of infected individuals [4]. With more options available for HIV prevention, the higher the potential for uptake. A 2-month long-acting injectable containing the integrase inhibitor cabotegravir has proved to be safe and highly effective and was recently approved for HIV prevention [5–7]. This eliminates the requirement of taking oral PrEP pills every day but requires painful 3 mL intramuscular (IM) injections every 60 days. As shown time and again, people want choices and there is no one-size-fits-all product that meets the needs of everyone at all stages in their life. Some individuals want user-controlled daily pills, some long-acting injectables, while others, especially those who engage in less frequent or clustered sex prefer short-acting on-demand products [8–10].

Currently, the only on-demand option proven to be effective in the MSM (men who have sex with men) population is TDF/FTC using a challenging 2-1-1 schedule (two pills 2–24 h before sex, one pill 24 h after the first dose, and another pill 24 h after the second dose). This regimen is not approved by the U.S. Food and Drug Administration (FDA). Nothing comparable exists for women or people who have vaginal sex. Topically applied mucosal HIV prevention products have a high potential for safety, ease of use, low cost, and privacy and bear a broad appeal among end-users [11]. The only topical product to prevent the vaginal transmission of HIV, which is recommended by The World Health Organization (WHO) and approved in some African countries, though not approved by the FDA, is the 1-month dapivirine vaginal ring [12]. This ring, however, is restricted to vaginal application only and must be worn daily, continuously, and replaced every month. Taken together, there exists a clear need for a product that can be applied topically, on-demand, and by both women and men and is forgiving (extended window of protection), safe, and highly effective. In the category of topical products, inserts can be applied vaginally or rectally and are a promising alternative to oral and parenteral formulations for HIV prevention.

This article brings together, for the first time, all of the latest preclinical efficacy testing and recently completed first-in-human (FIH) clinical studies on acceptability, safety, and pharmacokinetics (PK) evaluating CONRAD's dual-compartment TAF/EVG topical insert for on-demand HIV prevention. Ongoing clinical studies and future product development direction based on end-user feedback are also included. For more information on these inserts as drug delivery platforms, please see below and in the previous review [13].

2. Insert Platform as a Topical Drug Delivery System

Therapeutic products delivered topically via the vaginal or rectal mucosa have been around for years and used for multiple indications. Vaginal inserts on the U.S. market include, but are not limited to, Vagifem® (for atrophic vaginitis due to menopause), Cervidil® (for the initiation or continuation of cervical ripening in patients at term prior to labor induction), Encare® (contraceptive), and Mycelex-G (for vaginal yeast infections). Examples of rectal products on the market include CANASA® (suppository for the treatment of mildly to moderately active ulcerative proctitis), Diastat® (gel for the acute treatment of intermittent, stereotypic episodes of frequent seizure activity), and hydrocortisone-based suppositories for use in inflamed hemorrhoids and post-irradiation (factitial) proctitis. There are a number of research groups targeting the vaginal and rectal mucosa to topically deliver agents formulated in inserts, films, gels, rings, and enema for the prevention of HIV (a complete list can be found at AVAC Prevention Products in Pipeline; https://avac.org/resource/infographic/the-future-of-arv-based-prevention-and-more/, accessed on 6 November 2023). While there are no FDA or African regulatory-approved topical on-demand products for HIV prevention other than condoms, many in the form of enemas, films, and fast-dissolving inserts are showing promising preclinical and clinical findings [14–19]. The insert platform has also been explored for other drugs/indications to provide user-friendly, low-cost, on-demand product options, especially in lower- and middle-income countries (LMICs).

To fill the gap in the HIV prevention pipeline using topical ARV products, CONRAD is leading the development of an insert platform that can be applied using a finger both vaginally and rectally, before or after sex, is user-administered, discreet, and displays minimal side effects. This dual-compartment on-demand topical insert (Figure 1) contains TAF and EVG and is manufactured using a simple, robust, and highly scalable direct-compression process with a limited number of processing steps involved to produce inserts with high mechanical strength. The inserts comprise FDA-approved excipients from different categories, including lubricants, fillers, binders, and disintegrants in appropriate ratios to support the fast disintegration and high mechanical integrity of the final product (manuscript in preparation). The specific excipients include magnesium stearate, lactose monohydrate, poloxamers, povidone, mannitol, and polyethylene glycol [20]. When formulating inserts, multiple physicochemical characteristics are taken into consideration, such as hardness and friability, to provide a balance of fast disintegration (for immediate drug release) and appropriate mechanical integrity to withstand physical stress during the manufacturing, shipment, and usage.

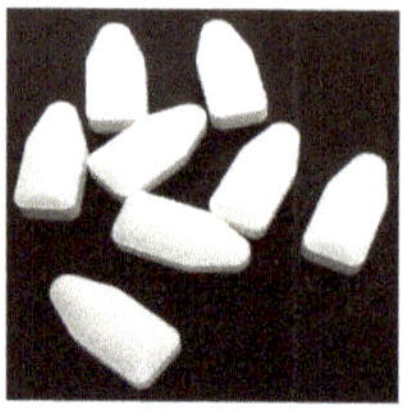

Figure 1. CONRAD's current TAF/EVG insert product (1.5 [L] × 0.7 [W] × 0.6 [H] cm).

3. Development of On-Demand Topical TAF/EVG Inserts for HIV Prevention

Earlier R&D efforts of topical inserts developed by CONRAD contained tenofovir (TFV, 40 mg) and FTC (40 mg), conceptually pioneering combination ARVs for topical HIV prevention. This drug combination was based on the approved oral PrEP tablet containing TDF and FTC and the TFV 1% gel under development but in a dosage form for topical delivery, intended to be more portable and discreet than the vaginal gel. In addition to being tested preclinically, these inserts were evaluated clinically, in CONRAD 117 (NCT01694407), where they demonstrated good safety and PK profile [21,22]. User input however highlighted excessive discharge and the presence of particulate materials as two negative features that needed to be corrected before moving forward. Four different placebo inserts were reformulated for further end-user feedback in the CONRAD 134 study (NCT02534779), leading to the selection of a bullet-shaped prototype with a shape different than what had been used before and more intuitive for vaginal insertion [13,23]. All placebo prototypes had improved user acceptability, including easy use, over the earlier TFV/FTC inserts and showed a superior dissolution profile, with initial disintegration occurring within 10 min and complete dissolution as quickly as 30 min, without leaving any visible residue or triggering excessive discharge. The clinical evolution of the bullet-shaped insert, initiating with the circular insert used in CONRAD 117, is presented in Figure 2 below, which includes completed clinical studies with the placebo form (QUATRO and MTN 035) and the TAF/EVG bullet-shaped insert (CONRAD 146 and MTN 039) in recently completed and ongoing clinical studies and development.

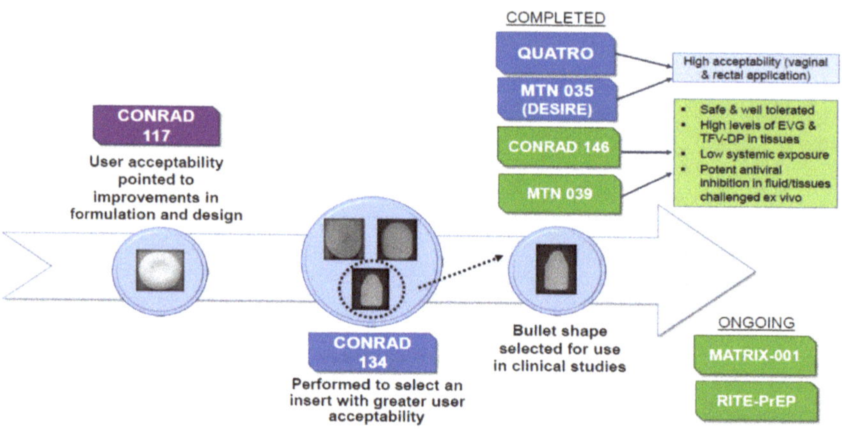

Figure 2. Sequence of clinical studies shaping the form and properties of the current TAF/EVG inserts, including recently completed and ongoing clinical studies.

While reformulating the platform, CONRAD explored the use of two more potent, synergistic, and mechanistically different ARVs, tenofovir alafenamide fumarate (TAF, equivalent to 20 mg of TAF free base) and elvitegravir (EVG, 16 mg), to expand the window of protection and dosing forgiveness. In vitro data using cell lines and cervicovaginal (CV) tissues supported activity under pre- and post-viral exposure conditions [24]. These new TAF/EVG inserts are manufactured using FDA-approved Generally Recognized as Safe (GRAS) materials and a simple direct-compression process, which enables low-cost [25,26], scalable, and readily transferable production of robust inserts, resulting in better handling and packaging feasibility of the final product. The accelerated and long-term stability data of these inserts provide a projected shelf life of >2 years if stored at or below 30 °C/65%RH, ensuring that the integrity, quality, and potency of the product is maintained for an extended duration. These inserts have undergone extensive acceptability, preclinical, and clinical testing to date supporting both vaginal and rectal use and a wide window of protection against HIV. The following sections discuss these aspects in detail.

4. Acceptability Assessment of Placebo and ARV-Containing Inserts

User acceptability is paramount to the uptake and adherence of HIV prevention interventions and their ultimate success. Acceptability assessments of products for HIV prevention following actual use have been limited, especially for rectal administration, with most being surveys inquiring about theoretical preferred use [27–31]. CONRAD's topical insert in placebo form has had the benefit of undergoing acceptability testing following vaginal use [23,32] and rectal use [27] to guide product development and ensure suitability for its users. Early on in insert development, the CONRAD 134 study (NCT02534779) was conducted to evaluate variations in the composition, technology type (direct-compressed versus freeze-dried), shape, and size of different prototypes for their disintegration time, residue/leakage, and acceptability after a single use to select a lead insert to advance into clinical development (reviewed in Peet 2019, [13]). The placebo form of the lead insert has also been evaluated for acceptability in comparison to other topical vaginal and rectal placebo products [27,32], as well as in Phase 1 studies following vaginal and rectal use of the TAF/EVG inserts [33,34], as summarized below.

The placebo insert was evaluated for its acceptability after vaginal application along with three other vaginal placebo products (film, gel, and ring) in QUATRO, a mixed-method crossover clinical study performed from June 2016 to 2017 in 178 HIV prevention product naïve, sexually active women living in Zimbabwe and South Africa [32]. Products were first assessed for preference prior to actual use, resulting in 41% preferring the gel, 19% the film, 15% the ring, and 25% the insert. After a period of using each of the products, in a randomized way, at least once weekly for one month prior to or independent of sex, the film, ring, and insert were preferred by 29%, 28%, and 26%, respectively, and the initially preferred gel by only 16% after use. At the end of the study, product choice varied by country with 45% in Zimbabwe selecting the film as most preferred and 35% in South Africa selecting the insert. A qualitative component was included in the QUATRO study to gain additional perspectives from women participants (n = 40), their male partners (n = 17), and key informants (KIs; n = 24) [35]. Focus groups discussions were conducted with the women participants and their male partners, and in-depth interviews (IDIs) were conducted with KIs. Key findings revealed that the insert was one of the two (film noted as the other) most preferred products for women in South Africa, men in Zimbabwe, and KIs in both countries. Noteworthy positive attributes related to the insert included female-initiated application, small and easy to use, tightened vagina/enhanced sexual pleasure, familiarity (similar looking to pills), discreet, easily dissolves, shape shows how to insert, and allowance for barrier-free sex. Negative attributes were wettened vagina/reduced sexual pleasure, short-term use, stigma (package and pill-like form similar to antiretroviral therapy), method of insertion may cause infection, and too small to insert properly. It should be noted that while short-term use was listed as a negative attribute, overall, the on-demand products (insert, film, and gel) were most liked.

These inserts were also included as one of the three placebo products (insert, suppository, enema) evaluated for rectal use prior to receptive anal intercourse (RAI) for the prevention of HIV in MTN-035 (NCT03671239), a randomized crossover trial for assessment of safety, acceptability, and adherence [36]. Participants (n = 217) included HIV-uninfected transgender men, transgender women, and cisgender MSM from five different countries (Malawi, Peru, South Africa, Thailand, and the U.S.). Inserts were reported as safe with high acceptability and adherence in general and when compared to the other product modalities. High acceptability was reported by 72% of participants for inserts, 66% for suppositories, and 73% for the enema. The percentage of participants fully adhering to the protocol (used at least once per week) was 75%, 74%, and 83% for the inserts, suppositories, and enema, respectively. Full adherence per RAI act was comparable across products: 58.9% for the insert, 58.0% for the suppository, and 58.8% for the enema.

Acceptability of the clinical form of the TAF/EVG insert for vaginal use was assessed in 16 healthy, HIV-uninfected women in the U.S. at low risk of HIV acquisition in CONRAD 146 (NCT03762772), a Phase 1 open-label randomized, parallel-group study [33]. Following a single intravaginal administration of the insert, participants completed a quantitative survey to assess acceptability across multiple attributes of the insert at 4 or 24 h post insertion. Size, ease of insertion, and color were considered very or somewhat acceptable in 100% of participants. Comfort after insertion, dissolvability, and discreetness were also highly acceptable for 15 of 16 participants (93.8%). Acceptability of residue and leakage ranked lower (81.3% and 81.5%, respectively) but remained acceptable in over half of the participants. A few participants described leakage as a small amount of whitish, odorless discharge. A total of 7 of 16 participants (43.8%) noted that using an applicator would make it easy to administer.

The clinical formulation of the TAF/EVG insert for rectal use was also assessed for acceptability in MTN 039 (NCT04047420), a Phase 1 open-label, multi-site, single-arm, two-period study in 23 healthy, HIV-uninfected male and female participants at two sites in the U.S. Although this study was focused on safety, PK, and pharmacodynamics (PD), acceptability was evaluated using computer-administered self-interviews (CASIs) prior to and after rectal administration of one or two inserts and a concluding IDI after final use of the product. Nearly all participants found the insert easy to use and comfortable. This clinical trial was recently completed, and a manuscript is pending review [34].

5. Preclinical Testing of TAF/EVG Inserts Shows Safety and High Efficacy When Applied before and after Virus Challenge

TAF/EVG inserts have undergone rigorous preclinical testing to verify local mucosal safety, characterize multi-compartment PK, and establish a predicted window of efficacy for advancement to clinical testing. Using the rabbit, a sensitive and preferred model to assess mucosal irritation and local tolerability and safety [37,38], insert safety was confirmed after several weeks of repeated daily intravaginal and intrarectal doses of pre-dissolved inserts using immunohistochemical and microscopic evaluations of mucosal tissues [13,21]. Ex vivo testing was performed to aid in defining the window of protection prior to testing in a primate model. CV explant tissue treated with TAF and EVG pre-exposure, during, or post-exposure to HIV-1BaL reduced p24 antigen production at all timepoints tested (up to 24 h pre-exposure and 48 h post-exposure) [24].

Using the repeated low-dose simian HIV hybrid (SHIV) challenge model along with multi-compartment PK assessments, TAF/EVG inserts have been evaluated to predict the clinical window of protection when used under PrEP and/or post-exposure prophylaxis (PEP) conditions and correlate drug levels with efficacy. This model has been used to predict clinical efficacy of FDA-approved ARVs for the prevention of HIV [39]. Through a collaboration with the Centers for Disease Control and Prevention (CDC), the prophylactic potential of inserts was tested against vaginal or rectal transmission of SHIV. TAF/EVG inserts at the clinical dose (20/16 mg) and administered vaginally to NHPs 4 h pre- or 4 h post-repeated weekly vaginal SHIV exposures provided protection in 5/6 animals pre-exposure and 6/6 animals post-exposure resulting in an efficacy of 91% and 100% for PrEP and PEP, respectively [18] (Figure 3a). PK evaluations confirmed rapid penetration of both drugs in vaginal tissue with EVG tissue levels above target at 2, 4, and 24 h and rising TFV-DP reaching target levels at 24 h and remaining high for at least 72 h. The efficacy of TAF/EVG inserts at 8 and 24 h post-SHIV exposure was also tested using the same challenge model, registering 94% and 72% efficacy at 8 h and 24 h post-viral exposure, respectively [40] (Figure 3a). Time to infection was delayed for animals infected with the TAF/EVG insert compared to placebo controls. Two of the animals which became infected tested positive only after 11 challenges. Using a similar methodology, the inserts were also tested for their efficacy against rectal SHIV transmission when administered 4 h pre-exposure followed by weekly repeated challenges [19]. In order to achieve better coverage of the recto-sigmoid compartment and greater than partial protection, we found that two inserts were required. Similar to what has been observed with topical TFV [41], higher drug concentrations were required in rectal tissue compared to vaginal tissue to provide protection with the inserts. When compared to single-insert application, the two-insert dose increased drug levels in rectal tissues by 10-fold, with an efficacy of 93.1% (Figure 3b).

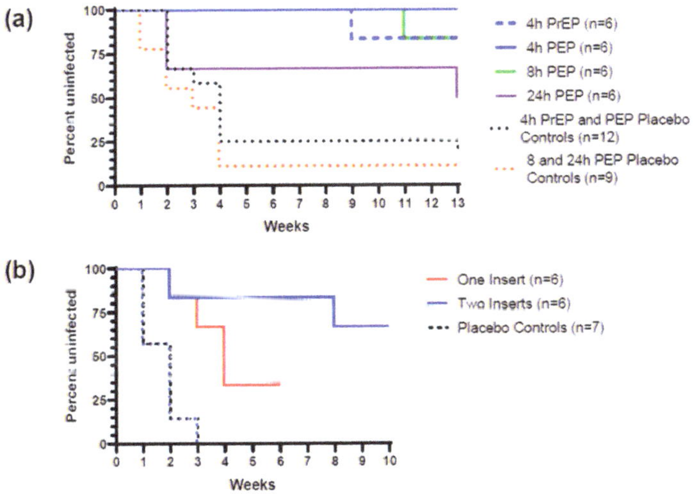

Figure 3. (**a**) Percent of uninfected macaques administered TAF/EVG inserts vaginally at 4 h prior to and 4, 8, and 24 h post-SHIV exposure (reproduced and modified with permission from [18,40]). (**b**) Rectal efficacy (percent uninfected) of one TAF/EVG insert vs. two TAF/EVG inserts administered 4 h before SHIV exposure in macaques (reproduced and modified with permission from [19]).

6. Clinical Testing of TAF/EVG Inserts Demonstrates Safety, Pharmacokinetics, and Pharmacodynamics Profiles Compatible with Protection against HIV

TAF/EVG inserts have been evaluated in two Phase 1 studies following vaginal (CONRAD 146) or rectal (MTN-039) administration. The clinical manufacturing of inserts under cGMP (Current Good Manufacturing Practice) conditions was conducted at Patheon, Inc., (Whitby, ON, Canada). CONRAD 146's primary objective was to assess safety and tolerability, multi-compartment PK, and PD out to 7 days following a single vaginal application of the TAF/EVG insert in 16 healthy women in the U.S. [33]. Safety was confirmed using treatment-emergent adverse events (TEAEs); no TEAEs were observed after a single dose of the insert. Vaginal administration was well tolerated and resulted in low plasma levels and no systemic side effects. PK and PD evaluated in vaginal fluid and tissue correlated well and supported an extended window of protection, revealing rapid delivery of both drugs at the earliest timepoint of collection, 4 h, and high levels out to 72 h post dose. The combination of TAF and EVG works well as EVG has a fast onset with rapid uptake into tissue, while TAF lasts longer, requiring conversion to TFV and TFV-DP. EVG vaginal tissue concentrations were above target protection levels (1000 ng/g; more than 20× EVG PA-IC95 [45 ng/mL] [42]) at 4 h and maintained out to 24 h, similar to what was seen in NHPs. Also, similarly to NHPs, TFV-DP levels were detected at 4 h post dose and were above the target benchmark (1000 fmol/mg) [43] at the 24, 48, and 72 h collection timepoints (Figure 4).

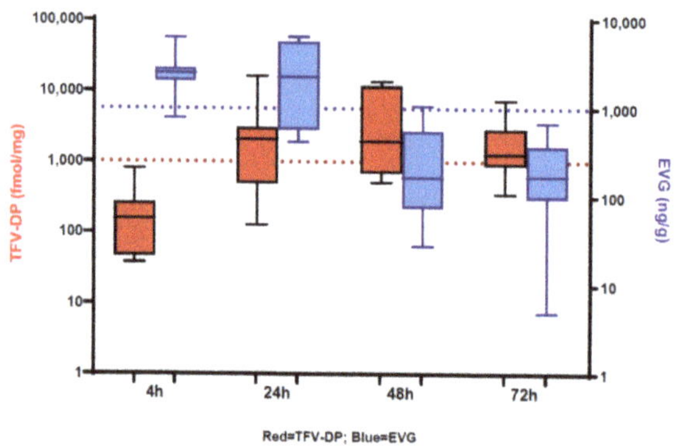

Figure 4. Concentrations of TFV-DP (red) and EVG (blue) in vaginal tissue at multiple timepoints post dose (reproduced and modified with permission from [33] under the terms and conditions of the Creative Commons Attribution (CC BY) license). Red dotted line represents TFV-DP target levels of 1000 fmol/mg, and blue dotted line represents target EVG levels of 1000 ng/g.

Cell-based and CV tissue assays revealed that the combination of TAF and EVG provided extended pre- and post-exposure inhibition of HIV replication. TAF demonstrated inhibitory activity against both HIV-1, its primary indication, and HSV-2, an additional exploratory indication, at 4 and 24 h post dose in vaginal fluid (Figure 5) [33]. TFV has been shown to inhibit HSV-2 infection in vitro, in animal models, and in clinical studies [44].

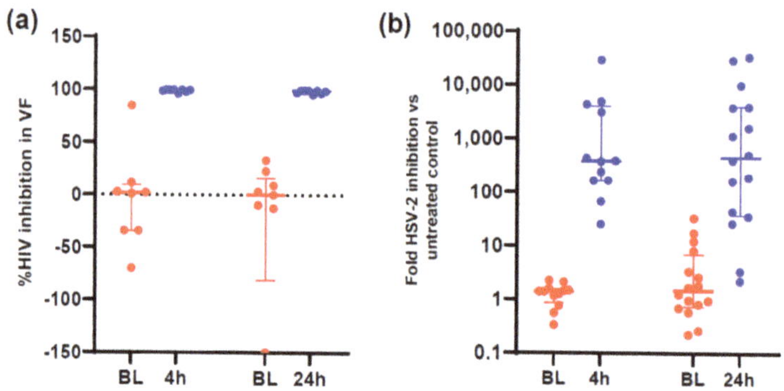

Figure 5. Ex vivo PD modeling of anti-viral effects in vaginal fluid. (**a**) Vaginal fluid HIV-1 inhibition. (**b**) Vaginal fluid HSV-2 inhibition. Red = baseline (BL) and blue = 4 or 24 h (reproduced and modified with permission from [33] under the terms and conditions of the Creative Commons Attribution (CC BY) license).

MTN-039 had a similar objective of assessing the safety and tolerability, multi-compartment PK out to 72 h post application, and PD of inserts following rectal administration in 23 participants at two U.S. sites [34]. Assessments were performed after rectal administration of one versus two inserts. Preliminary findings from this study have been reported [34], confirming that the rectal administration of TAF/EVG inserts is safe and well tolerated. Adverse effects were limited to mild anal erythema following one insert in a single participant. Rectal application also resulted in low plasma levels and no systemic side effects. EVG levels in rectal tissue were high at 2 h post application (exceeding 1 ng/mg) and showed a decline by the 24 h timepoint. TFV-DP levels were maintained markedly high, especially with two inserts out to 72 h post dose. Using the ex vivo rectal challenge model, HIV-1 p24 levels were significantly lower than the baseline with both one and two inserts out to 72 h. Further details are pending the publication of the manuscript.

7. Summary and Future Directions for TAF/EVG Inserts

The TAF/EVG inserts described in this review were rationally designed and developed by CONRAD following in vitro, animal model, acceptability/end-user input, and clinical testing. At present, these inserts are the most clinically advanced topical HIV prevention product for women and/or men with potential for flexible pre- or post-exposure use. Overall, the inserts have proven to be safe and efficacious preclinically, further displaying clinical PK/PD profiles compatible with HIV protection. They have also proven to be stable, easy to manufacture, low-cost, and potentially transferable to large-scale cGMP production in LMICs. It is worth noting that the concept of combining two synergistic ARVs with different physicochemical properties and PK profiles supports a highly effective, lasting, and forgiving (i.e., with a wider window of protection) dosing regimen, all properties that have been verified in both animal and clinical studies and represent unique characteristics of this novel method of on-demand HIV prevention for all genders. The development of TAF/EVG inserts has been and continues to be an iterative process. With a strong focus on end-user preferences, other design (shape/size) options are being explored that may be more intuitive and easier for application, along with different packaging options to maintain product discreetness and portability for on-demand use. End-user feedback is ongoing to determine whether an evolution of the bullet-shaped insert design, under continued clinical development for the TAF/EVG insert, to an almond shape with a dimple indent for the placement of a finger and ease of administration, for potential future insert products in preclinical development, is preferred (Figure 6).

Figure 6. Almond-shaped insert prototype with dimple indent (1.5 [L] × 1 [W] × 0.4 [H] cm).

Exploration of individual insert packaging and packaging for multiple inserts in one is also ongoing with the goal of producing a portable, functional, and discreet product package (Figure 7).

TAF/EVG inserts will be tested vaginally for the first time in Africa in a clinical study initiating at the end of 2023/beginning of 2024 (MATRIX-001, NCT06087913). This study will enroll approximately 60 healthy, non-pregnant, HIV-uninfected women at three clinical sites in South Africa, Kenya, and the U.S. MATRIX-001 will examine the safety, PK, modeled PD, disintegration, and acceptability of inserts. Participants will be randomized (1:1) to apply either a placebo or TAF/EVG insert vaginally once daily for 3 consecutive days and then every other day (QOD) for 14 days. Another clinical study planned, but assessing rectal application, is a placebo-controlled, randomized Phase 1 clinical trial in the U.S. (RITE PrEP) to evaluate the safety and PK of two TAF/EVG inserts administered rectally for 3 consecutive days and after seven doses on alternating days in up to 24 healthy, non-pregnant, HIV-uninfected participants.

Figure 7. Prototype packaging designs, informed by human-centered design process, intended for discreet, portable, and on-demand use.

Taken together, the current findings support the continued development and potential use of TAF/EVG inserts as a desired, gap-filling, and affordable option for on-demand/event-driven, discreet, flexible, and topical PrEP and PEP regimens against vaginal and/or rectal HIV transmission.

8. Patents

The authors are named in a patent application on "Pharmaceutical compositions and methods of making on demand solid dosage formulations" (PCT No.: PCT/US19/57645; Pub No. US 2021/0379089 A1), an invention developed under a U.S. Agency for International Development (USAID)-funded cooperative agreement.

Author Contributions: Conceptualization, M.M.P., V.A., M.R.C. and G.F.D.; writing—original draft preparation, M.M.P. and V.A.; writing—review and editing, M.M.P., V.A., M.R.C. and G.F.D.; visualization, M.M.P. and V.A.; supervision, G.F.D.; funding acquisition, M.R.C. and G.F.D. All authors have read and agreed to the published version of the manuscript.

Funding: CONRAD's R&D program advancing the TAF/EVG topical inserts has been funded primarily by the U.S. President's Emergency Plan for AIDS Relief (PEPFAR) through the U.S. Agency for International Development (USAID) under MAPS2: Development of Novel On-Demand and Longer-Acting Microbicide Product Leads a Cooperative Agreement #AID-OAA-A-14-00010 with CONRAD/Eastern Virginia Medical School and MATRIX: A USAID Project to Advance the Research and Development of Innovative HIV Prevention Products for Women, under Cooperative Agreement #7200AA22CA00002 with Magee Women's Research Institute. Additional financial support for the QUATRO study was provided by the Bill & Melinda Gates Foundation, and support for the Microbicide Trial Network's clinical studies (MTN 039 and MTN 035) of the TAF/EVG insert platform was provided by the National Institutes of Health, National Institute of Allergy and Infectious Diseases, Division of AIDS. CONRAD's initial development of the first-generation (TFV/FTC) insert was funded by USAID through Cooperative Agreement GPO-A-00-08-00005-00. Testing in NHPs at CDC was funded by CDC intramural funds and an interagency agreement between CDC and USAID (USAID/CDC IAA AID-GH-T-15-00002). The conclusions of this manuscript are those of the authors and do not necessarily reflect the views of USAID, PEPFAR, CDC, NIH, or the United States Government.

Acknowledgments: The authors thank Gilead Sciences, Inc., for the gracious provision of drug substances for CONRAD's development of the topical inserts used in the cited studies. The authors are thankful to Andrew P. Wong, Science Writer at CONRAD, Eastern Virginia Medical School, Norfolk, VA, for carefully reading and editing this manuscript.

Conflicts of Interest: The authors declare no conflicts of interest.

References

1. UNAIDS HIV Prevention 2025 Roadmap. Available online: https://www.unaids.org/en/resources/documents/2022/prevention-2025-roadmap (accessed on 7 December 2023).
2. Stover, J.; Rosen, J.E.; Carvalho, M.N.; Korenromp, E.L.; Friedman, H.S.; Cogan, M.; Deperthes, B. The case for investing in the male condom. *PLoS ONE* **2017**, *12*, e0177108. [CrossRef]
3. Hanscom, B.; Janes, H.E.; Guarino, P.D.; Huang, Y.; Brown, E.R.; Chen, Y.Q.; Hammer, S.M.; Gilbert, P.B.; Donnell, D.J. Brief Report: Preventing HIV-1 Infection in Women Using Oral Preexposure Prophylaxis: A Meta-analysis of Current Evidence. *J. Acquir. Immune Defic. Syndr.* **2016**, *73*, 606–608. [CrossRef]
4. Haire, B.G. Preexposure prophylaxis-related stigma: Strategies to improve uptake and adherence—A narrative review. *HIV/AIDS-Res. Palliat. Care* **2015**, *7*, 241–249. [CrossRef]
5. Prather, C.; Jeon, C. Cabotegravir: The first long-acting injectable for HIV pre-exposure prophylaxis. *Am. J. Health Syst. Pharm.* **2022**, *79*, 1898–1905. [CrossRef]
6. Bares, S.H.; Scarsi, K.K. A new paradigm for antiretroviral delivery: Long-acting cabotegravir and rilpivirine for the treatment and prevention of HIV. *Curr. Opin. HIV AIDS* **2022**, *17*, 22–31. [CrossRef] [PubMed]
7. Andrews, C.D.; Heneine, W. Cabotegravir long-acting for HIV-1 prevention. *Curr. Opin. HIV AIDS* **2015**, *10*, 258–263. [CrossRef] [PubMed]
8. Noret, M.; Balavoine, S.; Pintado, C.; Siguier, M.; Brun, A.; Bauer, R.; Loze, B.; Leplatois, A.; Aslan, A.; Moudachirou, K.; et al. Daily or on-demand oral tenofovir disoproxil fumarate/emtricitabine for HIV pre-exposure prophylaxis: Experience from a hospital-based clinic in France. *AIDS* **2018**, *32*, 2161–2169. [CrossRef] [PubMed]
9. Reyniers, T.; Nostlinger, C.; Laga, M.; De Baetselier, I.; Crucitti, T.; Wouters, K.; Smekens, B.; Buyze, J.; Vuylsteke, B. Choosing Between Daily and Event-Driven Pre-exposure Prophylaxis: Results of a Belgian PrEP Demonstration Project. *J. Acquir. Immune Defic. Syndr.* **2018**, *79*, 186–194. [CrossRef]
10. Vaccher, S.J.; Gianacas, C.; Templeton, D.J.; Poynten, I.M.; Haire, B.G.; Ooi, C.; Foster, R.; McNulty, A.; Grulich, A.E.; Zablotska, I.B.; et al. Baseline Preferences for Daily, Event-Driven, or Periodic HIV Pre-Exposure Prophylaxis among Gay and Bisexual Men in the PRELUDE Demonstration Project. *Front. Public Health* **2017**, *5*, 341. [CrossRef] [PubMed]
11. Baeten, J.M.; Hendrix, C.W.; Hillier, S.L. Topical Microbicides in HIV Prevention: State of the Promise. *Annu. Rev. Med.* **2020**, *71*, 361–377. [CrossRef]
12. Nel, A.; van Niekerk, N.; Kapiga, S.; Bekker, L.G.; Gama, C.; Gill, K.; Kamali, A.; Kotze, P.; Louw, C.; Mabude, Z.; et al. Safety and Efficacy of a Dapivirine Vaginal Ring for HIV Prevention in Women. *N. Engl. J. Med.* **2016**, *375*, 2133–2143. [CrossRef]
13. Peet, M.M.; Agrahari, V.; Anderson, S.M.; Hanif, H.; Singh, O.N.; Thurman, A.R.; Doncel, G.F.; Clark, M.R. Topical Inserts: A Versatile Delivery Form for HIV Prevention. *Pharmaceutics* **2019**, *11*, 374. [CrossRef] [PubMed]
14. Derby, N.; Lal, M.; Aravantinou, M.; Kizima, L.; Barnable, P.; Rodriguez, A.; Lai, M.; Wesenberg, A.; Ugaonkar, S.; Levendosky, K.; et al. Griffithsin carrageenan fast dissolving inserts prevent SHIV HSV-2 and HPV infections in vivo. *Nat. Commun.* **2018**, *9*, 3881. [CrossRef]
15. Hoang, T.; Date, A.A.; Ortiz, J.O.; Young, T.W.; Bensouda, S.; Xiao, P.; Marzinke, M.; Rohan, L.; Fuchs, E.J.; Hendrix, C.; et al. Development of rectal enema as microbicide (DREAM): Preclinical progressive selection of a tenofovir prodrug enema. *Eur. J. Pharm. Biopharm.* **2019**, *138*, 23–29. [CrossRef]
16. Li, J.; Regev, G.; Patel, S.K.; Patton, D.; Sweeney, Y.; Graebing, P.; Grab, S.; Wang, L.; Sant, V.; Rohan, L.C. Rational Design of a Multipurpose Bioadhesive Vaginal Film for Co-Delivery of Dapivirine and Levonorgestrel. *Pharmaceutics* **2019**, *12*, 1. [CrossRef] [PubMed]
17. Patel, S.K.; Agashe, H.; Patton, D.L.; Sweeney, Y.; Beamer, M.A.; Hendrix, C.W.; Hillier, S.L.; Rohan, L.C. Tenofovir vaginal film as a potential MPT product against HIV-1 and HSV-2 acquisition: Formulation development and preclinical assessment in non-human primates. *Front. Reprod. Health* **2023**, *5*, 1217835. [CrossRef]
18. Dobard, C.W.; Peet, M.M.; Nishiura, K.; Holder, A.; Dinh, C.; Mitchell, J.; Khalil, G.; Pan, Y.; Singh, O.N.; McCormick, T.J.; et al. Single dose topical inserts containing tenofovir alafenamide fumarate and elvitegravir provide pre- and post-exposure protection against vaginal SHIV infection in macaques. *EBioMedicine* **2022**, *86*, 104361. [CrossRef] [PubMed]
19. Makarova, N.; Singletary, T.; Peet, M.M.; Mitchell, J.; Holder, A.; Dinh, C.; Agrahari, V.; Mendoza, M.; Pan, Y.; Heneine, W.; et al. Pharmacokinetics and efficacy of topical inserts containing tenofovir alafenamide fumarate and elvitegravir administered rectally in macaques. *EBioMedicine* **2022**, *86*, 104338. [CrossRef]
20. Singh, O.N.; Mccormick, T.J.; Agrahari, V.; Peet, M.M.; Clark, M.R.; Schwartz, J.; Doncel, G.F. Pharmaceutical Compositions and Methods of Making on Demand Solid Dosage Formulations. U.S. Patent 20210379089 A1, 9 December 2021.
21. Clark, M.R.; Peet, M.M.; Davis, S.; Doncel, G.F.; Friend, D.R. Evaluation of Rapidly Disintegrating Vaginal Tablets of Tenofovir, Emtricitabine and Their Combination for HIV-1 Prevention. *Pharmaceutics* **2014**, *6*, 616–631. [CrossRef]
22. Mauck, C.K.; Thurman, A.; Keller, M.; Archer, D.F.; Kimble, T.; Kashuba, A.; Sykes, C.; Friend, D.; Schwartz, J.L.; Doncel, G.F. Pharmacokinetics of Tenofovir and Emtricitabine Delivered by Vaginal Tablets. Presented at the Conference on Retroviruses and Opportunitis Infections (CROI), Seattle, WA, USA, 23–26 February 2015.
23. Littlefield, S.; Zhang, W.; Schwartz, J.; Clark, M.; Thurman, A.; Archer, D.; Mauck, C.; Linton, K.; McCormick, T.; Jacot, T.; et al. Clinical performance, acceptability and optimization of fast dissolve vaginal inserts designed for HIV prevention: Results from two clinical studies. In Proceedings of the in Conference Presentation at HIV Research for Prevention (HIV R4P), Chicago, IL, USA, 18 October 2016.

24. Ouattara, A.; Kerr, B.; Abdelmawla, F.; Anderson, P.; Clark, M.; Doncel, G.F. Time of addition studies of elvitegravir/TFV prodrug combinations demonstrate extended window of prophylactic activity in cells and tissue models. In Proceedings of the In Conference Presentation at HIV Research for Prevention (HIV R4P), Chicago, IL, USA, 17–21 October 2016.
25. Chen, H.; Aburub, A.; Sun, C.C. Direct Compression Tablet Containing 99% Active Ingredient-A Tale of Spherical Crystallization. *J. Pharm. Sci.* **2019**, *108*, 1396–1400. [CrossRef]
26. Hill, A.M.; Barber, M.J.; Gotham, D. Estimated costs of production and potential prices for the WHO Essential Medicines List. *BMJ Glob. Health* **2018**, *3*, e000571. [CrossRef] [PubMed]
27. Bauermeister, J.A.; Downs, J.S.; Krakower, D.S. PrEP Product Acceptability and Dual Process Decision-Making Among Men Who Have Sex with Men. *Curr. HIV/AIDS Rep.* **2020**, *17*, 161–170. [CrossRef] [PubMed]
28. Carballo-Dieguez, A.; Giguere, R.; Lentz, C.; Dolezal, C.; Fuchs, E.J.; Hendrix, C.W. Rectal Douching Practices Associated with Anal Intercourse: Implications for the Development of a Behaviorally Congruent HIV-Prevention Rectal Microbicide Douche. *AIDS Behav.* **2019**, *23*, 1484–1493. [CrossRef] [PubMed]
29. Luecke, E.H.; Cheng, H.; Woeber, K.; Nakyanzi, T.; Mudekunye-Mahaka, I.C.; van der Straten, A.; Team, M.-D.S. Stated product formulation preferences for HIV pre-exposure prophylaxis among women in the VOICE-D (MTN-003D) study. *J. Int. AIDS Soc.* **2016**, *19*, 20875. [CrossRef]
30. Pines, H.A.; Gorbach, P.M.; Weiss, R.E.; Hess, K.; Murphy, R.; Saunders, T.; Brown, J.; Anton, P.A.; Cranston, R.D. Acceptability of potential rectal microbicide delivery systems for HIV prevention: A randomized crossover trial. *AIDS Behav.* **2013**, *17*, 1002–1015. [CrossRef]
31. Tingler, R.C.; Connochie, D.; Bauermeister, J.A. Rectal Douching and Microbicide Acceptability among Young Men who Have Sex with Men. *AIDS Behav.* **2020**, *24*, 1414–1421. [CrossRef]
32. Montgomery, E.T.; Beksinska, M.; Mgodi, N.; Schwartz, J.; Weinrib, R.; Browne, E.N.; Mphili, N.; Musara, P.; Jaggernath, M.; Ju, S.; et al. End-user preference for and choice of four vaginally delivered HIV prevention methods among young women in South Africa and Zimbabwe: The Quatro Clinical Crossover Study. *J. Int. AIDS Soc.* **2019**, *22*, e25283. [CrossRef]
33. Thurman, A.R.; Ouattara, L.A.; Yousefieh, N.; Anderson, P.L.; Bushman, L.R.; Fang, X.; Hanif, H.; Clark, M.; Singh, O.; Doncel, G.F. A phase I study to assess safety, pharmacokinetics, and pharmacodynamics of a vaginal insert containing tenofovir alafenamide and elvitegravir. *Front. Cell Infect. Microbiol.* **2023**, *13*, 1130101. [CrossRef]
34. Riddler, S.A.; Kelly, C.; Hoesley, C.; Ho, K.; Piper, J.; Edick, S.; Heard, F.; Jiao, Y.; Doncel, G.F.; Johnson, S.; et al. Safety and PK/PD of a Tenofovir alafenamide/Elvitegravir insert administrated rectally. In Proceedings of the Conference on Retroviruses and Opportunistic Infections (CROI), Seattle, WA USA, 19–22 February 2023.
35. Musara, P.; Milford, C.; Shapley-Quinn, M.K.; Weinrib, R.; Mutero, P.; Odoom, E.; Mgodi, N.M.; Chirenje, Z.M.; Hanif, H.; Clark, M.R.; et al. Preferences and Acceptability of Vaginal Delivery Forms for HIV Prevention Among Women, Male Partners and Key Informants in South Africa and Zimbabwe: Qualitative Findings. *AIDS Behav.* **2021**, *25*, 124–138. [CrossRef] [PubMed]
36. Bauermeister, J.A.; Dominguez Islas, C.; Jiao, Y.; Tingler, R.; Brown, E.; Zemanek, J.; Giguere, R.; Balan, I.; Johnson, S.; Macagna, N.; et al. A randomized trial of safety, acceptability and adherence of three rectal microbicide placebo formulations among young sexual and gender minorities who engage in receptive anal intercourse (MTN-035). *PLoS ONE* **2023**, *18*, e0284339. [CrossRef] [PubMed]
37. Doncel, G.F.; Chandra, N.; Fichorova, R.N. Preclinical assessment of the proinflammatory potential of microbicide candidates. *J. Acquir. Immune Defic. Syndr.* **2004**, *37* (Suppl. S3), S174–S180. [CrossRef] [PubMed]
38. Eckstein, P.; Jackson, M.C.; Millman, N.; Sobrero, A.J. Comparison of vaginal tolerance tests of spermicidal preparations in rabbits and monkeys. *J. Reprod. Fertil.* **1969**, *20*, 85–93. [CrossRef] [PubMed]
39. Garcia-Lerma, J.G.; McNicholl, J.M.; Heneine, W. The predictive value of macaque models of preexposure prophylaxis for HIV prevention. *Curr. Opin. HIV AIDS* **2022**, *17*, 179–185. [CrossRef] [PubMed]
40. Makarova, N.; Singletary, T.; Peet, M.M.; Mitchell, J.; Bachman, S.; Holder, A.; Dinh, C.; Lipscomb, J.; Agrahari, V.; Mendoza, M.; et al. Extended post-exposure protection against vaginal SHIV infection with tenofovir alafenamide fumarate/elvitegravir inserts in macaques. *J. Infect. Dis.* **2023**, jiad599. [CrossRef] [PubMed]
41. Dobard, C.W.; Sharma, S.; Cong, M.E.; West, R.; Makarova, N.; Holder, A.; Pau, C.P.; Hanson, D.L.; Novembre, F.J.; Garcia-Lerma, J.G.; et al. Efficacy of topical tenofovir against transmission of a tenofovir-resistant SHIV in macaques. *Retrovirology* **2015**, *12*, 69. [CrossRef]
42. Ramanathan, S.; Mathias, A.A.; German, P.; Kearney, B.P. Clinical pharmacokinetic and pharmacodynamic profile of the HIV integrase inhibitor elvitegravir. *Clin. Pharmacokinet.* **2011**, *50*, 229–244. [CrossRef]
43. Dobard, C.; Sharma, S.; Martin, A.; Pau, C.P.; Holder, A.; Kuklenyik, Z.; Lipscomb, J.; Hanson, D.L.; Smith, J.; Novembre, F.J.; et al. Durable protection from vaginal simian-human immunodeficiency virus infection in macaques by tenofovir gel and its relationship to drug levels in tissue. *J. Virol.* **2012**, *86*, 718–725. [CrossRef]
44. Andrei, G.; Gillemot, S.; Topalis, D.; Snoeck, R. The Anti-Human Immunodeficiency Virus Drug Tenofovir, a Reverse Transcriptase Inhibitor, Also Targets the Herpes Simplex Virus DNA Polymerase. *J. Infect. Dis.* **2018**, *217*, 790–801. [CrossRef]

pharmaceutics

Article

A Bilayer Microarray Patch (MAP) for HIV Pre-Exposure Prophylaxis: The Role of MAP Designs and Formulation Composition in Enhancing Long-Acting Drug Delivery

Lalitkumar K. Vora [1], Ismaiel A. Tekko [1], Fabiana Volpe Zanutto [1], Akmal Sabri [1], Robert K. M. Choy [2], Jessica Mistilis [2], Priscilla Kwarteng [2], Courtney Jarrahian [2], Helen O. McCarthy [1] and Ryan F. Donnelly [1,*]

1 Medical Biology Centre, School of Pharmacy, Queen's University Belfast, 97 Lisburn Road, Belfast BT9 7BL, UK; l.vora@qub.ac.uk (L.K.V.); i.tekko@qub.ac.uk (I.A.T.); f.volpezanutto@qub.ac.uk (F.V.Z.); a.binsabri@qub.ac.uk (A.S.)

2 PATH, 2201 Westlake Avenue, Seattle, DC 98121, USA; rchoy@path.org (R.K.M.C.); jmistilis@path.org (J.M.)

* Correspondence: r.donnelly@qub.ac.uk

Abstract: Microarray patches (MAPs) have shown great potential for efficient and patient-friendly drug delivery through the skin; however, improving their delivery efficiency for long-acting drug release remains a significant challenge. This research provides an overview of novel strategies aimed at enhancing the efficiency of MAP delivery of micronized cabotegravir sodium (CAB Na) for HIV pre-exposure prophylaxis (PrEP). The refinement of microneedle design parameters, including needle length, shape, density, and arrangement, and the formulation properties, such as solubility, viscosity, polymer molecular weight, and stability, are crucial for improving penetration and release profiles. Additionally, a bilayer MAP optimization step was conducted by diluting the CAB Na polymeric mixture to localize the drug into the tips of the needles to enable rapid drug deposition into the skin following MAP application. Six MAP designs were analyzed and investigated with regard to delivery efficiency into the skin in ex vivo and in vivo studies. The improved MAP design and formulations were found to be robust and had more than 30% in vivo delivery efficiency, with plasma levels several-fold above the therapeutic concentration over a month. Repeated weekly dosing demonstrated the robustness of MAPs in delivering a consistent and sustained dose of CAB. In summary, CAB Na MAPs were able to deliver therapeutically relevant levels of drug.

Keywords: cabotegravir; dissolving microarray patch; microneedles; HIV; long-acting drugs; pre-exposure prophylaxis; intradermal delivery

1. Introduction

HIV continues to be a major cause of morbidity and mortality worldwide; more than 76 million people have been infected, culminating in more than 33 million deaths globally since the start of this epidemic in the 1980s [1–4]. A significant tool to reduce HIV infections is the use of pre-exposure prophylaxis (PrEP) to deliver an antiretroviral (ARV) to individuals at substantial risk of HIV infection. In 2021, an estimated 1.6 million people received oral PrEP (daily dose) at least once [5].

Despite the effectiveness of oral PrEP, daily administration may impose a considerable burden on patients, resulting in poor compliance and therefore a persistent risk of HIV acquisition [6]. Because of this limitation, there is impetus to develop a long-acting ARV formulation to simplify the overall prevention regimen and thus increase overall patient compliance. Dedicated research within the field of pharmaceutical science led to the approval of Apretude® (ViiV Healthcare), a long-acting injectable formulation of the integrase strand transfer inhibitor cabotegravir (CAB). The intramuscular (IM) injection is given in-clinic every 2 months to reduce the risk of HIV infection [7]. Additionally, a monthly dapivirine vaginal ring has been approved for the prevention of HIV in several countries [8], and other regimens, such as a twice-yearly subcutaneous injection of lenacapavir, have been recently approved [9]. Long-acting drug delivery technologies such as implants are also under development for HIV PrEP [10–13].

Injectable formulations must be delivered by skilled health care professionals, adding to the cost of PrEP and increasing the burden on health care programs, particularly in areas with fewer health care workers. This is a major global health issue, as the HIV incidence rate is exceptionally high in sub-Saharan Africa relative to other parts of the world [14]. These routine, painful injections may also have a negative effect on patient compliance over the long term. Therefore, there is a need to refine the way PrEP is administered to increase accessibility. One approach is to reformulate ARVs for delivery via microarray patches (MAPs), also known as microneedle patches [15–17]. These are patches with micron-size projections on a flat baseplate, which upon application to the skin, in a manner similar to the use of a bandage, generate microchannels for drug delivery. In general, MAPs can be classified into three main categories based on their mode of delivering payload into the skin: dissolving, hydrogel-forming, and coated MAPs. Dissolving MAPs have been the most widely explored and studied with respect to the delivery of ARVs [18].

Compared to the conventional IM injection of ARV formulations, MAPs can be applied in a minimally invasive fashion to the skin to achieve systemic delivery of ARVs. Despite the ease and simplicity of applying these patches, the insertion and delivery of the payload via MAP involve complex factors such as design and needle tip composition, as well as the overall physiochemical properties of the delivered active pharmaceutical ingredient [19,20]. Therefore, it would be judicious to meticulously consider these crucial factors when designing a MAP formulation for the delivery of ARVs. In addition, these MAPs ought to be able to achieve high drug loading and be capable of delivering the active pharmaceutical ingredient with acceptable delivery efficiency into the skin to serve as an intradermal depot for sustained delivery of hydrophobic particulate payload overall for several weeks. Guided by these formulation goals and the purpose of developing a minimally invasive MAP strategy to deliver ARVs, this paper summarizes the different strategies that have been explored with the aim of augmenting the efficiency of MAPs for the long-acting delivery of CAB sodium (CAB Na) for HIV PrEP.

2. Materials and Methods

2.1. Materials

Long-acting cabotegravir (CAB LA) nanosuspension vials for IM injection with drug content (400 mg/2 mL) and micronized powders of CAB Na were supplied by ViiV Healthcare (Research Triangle Park, NC, USA). Poly(vinyl alcohol), or PVA, of molecular weight (MW) 9 to 10 kDa, 80% hydrolyzed (PVA 10K), and of MW 31 to 50 kDa, 87% to 89% hydrolyzed (PVA 50K), were purchased from Sigma–Aldrich (Dorset, UK). Poly(vinyl pyrrolidone) K29-32 gel (MW 58 kDa) was a gift from Ashland (Kidderminster, UK). Liquid silicone elastomer mix was purchased from Nusil Technology (Buckinghamshire, UK). Ultrapure water was obtained from a water purification system (Elga PURELAB® DV 25, Veolia Water Systems, Dublin, Ireland). All other chemicals and materials were of analytical reagent grade and supplied by Sigma–Aldrich.

2.2. Strategies to Improve MAP Delivery Efficiency via Refinement of MAP Designs

The MAP design parameters, including needle length, shape, density, and arrangement, are crucial for improving various aspects of microneedle performance, such as penetration efficiency, drug distribution within the skin, and drug release profiles. Six MAPs of various geometries and shapes (Table 1) were designed and manufactured for use in the investigation of various strategies to improve CAB intradermal delivery efficiency [21,22].

Table 1. Different MAP mold designs consisting of different geometries and array sizes.

Code	Design	Length (μm)	Base Width (μm)	Interspacing (μm)	Array Size (cm^2)
MAP-D1	16 × 16 (256) Cuboidal base/pyramidal tips	900	300	100	0.5
MAP-D2	11 × 11 (121) Cuboidal base/pyramidal tips	900	300	300	0.5
MAP-D3	16 × 16 (256) Full conical	900	300	100	0.5
MAP-D4	16 × 16 (256) Full pyramidal	900	300	100	0.5
MAP-D5	24 × 25 (600) Full pyramidal	700	300	50	0.75
MAP-D6	9 × 9 Cuboidal base/pyramidal tips	1150	500	300	0.5

Abbreviation: MAP, microarray patch.

Computer-aided design was employed to create the various MAP designs. These designs were then used to generate a primary template (Figure 1a) through a process called two-photon polymerization of IP-S resins, which are acrylate-based polymers. This process was carried out using a Photonic Professional GT 3D printer system from Nanoscribe GmbH (Leopoldshafen, Germany), as detailed in our previous research [23]. Next, each MAP design's primary template was affixed to a 3D-printed (UltiMaker B.V., Geldermalsen, Germany) holder made of polylactic acid and utilized to fabricate female molds for the MAPs. These molds were produced using a transparent silicone/elastomer mixture (Figure 1a). To eliminate any remaining air bubbles, the silicone-filled templates underwent centrifugation at 3500 rpm for 5 min, followed by curing at 60 °C for a period of 3 to 6 h. Once the primary template cooled down, the PDMS mold (Figure 1b) was extracted. This PDMS mold served as the main component for manufacturing the bilayer CAB Na MAPs and could be reused multiple times.

2.3. Strategies to Enhance MAP Delivery Efficiency via Fabrication of Improved Bilayer MAPs

Bilayer CAB Na MAPs were prepared from various CAB Na-loaded hydrogels (to form the first MAP layer) and drug-free hydrogels (to form the second MAP layer) using a positive pressure chamber by either two-step or three-step casting as needed, as depicted in Figure 2 and previously described [24].

Briefly, 100 mg of the finalized CAB Na-loaded hydrogel was dispensed onto the PDMS mold of the developed MAP and then placed in the positive pressure chamber at 5 bars for 3 min. The PDMS mold was removed from the chamber, excess drug-loaded hydrogel was carefully scraped using a stainless steel spatula, and the mold was immediately returned to the chamber at 5 bar pressure for 15 min. Subsequently, the PDMS mold was removed from the chamber, and the first MAP layer was dried overnight at room temperature. The second MAP layer was then prepared by pouring the optimized drug-free polymeric blend on top of the first layer and placing the mold in a positive pressure chamber at 5 bar for 15 min. Finally, the bilayer CAB Na MAP was dried at room temperature for 24 h and at 37 °C for 24 h before being gently removed from the mold and stored in a desiccator until further testing or use.

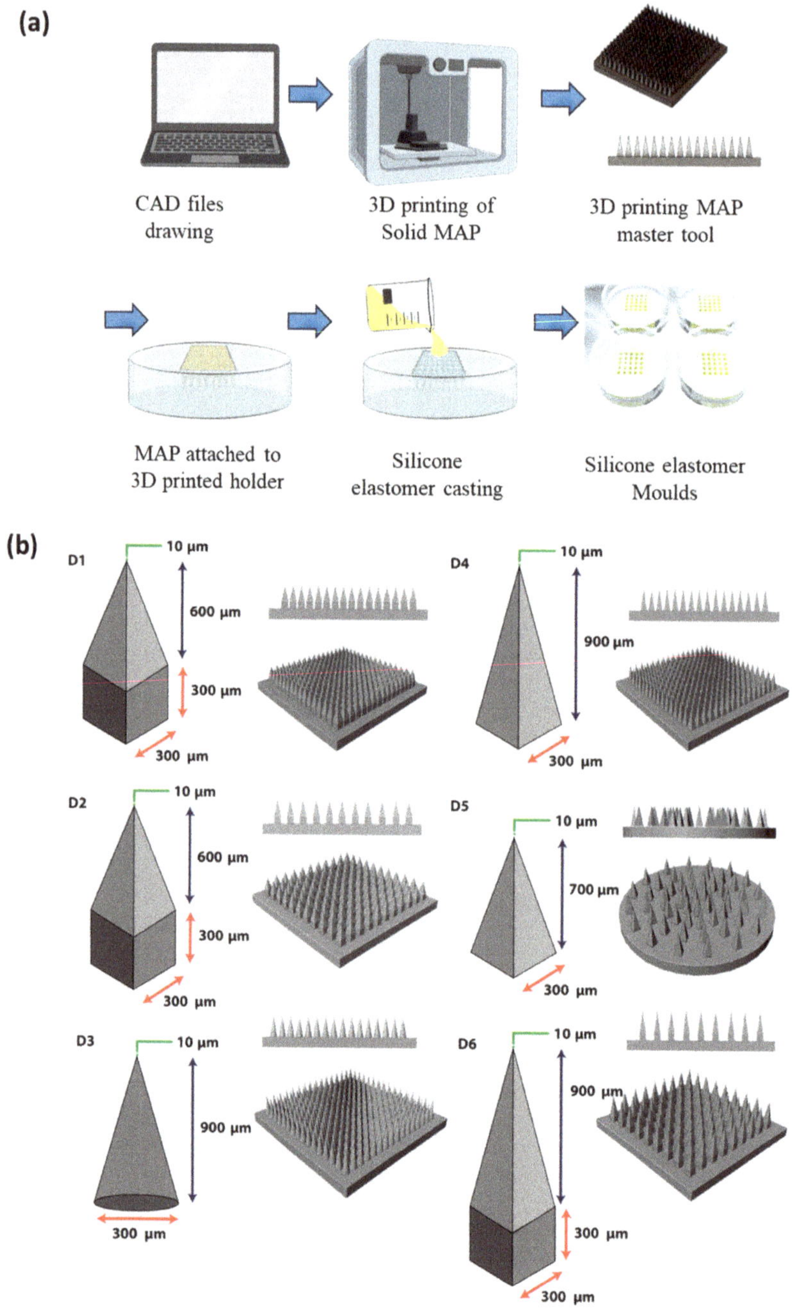

Figure 1. (**a**) Schematic illustrating the fabrication and manufacture of MAP molds via two-photon polymerization 3D printing. (**b**) MAP array designs developed for two-photon polymerization 3D printing (except D5).

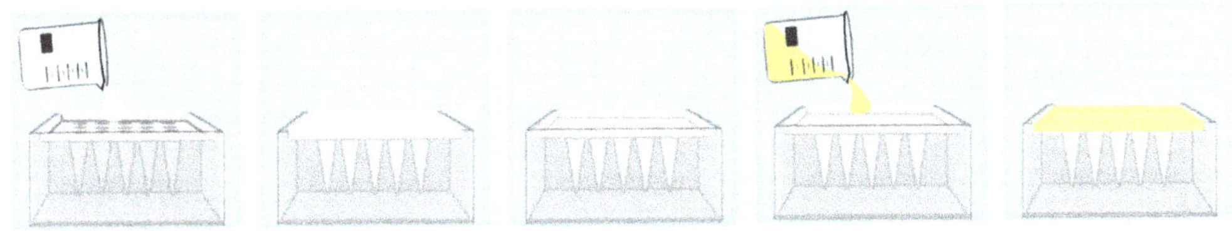

Figure 2. Schematic representation of the bilayer MAP preparation method.

To optimize loading the drug into the MAP tips, five CAB Na hydrogel formulations were developed and used to cast the MAP first layer. The composition of these hydrogel formulations is summarized in Table 2.

Table 2. Different hydrogel (PVA/PVP/CAB Na/water) compositions cast into the MAP mold.

	Formulation Code				
Composition	F1	F2 *	F3	F4	F5
CAB Na, g	0.84	0.84	0.84	0.84	0.84
PVA (10 kDa) gel (40%*w*/*w*), g	0.6	0.6	0.6	0.6	0.6
PVP K29-32 gel (40%*w*/*w*), g	0.6	0.6	0.6	0.6	0.6
Water, mL	1.0	1.0	1.5	2.0	2.5

Abbreviations: CAB Na, micronized cabotegravir sodium; PVA, poly(vinyl alcohol); PVP, poly(vinyl pyrrolidone). * F2: The same CAB Na-loaded hydrogel formulation as F1; however, following its casting into the MAP mold and removing the excess hydrogel, the top part of the hydrogel was removed by washing the first layer with deionized water.

2.4. *MAP Characterization*

2.4.1. Determination of CAB Na MAP Drug Content

The patches were first dissolved in 5 mL of deionized water to determine the drug content of the various bilayer CAB Na MAPs. Following complete dissolution, 100 µL of the resulting suspension was transferred to 1.5 mL tubes and mixed vigorously with 0.9 mL of acetonitrile. This process allowed the drug to be completely dissolved and led to the precipitation of water-soluble polymers. The resulting suspension was subsequently centrifuged at 14,800 rpm for 15 min, and 100 µL of the clear supernatant solution was collected and analyzed after making an appropriate dilution with acetonitrile by the validated high-performance liquid chromatography–ultraviolet (HPLC–UV) method [15].

2.4.2. Assessment of CAB Na MAP Mechanical Properties

A TA.XT2 Texture Analyser (Stable Micro Systems, Surrey, UK) in compression mode was used to determine the mechanical properties of the fabricated MAPs, as previously described [24]. Briefly, each MAP was attached to a cylindrical probe (cross-sectional area 1.5 cm^2), and the texture analyzer arm moved vertically downward at a speed of 0.5 mm/s and the MAP compressed against a flat aluminum block. A force of 32 N was applied for 30 s before the probe moved upward again. The length of the individual needles in each MAP was measured before and after compression using a Leica EZ4 W stereo microscope (Leica Microsystems, Milton Keynes, UK). The percentage reduction of microneedle heights was calculated and reported.

2.5. *Ex Vivo Drug Deposition Studies*

Ex vivo skin deposition studies were conducted to evaluate the efficiency of the MAPs in delivering CAB Na into the skin following administration. These studies were performed using fully excised neonatal pig skin (Figure 3), as described previously [25]. Briefly, the ex vivo skin samples were placed on paper tissue sheets soaked in 10 mM phosphate-buffered saline (PBS) (pH 7.4) solution to maintain skin hydration and placed in a weighing boat. To mimic human skin temperature, the pig skin samples were placed in a preheated incubator for 30 min at 32 °C ± 1 °C. The selected MAPs were subsequently inserted into the skin manually using thumb pressure for 30 s. Afterward, a 13 g cylindrical stainless steel block was placed on top of each MAP to keep it in place, and the skin samples were incubated again at 32 °C ± 1 °C for 24 h. Another weighing boat containing 10 mM PBS (pH 7.4) was placed on top to maintain skin hydration. The inserted MAPs were removed from the skin following 24 h of application, and the skin surface was thoroughly cleaned by applying 3 × 1 mL of PBS (pH 7.4) solution and gently wiping with wet paper tissue. The skin at the MAP application site was then visualized using a Leica EZ4 W stereo microscope and stored in Microcentrifuge tubesat −20 °C until further processing and analysis using the validated HPLC–UV method.

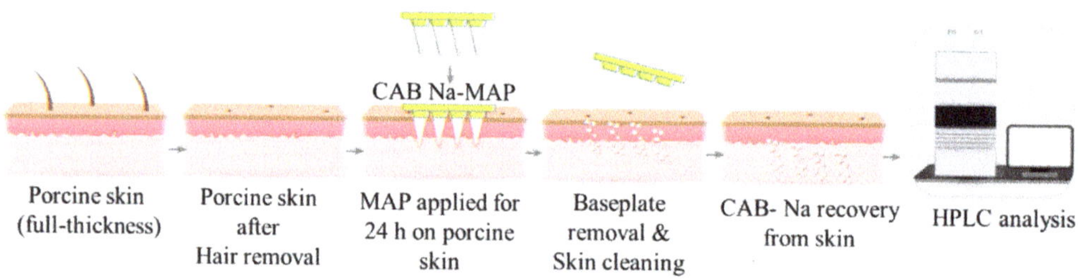

Figure 3. Schematic diagram illustrating the experimental procedures used to evaluate CAB Na deposition ex vivo.

2.6. In Vivo Drug Deposition and Pharmacokinetic Studies

In vivo studies were performed using a Sprague Dawley rat model to (i) evaluate MAP delivery efficiency by conducting drug deposition studies in vivo; (ii) evaluate drug pharmacokinetics from the finalized MAPs following single-dose administration; (iii) evaluate the advantage of potentially initiating therapy by administering a loading dose of CAB intramuscularly, followed by MAP application as a maintenance dose; and (iv) assess MAP tolerance and safety.

These in vivo studies were conducted according to the policy of the Federation of European Laboratory Animal Science Associations and the European Convention for the Protection of Vertebrate Animals Used for Experimental and Other Scientific Purposes, following the principles of the 3Rs (replacement, reduction, and refinement). The studies were performed under Institutional Animal Project License number 2903 and Personal License numbers 1747 and 1892 at the Queen's University Belfast Biological Services Unit after obtaining the required ethical permission from the university's Animal Welfare and Ethics Review Body.

2.6.1. Rat Dose Calculation

The clinical dose of CAB LA equates to 11.42 mg/kg (800 mg per month administered to a person with an average weight of 70 kg) for the initial dose, or 5.71 mg/kg (400 mg per month administered to a person with an average weight of 70 kg) for the maintenance dose. Here, we used a fixed dose per cohort that best approximated the allometric scaling-based calculated theoretical drug dose (as mg/kg) as required per cohort, rather than using a dose as per the rat's weight. This was carried out for three reasons: (i) a large number of animals were used in these comprehensive in vivo studies, and numerous blood samples per cohort were collected over prolonged periods, which necessitated conducting the experiments in several consecutive stages, leading to inevitable inter-cohort rat body weight variations; (ii) the difficulty of administering a rat body weight-tailored drug volume from a highly concentrated CAB LA nanosuspension (200 mg/mL), considering the minimum volume that can be accurately administered to a rat using commercially available syringes is 50 μL;

and (iii) the difficulty of tailoring MAP drug content per the body weight of each rat, as this could have affected the MAP properties. Therefore, we calculated the IM dose per cohort, assuming that the average weight of a rat aged 12 to 18 weeks is 0.25 kg. For MAP cohorts, four MAPs were applied with the maximum possible dose for the respective MAP design and formulation.

2.6.2. In Vivo Study Design and Experimental Procedure

Healthy female Sprague Dawley rats (n = 57) aged 12 to 18 weeks were purchased from Envigo Holdings, Inc. (Huntingdon, UK) and allowed to acclimatize to laboratory conditions for at least 7 days before experimental work was commenced. The rats were split randomly into 16 cohorts: 10 cohorts for the in vivo drug deposition study (cohorts 1 through 10; n = 3 rats per cohort) and 6 cohorts for the in vivo pharmacokinetic studies (cohorts 11 through 16; n = 6 rats per cohort). Cohorts 11 through 13 received a single dose of CAB via either MAP or IM injection, and the resulting plasma levels were measured for 28 days. Cohorts 14 through 16 were used in a multidose study, starting with a single loading dose by MAP or IM injection, followed by a maintenance dose either by MAP or IM injection every week for three weeks. A summary of the experimental details for each cohort is reported in Table 3.

For rats that received the drug by MAP application (cohorts 2 through 10, 12, 13, 15, and 16), to minimize interference with the rat's hair, an electric hair clipper was used to remove the bulk hair, and a depilatory cream was applied to remove any residual hair. To facilitate hair shaving and MAP application, rats were sedated using gas anesthesia (2% to 4% v/v isoflurane in oxygen). This was performed 24 h before MAP application to allow skin integrity to be restored in the event of being affected by the shaving or hair removal cream. The MAPs were then applied manually to the rat's back, followed by applying a pressure-sensitive adhesive tape (Microfoam™ Surgical Tape, 3M, Bracknell, UK) and an occlusive dressing film layer (Tegaderm™, 3M, St. Paul, MN, USA). Finally, kinesiology tape (ProWorks Corporation, Corvallis, OR, USA) was used to wrap the backs of the animals to hold the MAPs in place and provide occlusion. The MAPs were applied for 24 h even though the MAP dissolution time was much shorter because anesthesia was needed to remove the MAP, which could not be administered more than once within 24 h, as per the Institutional Animal Project Licence. Following application for 24 h, the adhesive was removed, and the MAP residuals were gently cleaned by washing the application sites with 1 mL of 10 mM PBS (pH 7.4) solution three times and wiping with prewetted tissue paper.

In the rats used for the in vivo drug deposition studies (cohorts 1 through 10), following 24 h of MAP application, the rats were euthanized, and a blood sample (approximately 200 μL) was collected from each rat in heparin-containing Eppendorf Tubes via cardiac puncture. Additionally, the skin at each MAP application site was harvested individually and stored in Eppendorf Tubes at −80 °C until processing and analysis using the high-performance liquid chromatography-mass spectrometry (HPLC-MS) method [15]. Drug deposition into the in vivo skin was considered cumulatively from the skin-deposited drug and the calculated drug amount from the plasma within 24 h.

In the rats used for in vivo pharmacokinetic studies, a blood sample (approximately 200 μL) was collected from each rat in Eppendorf Tubes (containing 10 μL of heparin) via tail vein bleeds at predefined time intervals over the study periods as per cohort. Plasma was separated by centrifuging the blood at 5000 rpm for 15 min at 4 °C in a refrigerated centrifuge (Sigma® 2-16 K benchtop refrigerated centrifuge, SciQuip Ltd., Shropshire, UK). Plasma samples were collected into microtubes and stored at −80 °C until processing and analysis using the validated HPLC-MS. The obtained data were used to construct CAB concentration versus time profiles (pharmacokinetic profiles).

Table 3. Summary of experimental details for each cohort, including rat weight, drug form, administration route, array design, dosing regimen, administered dose per rat, and study period.

Cohort	Administration Route/Array Design/Dosing Regimen	Study Objective	Rat Weight	Number of MAPs or Injection/Rat	Total Dose/Rat (mg)	Study Period (Days)
1	Intradermal/Single dose	Drug deposition	228 ± 8	2 × 50 µL	2.6 mg	1
2	MAP-D1-F1/Single dose		236 ± 4	4 MAP	10.9 mg	1
3	MAP-D2-F1/Single dose		222 ± 4	4 MAP	4.4 mg	1
4	MAP-D3-F1/Single dose		224 ± 9	4 MAP	7.2 mg	1
5	MAP-D4-F1/Single dose		215 ± 3	4 MAP	7.04 mg	1
6	MAP-D5-F1/Single dose		217 ± 7	4 MAP	13.1 mg	1
7	MAP-D6-F1/Single dose		215 ± 6	4 MAP	11.5 mg	1
8	MAP-D1-F3/Single dose		207 ± 5	4 MAP	4.3 mg	1
9	MAP-D4-F3/Single dose		209 ± 21	4 MAP	2.9 mg	1
10	MAP-D5-F3/Single dose		210 ± 12	4 MAP	4.6 mg	1
11	IM/Single dose	Single dose PK study	218 ± 11	50 µL	2.5 mg	28
12	MAP-D1-F3/Single dose		212 ± 8	4 MAP	4.3 mg	28
13	MAP-D5-F3/Single dose		200 ± 6	4 MAP	4.6 mg	28
14	LD: IM MD: IM	Repeated dose PK study	259 ± 24	LD: 50 µL MD: 50 µL	LD: 2.5 mg MD: weekly 1.25 mg	42
15	LD: IM MD: MAP		274 ± 29	LD: 50 µL MD: 4 MAP	LD: 2.5 mg MD: weekly 4.3 mg	42
16	LD: MAP MD: MAP		269 ±30	LD: 4 MAP MD: 4 MAP	LD: 4.3 mg MD: weekly 4.3 mg	42

Abbreviations: IM, intramuscular; LD, loading dose; MAP, microarray patch; MD, maintenance dose; PK, pharmacokinetic.

2.7. Pharmaceutical Analysis

The samples collected from the above-mentioned studies were processed and analyzed as described in the following subsections.

2.7.1. Preparation of Drug-Containing Samples from the In Vitro Studies for Drug Analysis

Samples were processed as previously described [15]. Briefly, samples collected from the in vitro studies were diluted with an appropriate volume of acetonitrile and vortexed (IKA Vortex 2) for 5 min to ensure complete solubilization of the drug, and then centrifuged at 14,800 rpm for 10 min using Eppendorf MiniSpin™ benchtop centrifuges (Fisher Scientific, Leicestershire, UK). The clear supernatant solution was collected and analyzed for drug content using the HPLC–UV method described in Section 2.4.1.

2.7.2. Preparation of Skin Tissue Samples from Ex Vivo Studies for Drug Analysis

To quantify the amount of drug present in biopsied skin samples from the ex vivo drug deposition studies, two stainless steel beads and 500 µL of deionized water were added to each skin sample after being shredded into small pieces, followed by homogenization for 15 min using a TissueLyser LT homogenizer (Qiagen, Hilden, Germany) to solubilize the remaining CAB MAP shafts deposited in the skin. Subsequently, 1 mL of acetonitrile was added to each sample, and the mixture was homogenized for another 15 min to solubilize the drug again. Each sample was then diluted with 3.5 mL of acetonitrile or acetonitrile/water (1:1, v/v) and vortexed for 5 min. An aliquot of 100 µL of the final skin homogenate was diluted with 900 µL of acetonitrile, vortexed for 30 s, and then centrifuged at 14,800 rpm using an Eppendorf MiniSpin benchtop centrifuge for 10 min. The clear supernatant solution was collected and analyzed for drug content using the validated HPLC–UV method [15].

2.7.3. Preparation of Plasma Samples from In Vivo Studies for Drug Analysis

A total of 300 μL of acetonitrile was added to an aliquot of 100 μL of each plasma sample collected from the in vivo studies (for blood protein precipitation) and vortexed (IKA Vortex 2, at level 6) for 15 min. Samples were then centrifuged at 14,800 rpm for 15 min. Then, 100 μL of supernatant was transferred into an HPLC vial and analyzed for drug content using the previously reported HPLC-MS method [26].

2.8. Pharmacokinetic Parameter Calculation

Data collected from analyzing the blood samples from the in vivo studies were used to construct drug concentration versus time profiles for rats in each cohort individually and then as a cohort after calculating the mean ± standard deviation (SD) of drug concentration for each time point. Noncompartmental pharmacokinetic analysis was carried out for each cohort using Microsoft Excel® 2016 (Microsoft Corporation, Washington, DC, USA) with the PKSolver 2 add-in [27], and key pharmacokinetic parameters (as required per cohort) were calculated.

The relative bioavailability (FR) of CAB from MAPs in comparison with IM injection was calculated using the following equation:

$$\mathrm{FR} = \frac{AUC_{0\text{–}last}(\mathrm{MAP}) \times \mathrm{Dose(IM)}}{AUC_{0\text{–}last}(\mathrm{IM}) \times \mathrm{Dose(MAP)}} \quad (1)$$

where $\mathrm{AUC}_{0\text{–last}}$ (MAP) is the area under the curve ($\mathrm{AUC}_{0\text{–last}}$) obtained following the finalized bilayer CAB MAP application; $\mathrm{AUC}_{0\text{–last}}$ (IM) is the area under the curve ($\mathrm{AUC}_{0\text{–last}}$) obtained following CAB LA administration by IM injection; Dose(IM) is the drug dose administered by IM injection; and Dose(MAP) is the dose administered by MAP.

2.9. Statistical Analysis

All results are expressed as the mean ± SD (n ≥ 3), calculated using Microsoft Excel 2013. Statistical analysis was carried out using GraphPad Prism version 6 (GraphPad Software, La Jolla, CA, USA), followed by Tukey's multiple comparisons post hoc test and *t* test for two-group comparisons. A value of $p < 0.05$ was considered statistically significant.

3. Results and Discussion

3.1. Characterization of MAP Designs for Improved Insertion and Drug Delivery Performance

The objective of this work was to evaluate various strategies to increase MAP delivery efficiency. These strategies included (i) increasing MAP insertion capability by reducing the MAP needle density from 256 needles per array (D1) to 121 needles per array (D2); (ii) changing the shape of the MAP needles to a full conical shape (D3) or to a full pyramidal shape (D4); (iii) reducing the needle height (D5); and (iv) increasing the length and width of the needles themselves (D6). All the CAB Na-loaded dissolving MAPs displayed sharp microprojections on a clean and smooth baseplate following the outlined fabrication procedure.

The drug content of each MAP was evaluated following microscopy characterization of the MAPs. Figure 4a shows that the drug content of the patches varied based on the overall design of the patches, as expected. Patch MAP-D2, which had the second lowest number of needles per array (121 per patch), had the lowest drug loading of all the MAP designs: an overall drug loading of 1100 μg per patch. Doubling the number of needles per patch from 121 (MAP-D2) to 256 (MAP-D1) resulted in an increase in the drug loading of the formulation up to approximately 2720 μg per patch. It was observed that the geometry of the needle (full conical MAP-D3 versus full pyramidal MAP-D4) had no significant impact on the overall drug loading of the patch, as the amount of CAB Na loaded into these patches was similar at approximately 1800 μg per patch. It should be noted that increasing the base width of the needles, from 300 μm (MAP-D2) to 500 μm (MAP-D6) also resulted in an increase in MAP drug loading from 1104 μg to 2870 μg per patch, despite having a lower number of needles per patch (81 needles/patch). Next, an ex vivo skin deposition study was conducted to evaluate the ability of these MAPs with different needle geometries and array sizes to deposit CAB Na into the skin. The appearance of the ex vivo neonatal porcine skin following MAP application is shown in Figure S1. Following skin application, the tip of the needle dissolved, resulting in the deposition of the drug into the skin and the formation of white microchannels that were filled with CAB Na drug particles.

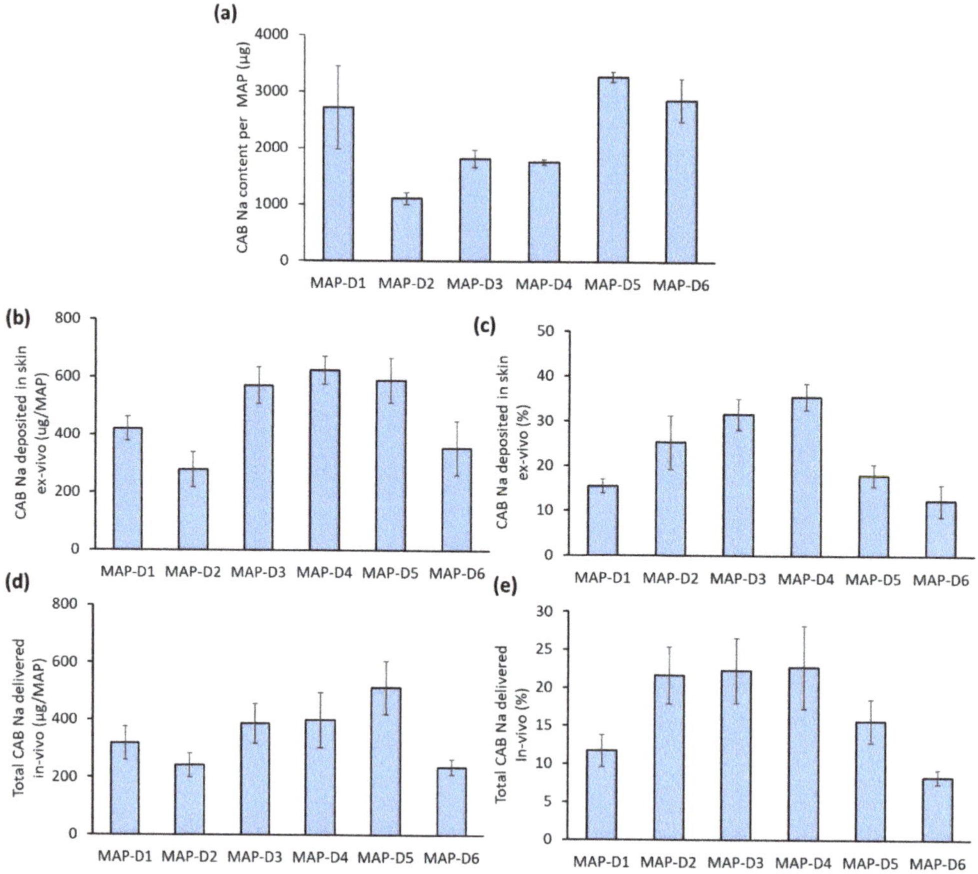

Figure 4. (**a**) Drug loading per patch (µg) for MAPs of different designs and geometries. Data are shown as the mean ± SD ($n \geq 3$). (**b**) Amount and (**c**) delivery efficiency of CAB Na into the ex vivo porcine skin following MAP application. Data are shown as the mean ± SD ($n \geq 4$). (**d**) Amount and (**e**) delivery efficiency of CAB Na into female Sprague Dawley rats. Data are shown as the mean ± SD ($n \geq 8$).

Subsequently, the skin samples were harvested, and the amount of drug deposited was quantified. It can be seen in Figure 4b that formulations D3 through D5 had the highest amount of drug deposited into the skin, resulting in an overall deposition of approximately 600 µg of CAB Na into the skin, while formulations D1 and D6 achieved a skin deposition of approximately 400 µg. In contrast, formulation D2 had the lowest skin deposition, amounting to only 280 µg following a single patch application. When this was viewed in terms of delivery efficiency, as shown in Figure 4c, formulations D3 and D4 had the highest percentage delivery efficiencies (approximately 33%) relative to the other formulations. When the amount of drug delivered ex vivo is viewed in terms of amount (µg) and delivery efficiency (%), it can be seen that overall, MAP designs D3 and D4, which consisted of 256 needles per patch and employed either a full conical or a full pyramidal design with a needle length of 850 μm, resulted in the best deposition of CAB Na into the skin. While the amount of drug delivered from D5 was comparable to D3 and D4, the calculated percentage delivery was low due to higher drug loading and the large patch size (0.75 cm^2) compared to the other designs (0.5 cm^2).

Nevertheless, considering the physiological and mechanical differences between ex vivo and in vivo skin, it is possible to expect discrepancies in the delivery amounts and efficiencies [28]. Anticipating these discrepancies, we proceeded to evaluate formulations D1 through D6 in vivo using female Sprague Dawley rats. Figure 4d shows that overall, the amount of drug deposited into the skin in vivo followed the same trends but was generally lower for all MAP designs relative to ex vivo skin. Nevertheless, we still observed that MAPs D3 through D5 deposited the highest amount of drug into the rat skin: approximately 400 μg/patch. In contrast, D2 and D6 deposited the lowest amount of drug into the skin, achieving only approximately 240 μg/patch. When the data are viewed in terms of delivery efficiencies (Figure 4e), compared to the other MAP designs, D2, D3, and D4 achieved the highest delivery efficiencies: approximately 22%. The lower amount delivered and delivery efficiencies observed in vivo may be attributed to the reduced insertion depth that was able to be achieved in live animals relative to ex vivo skin samples. In a live animal, the skin is still connected to additional adipose and muscle tissues, which may present some level of resistance (as well as inconsistency in texture/firmness) to the overall insertion of the needle in the array [29]. This reduced insertion depth would result in less drug-containing needle length being inserted into the skin, resulting in a reduction in the amount of drug deposited, which is then reflected in an overall reduced delivery efficiency.

It is worth noting that MAP D6 not only exhibited the lowest amount of CAB Na delivered but also had the lowest delivery efficiency of all six MAP designs. This may be attributed to the large width of the needles, which provided more surface area per needle, resulting in more area of contact with the skin during insertion, which would have caused each needle to encounter more frictional resistance during insertion, resulting in a lower insertion depth. Ultimately, this culminated in lower needle dissolution and drug deposition into the skin.

MAP D5 exhibited the highest amount of CAB Na delivered into the rat skin in vivo (500 μg/patch) relative to the other patches. This observation was attributed to the fact that there were more needles per patch, enabling more tips to be inserted into the skin, leading to greater deposition of the drug into the skin. Overall, this first part of the study clearly shows that overall MAP geometry as well as the number of needles per array have a profound impact on the delivery of CAB Na into ex vivo and in vivo skin. Moving forward, designs D1, D3, D4, and D5 were subjected to further refinement to enhance delivery efficiency. In contrast, we decided to exclude D2 from further optimization work due to its low needle per array count, which led to the lowest drug loading (Figure 4a). In addition, design D6 was omitted, as this design presented the lowest drug delivery efficiency during the initial screening study.

3.2. *Improving the MAP Formulation to Enhance MAP Delivery Efficiency*

We then evaluated whether we could optimize the drug composition of the needle for each MAP design to localize the drug into the tip of the patch in an attempt to further improve the delivery efficiency of the formulation. One approach was to alter the method by which the drug-gel mixture was cast into the mold (F1 to F2 in Table 2). In F2, the same CAB Na-loaded hydrogel as in F1 was cast into the molds. However, after positive pressure casting the first layer mixture into the molds and removing the excess gel from the molds, the top part of the first layer hydrogel was removed by washing with deionized water. This was undertaken to localize the drug into the tip of the formulation. This would enable only the drug-loaded portion of the tip to be inserted into the skin during application, thereby enhancing the overall delivery efficiency of the formulation [30]. However, it was observed that this method resulted in significant variability in drug loading between patches, as evidenced by microscopic images (Figure 5a) and the large error bar shown in Figure 5b. Given the inconsistency of this approach to localizing the drug into the tip, we explored whether diluting the gel mixture would help to promote drug localization into the tip of the MAP. When the water content of the drug/polymer gel mixture was increased from 1.0 mL to 1.5 mL, the drug content was reduced from 2895 µg to 1085 µg (F1 versus F3) (Figure 5b). However, when the mixture was diluted further (F4 and F5), we observed that the drug content in the patches plateaued at approximately 750 µg/patch. The mechanical characteristics in Figure 5c demonstrate a decrease in the mechanical strength of the MAP when the hydrogel formulation was used with increasing water content. The reduction in strength was directly proportional to the dilution factor. This can be attributed to a lower amount of polymer filling the needle cavity during the first layer casting, caused by the higher water content compared to hydrogel formulation F1. However, it is important to note that all MAP-D1 samples prepared from different formulations still exhibited good mechanical strength, with a height reduction of less than 11%. Among the formulations, F3 displayed an intermediate drug content (37% of F1) and maintained strong mechanical properties (height reduction < 8%). As a result, F3 was chosen as the preferred formulation to proceed with the preparation of other MAP designs. Based on the approach of localizing CAB Na at the tips of the MAPs, formulation F3 yielded the highest drug loading, as depicted in Figure 5b. Conversely, formulations F2, F4, and F5 showed uneven drug distribution in the MAP needle tips, making them unsuitable for reproducible MAP production (Figure 5a). Hence, formulation F3 is considered the most suitable for further MAP preparation and drug deposition studies.

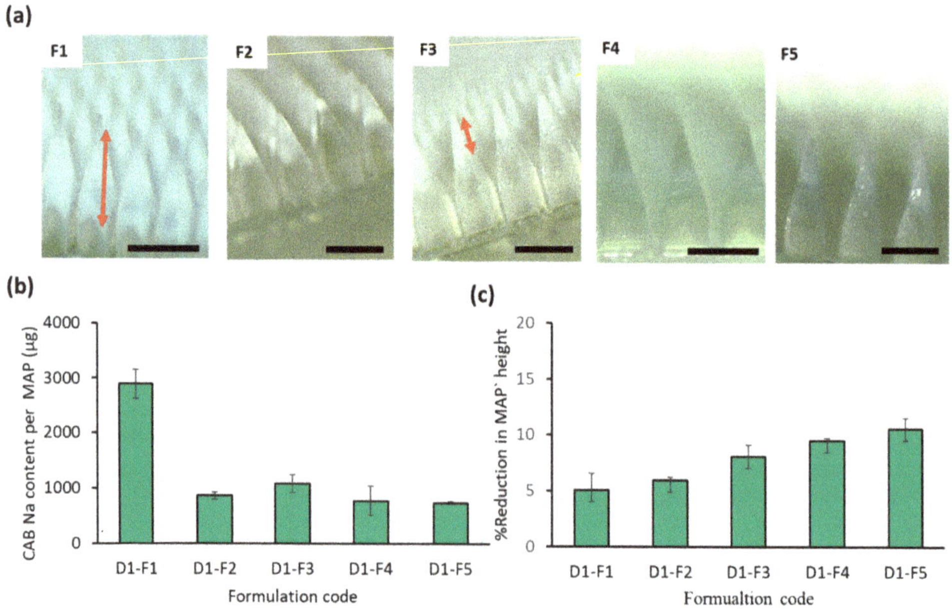

Figure 5. (**a**) Representative light microscope images of MAPs manufactured using diluted CAB Na-hydrogel formulations F1 through F5, with MAP design D1 showing the localization of the drug into the tip (scale bar = 500 µm). Red arrow shows the drug loaded part in the MAP tips (**b**) Drug loading per patch (µg) for MAP-D1 with different hydrogel compositions. Data are shown as the mean ± SD (n ≥ 3). (**c**) Mechanical strength of MAP-D1 prepared by using formulations F1 through F5 (reported as %MAP shaft height reduction after applying a compression force of 32 N against aluminum block following a standardized test using a texture analyzer. Data are reported as mean ± SD (n = 3).

We then proceeded to apply this formulation approach (F3) to the other MAP designs (D4 and D5). The microscopic images in Figure 6a clearly show that selected MAP designs D1, D4, and D5, with hydrogel formulation F1, exhibited complete formation and consistent drug loading across the MAP tips. However, upon changing the formulation from F1 to F3 (Figure 6b), a notable decrease in drug loading within the MAPs was observed for all three selected designs. In general, all MAPs had a drug loading of approximately 0.7 mg to 1.1 mg per patch when F3 was cast into these MAP designs. We did not evaluate MAP-D3 (full conical) any further, as we observed that D3 resulted in poor needle fidelity and integrity post demolding when F3 was cast into this design, as shown in Figure S2.

(a)

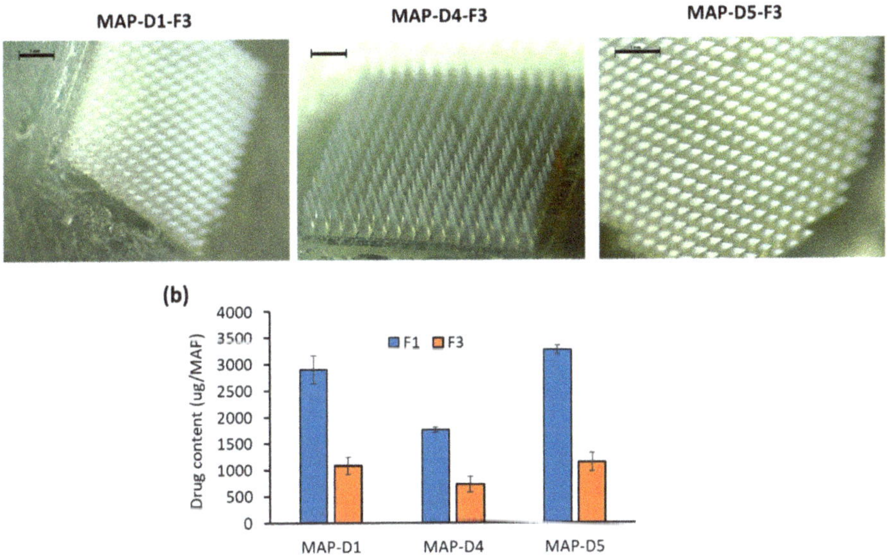

Figure 6. (**a**) Representative light microscope images for the three MAP designs prepared by using formulation F3. (**b**) Comparison of drug content between F1 and F3 for different MAP designs. Data are shown as the mean ± SD (n ≥ 3).

When the formulation was taken forward into a skin deposition study, it can be seen that, overall, formulation F3 resulted in significantly lower ($p < 0.05$) drug deposition into the skin, as shown in Figure 7a. This was anticipated, as the localization of the drug into the tip resulted in an overall reduction in the drug loading of the patches, which ultimately reduced the amount of drug that could be delivered into the skin. Overall, designs D1, D4, and D5 using formulation F3 resulted in a similar amount of drug deposited into the ex vivo skin: 280 µg to 300 µg. However, when these data were viewed in terms of delivery efficiency, as shown in Figure 7b, we could see that diluting the casted hydrogel via F3 augmented the delivery efficiency of the MAPs. This was most apparent for designs D1 and D5. However, it should be noted that the delivery efficiency of MAP D4 was similar between F1 and F3.

In contrast, when the formulations were taken forward in vivo, we observed that the amount of drug delivered using formulations F3 and F1 was similar ($p > 0.05$), which contradicts the finding that we observed using ex vivo skin. Such discrepancies in results may be attributed to several factors. One is the overhydration of ex vivo skin relative to in vivo skin, which may augment the faster dissolution of MAPs when inserted into ex vivo skin that leads to more expulsion of the drug to the skin surface [31,32].

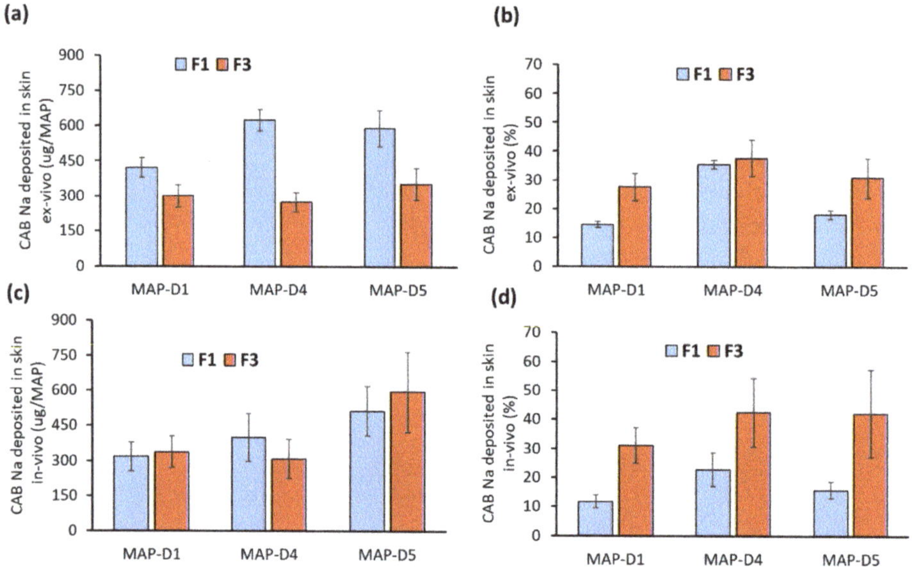

Figure 7. (**a**) Amount and (**b**) delivery efficiency of CAB Na into the ex vivo skin following MAP application. Data are shown as the mean ± SD (n = 5). (**c**) Amount and (**d**) delivery efficiency of CAB Na into female Sprague Dawley rats. Data are shown as the mean ± SD (n = 8).

In this instance, formulation F1, which provided higher drug loading due to the presence of more CAB Na along the needle length, may have undergone partial needle dissolution, mostly in the tip, due to incomplete insertion in vivo. In contrast, as most of the drug was localized in the tips with formulation F3, this portion of the MAP underwent dissolution, which resulted in a similar amount of drug being deposited in vivo. Therefore, when these data are viewed in terms of delivery efficiency, we notice that, overall, F3 had a significantly higher ($p < 0.05$) drug delivery efficiency relative to F1 for designs D1, D4, and D5. Overall, this portion of the work highlights that by diluting the drug/polymer blend that was cast into the molds, we were able to localize the drug into the tip of the MAPs and therefore augment the percentage delivery efficiency of the system, allowing us to reduce the wastage of the undelivered drug.

3.3. *In Vivo Pharmacokinetic Studies: Combination MAP Designs/Hydrogel Formulation to Enhance MAP Delivery Efficiency*

Guided by this series of work, we proceeded to evaluate the combinatorial effect of formulation F3 with two different MAP designs, D1 and D5, on the pharmacokinetic profile of CAB Na when delivered into Sprague Dawley rats in vivo. In addition, another group of rats, administered an IM injection of CAB Na at a dose of 2.5 mg/animal, was used as a control arm, as illustrated in Figure 8a. It should be noted that although design D4 in combination with formulation F3 exhibited a high delivery efficiency of approximately 40%, this formulation, similar to D3, suffered poor MAP fidelity post demolding. Therefore, sufficient quantities of MAP-D4-F3 could not be manufactured for the pharmacokinetic study; only designs D1 and D5 were taken forward for further formulation characterization in vivo.

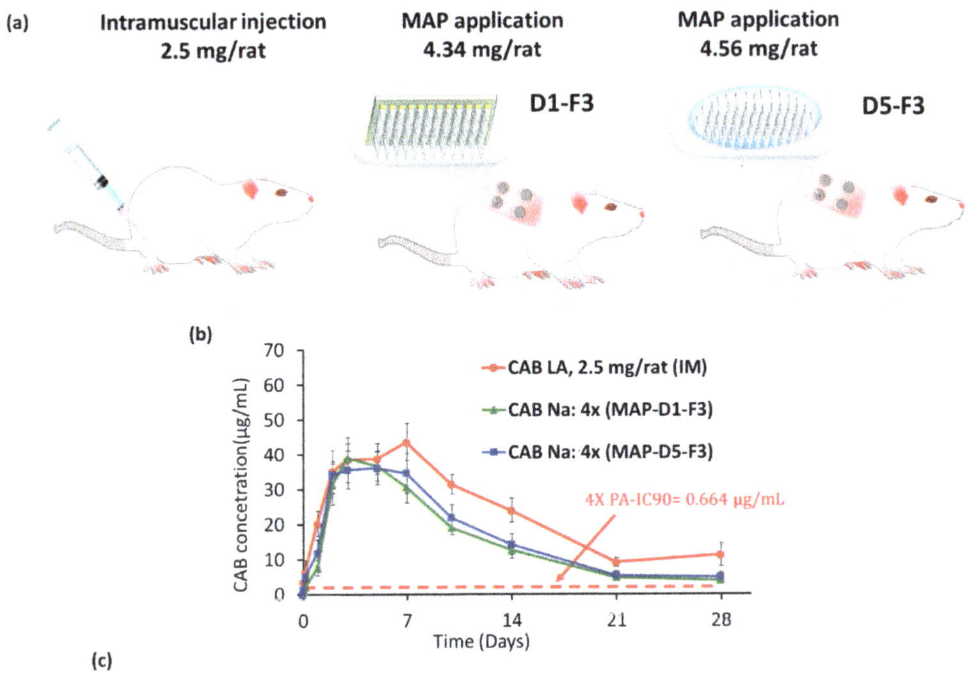

(c)

PK parameters	CAB LA: 2.5 mg/rat (IM)	CAB Na: 4x (MAP-D1-F3)	CAB Na: 4x (MAP-D5-F3)
Dose(mg/rat)	2.5	4.34	4.56
C_{max} (µg/mL)	43.6 ± 5.3	38.95 ± 4.3	36.1 ± 5.5
T_{max} (days)	7	3	5
C_{28days} (µg/mL)	11.2 ± 3.3	3.9 ± 0.8	4.9 ± 1.3
$AUC_{0\text{-}28}$	645 ± 55	430.3 ± 51.4	469 ± 67
Relative bioavailability, FR	1.00	0.34	0.36

Figure 8. (**a**) Schematic of the pharmacokinetic study design consisting of three different experimental cohorts: (i) rats administered an IM injection at a dose of 2.5 mg/rat; (ii) rats administered MAP-D1-F3 at a dose of 4.34 mg/rat; and (iii) rats administered MAP-D5-F3 at a dose of 4.56 mg/rat. (**b**) Plasma profile of CAB following different experimental strategies. (**c**) Pharmacokinetic parameters for the different experimental groups. Data are shown as the mean ± SD (n ≥ 6).

When the MAPs were applied as a single dose in vivo, we observed that the plasma level of CAB Na followed a similar trend to that of the IM injection arm. Although the total dose administered to the animal in the MAP experimental groups was much higher (approximately 4 mg/rat) (the delivered dose based on the in vivo skin deposition study was 1.3 mg and 2.4 mg per rat from MAP-D1-F3 and MAP-D5-F3, respectively) than that in the IM injection group (approximately 2.5 mg/rat), the C_{max} values achieved by all the experimental groups were similar ($p > 0.05$) at approximately 40 µg/mL. Nevertheless, we observed that the T_{max} for the MAP experimental group was much earlier relative to the IM injection group, with MAP-D1-F3 displaying a T_{max} of 3 days. This observation may be attributed to the physiochemical properties of the drug injected or applied to the animal. Regarding the IM injection group, CAB was administered as a free acid drug in the form of a nanosuspension. Injection into the highly vascularized muscle tissue allowed the drug to slowly dissolve and diffuse into the rich muscular vasculature, enabling it to reach systemic circulation [33]. However, the drug is a BCS Class II drug, exhibiting good permeability but poor aqueous solubility, and displays a solubility profile that decreases when the surrounding pH is below pH 10 [34]. Given that the pH of normal resting muscle tissue is 5.99 [35], it can be expected that the injected CAB would exhibit very low solubility, causing the drug particle to slowly dissolve from the injection site before reaching systemic circulation. In contrast, when CAB was delivered into the skin via MAPs, the drug was delivered as a sodium salt (CAB Na), which exhibits higher aqueous solubility relative to the free acid, enabling more rapid dissolution from the application site into systemic circulation [26]. Furthermore, this form of administration would generate multiple micron-sized depots from the microneedles inserted. Relative to one large IM depot, the presence of separate and individual micro-depots generated by MAPs would exhibit a larger total surface area, enabling more rapid drug dissolution into systemic circulation [16]. The combination of these effects may serve as a possible explanation for why the MAP group exhibited a more rapid onset of C_{max} relative to the IM injection group.

When examining the in vivo skin deposition, the amount of drug delivered by MAP-D1-F3 was 31% and by MAP-D5-F3 was 42%, which is a percentage of the total dose loaded into the MAP. This pattern was also reflected in the in vivo pharmacokinetic studies. Notably, the area under the curve (AUC) for both MAP groups was significantly lower ($p < 0.05$) when compared to the IM injection group. This reduction in AUC corresponds to the relative bioavailability; upon comparison, it was found that the bioavailability of both MAP groups was 66% less than that of the IM injection group.

This observation was attributed to the incomplete insertion of the MAPs into the skin, as discussed in Section 3.2, which resulted in the incomplete delivery of the dose applied to the skin. This is an occurrence frequently observed when MAPs are applied to the skin in vivo in animals [36,37]. Such an observation is frequently attributed to the elastic nature of the skin, which resists insertion and complete penetration of the MAPs into the skin [38]. Despite this inherent drawback with MAP application, we still observed the capability of MAPs to deliver CAB concentrations to levels exceeding the IC_{90} values. This is of great importance, as therapeutically relevant levels of CAB ought to be greater than IC_{90} to ensure the drug is capable of exerting an efficacious effect in vivo. In addition, when extrapolating these concentrations from a nonhuman model, $4\times$ IC_{90} is usually preferred to account for any physiological differences between species [38]. As shown in Figure 8b, the plasma levels achieved by both MAP groups were well above $4\times$ IC_{90} at 28 days after a single application.

3.4. In Vivo Pharmacokinetic Studies: Repeated Dosing of D1 and D5 Using F3

Finally, we evaluated the effect of a repeated maintenance dose of the final optimized dissolving MAP on CAB plasma levels relative to IM injection. In this part of the study, we evaluated only MAP-D1-F3. Although MAP-D1-F3 and MAP-D5-F3 had similar drug loading and achieved similar pharmacokinetic profiles following a single MAP application, as shown in Figures 6c and 8b, respectively, formulation MAP-D1-F3 displayed more reproducibility in delivery efficiency, as evidenced by the smaller error bar shown in Figure 7d. From a design perspective, MAP-D5-F3 had 600 needles/patch, which is far greater than MAP-D1-F3, which had only 256 needles/patch. Having a greater number of needles per patch may be advantageous in the case of increasing the drug loading of the system. However, in this case, as the drug loading for both systems was similar when casted with F3, having more needles per array would have been disadvantageous. This can be attributed to the system being more difficult to insert into the skin due to the bed of nails effect with more needles per patch [39]. Additionally, from an end-user perspective, the use of a smaller patch design may be advantageous, as this would enable patients to easily apply and insert the patch into the skin with minimal effort and thus provide reassurance that they applied the formulation correctly. Nevertheless, the ease and usability of these different designs from an end-user perspective would warrant a separate human factors evaluation to identify which design would be most preferable for patients.

In the repeated dosing study, MAP-D1-F3 was given once weekly over the course of one month, and the plasma concentration of CAB Na was monitored over the course of 42 days. In this study design, three cohorts were tested: (i) IM injection at a dose of 2.5 mg/rat as both a loading dose and a maintenance dose once weekly; (ii) IM injection at a dose of 2.5 mg/rat as a loading dose, followed by once-weekly MAP application as a maintenance dose; and (iii) MAP application as a loading dose, followed by once-weekly MAP application as a maintenance dose, as shown in Figure 9a. In cohorts 1 and 2, a similar $C_{max\text{-}1}$ of approximately 60 µg/mL was achieved on day 7. In contrast, cohort 3 exhibited a more rapid $C_{max\text{-}1}$ on day 4 at a lower plasma concentration of approximately 50 µg/mL. When the formulations were administered on a weekly basis, we observed that the plasma concentration of CAB Na managed to achieve pseudosteady-state concentrations after the second dose. Cohorts 1 and 2 achieved pseudosteady-state concentrations of approximately 60 µg/mL, while cohort 3 achieved a much lower steady-state concentration of approximately 30 µg/mL. Although the steady-state concentration achieved by cohort 3 was only half that achieved by cohorts 1 and 2, this concentration was well above 4× IC_{90}. Overall, this repeated dosing pharmacokinetic study not only showed that the MAP-only group was able to achieve a more rapid C_{max}, but the pseudosteady-state concentrations achieved with repeated applications were well above the required plasma concentration needed to achieve viral suppression. Based on the results shown in Figure 9b, it can be postulated that the patches may need to be administered once a week to achieve physiologically relevant plasma concentrations of CAB. Alternatively, when this result is viewed in tandem with the data shown in Figure 8, it is possible to postulate that the CAB Na-loaded dissolving MAPs could also be administered as a monthly dose. However, the reduction in dosing frequency from once weekly to once monthly would necessitate a larger patch size in humans, which is likely to be larger than would be acceptable to potential users [40,41]. Additionally, the reduction in dosing frequency to once a month may also result in a large dynamic range of CAB plasma levels relative to a weekly patch application.

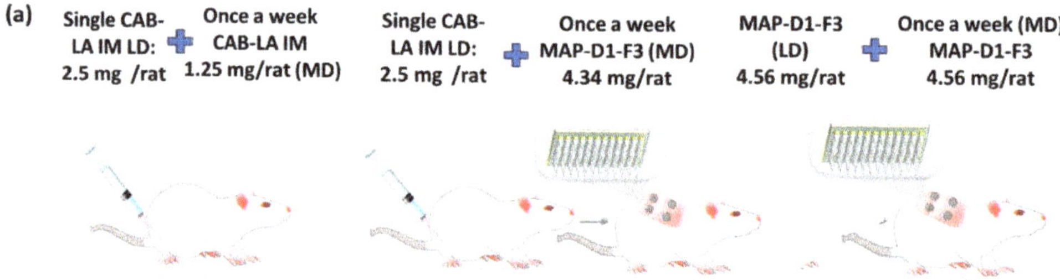

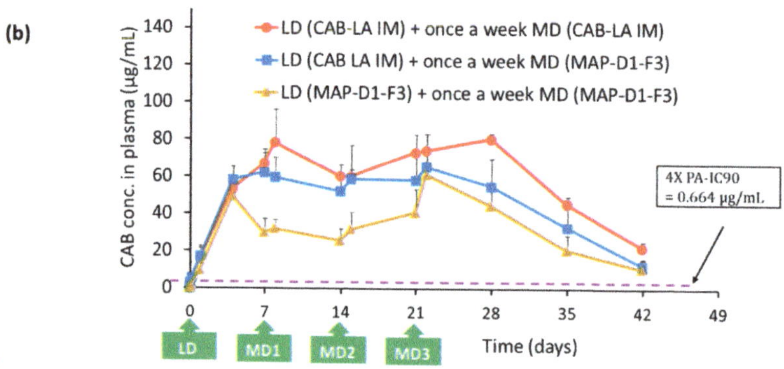

(c)

Recent study	Drug form/Dosage form		
PK parameters	**LD: CAB-LA(IM)+ MD: CAB-LA (IM) (Cohort 1)**	**LD: CAB-LA (IM)+ MD:MAP-D1-F3 (Cohort 2)**	**LD: MAP-D1-F3 MD:MAP-D1-F3 (Cohort 3)**
Applied dose(mg/rat)	LD: 2.5 mg MD: 1.25 mg	LD: 2.5 mg MD:4.34 mg	LD: 4.34 mg MD: 4.34 mg
Rats weight (kg)	0.259 ± 0.024	0.274 ± 0.029	0.269 ± 0.030
C_{max-1} (μg/mL)	67.10± 5.50	62.27 ± 12.45	49.6 ± 9.65
T_{max-1} (days)	7	7	4
C_{max-2} (μg/mL)	78.48± 17.59	59.70 ± 10.37	32.04 ± 4.62
T_{max-s} (days)	8	8	8
$C_{42\ days}$ (μg/mL)	22.48± 2.90	12.64± 3.35	10.92± 4.65

Figure 9. (**a**) Schematic of the pharmacokinetic study design for repeated dosing consisting of three different experimental cohorts: (i) rats administered an IM injection at a loading dose of 2.5 mg/rat, followed by a maintenance dose of 1.25 mg/rat via IM injection once a week; (ii) rats administered a loading dose of 2.5 mg/rat via IM injection, followed by a maintenance dose using a MAP at a dose of 4.34 mg/rat once a week; (iii) rats administered a MAP at a loading dose and a maintenance dose of 4.34 mg/rat once a week. (**b**) Plasma profile of CAB following different administration strategies. (**c**) Pharmacokinetic parameters for the different experimental groups. Data are shown as the mean ± SD (n ≥ 5). C_{max-1} and C_{max-2}: first and second maximum plasma concentration, respectively. T_{max-1} and T_{max-2}: first and second minimum time for C_{max}, respectively.

4. Conclusions

Overall, this work highlights the development of a range of dissolving MAPs of different designs and hydrogel compositions loaded with CAB Na. This library was evaluated both ex vivo and in vivo to develop a dissolving MAP that can efficiently deliver CAB Na for the prevention of HIV. This research shows that a combinatorial approach to design and formulation is needed to develop MAPs with tip-localized CAB Na that can aid in augmenting the delivery efficiency of the MAP. A limitation of this work is that when scaled to humans, a very large patch may be required to deliver the dose needed to elicit a protective effect; larger patches would not necessarily be easy to apply and may not be acceptable for end users. Nevertheless, the patches developed in this work could potentially be useful for pediatric patients who require a much lower dose than adults. Alternatively, the study design may be suitable for more potent ARVs that require a lower dose. Further studies are required to identify whether such a formulation strategy is applicable to the novel anti-HIV compounds that are emerging on the market. Based on the results, along with the advancements made within the MAP field, it could be postulated that future MAPs may co-deliver (side by side) several anti-HIV drugs in a single patch that could be used for ARV treatment of HIV-infected patients. Such a composite system would indubitably simplify the complex regimen to which these patients are strictly required to adhere. Nevertheless, before such a system could have a true impact in the clinic, suitable pharmacokinetics and safety must first be demonstrated in humans.

Supplementary Materials: The following supporting information can be downloaded at: https://www.mdpi.com/article/10.3390/pharmaceutics16010142/s1, Figure S1: Appearance of the skin following the application of the MAP on ex vivo neonatal porcine skin. Figure S2: Representative light microscope images for MAP design D4 manufactured using diluted CAB Na-hydrogel formulation F3. The images show the formation of MAPs with poor needle fidelity post demolding, highlighting the incompatibility of formulation F3 with this MAP design. Insufficient quantities of this MAP design with formulation F3 need to be taken forward into an in vivo pharmacokinetic study.

Author Contributions: Conceptualization, R.F.D., L.K.V. and I.A.T.; Methodology, L.K.V. and I.A.T.; Validation, L.K.V., I.A.T. and F.V.Z.; Formal analysis, L.K.V., I.A.T. and F.V.Z.; Investigation, L.K.V., I.A.T. and F.V.Z.; Resources, R.F.D. and H.O.M.; Data curation, I.A.T. and L.K.V.; Original draft preparation, L.K.V. and A.S.; Review and editing, L.K.V., A.S., R.K M.C., J.M., P.K. and C.J.; Visualization, L.K.V., I.A.T. and R.F.D.; Supervision, R.F.D.; Project administration, R.F.D.; Funding acquisition, R.F.D. All authors have read and agreed to the published version of the manuscript.

Funding: This work was made possible by the generous support of the American people through the United States Agency for International Development (USAID) through the United States President's Emergency Plan for AIDS Relief (PEPFAR) under the terms of Cooperative Agreement #AID-OAA-A-17-00015. The contents are the responsibility of Queen's University Belfast and PATH and do not necessarily reflect the views of USAID, PEPFAR, or the United States government. This project is also supported in part by EPSRC (grant number EP/S028919/1).

Institutional Review Board Statement: Not applicable.

Informed Consent Statement: Not applicable.

Data Availability Statement: All data are included in this manuscript and its Supplementary Files.

Acknowledgments: We thank ViiV Healthcare for supplying both long-acting cabotegravir injectable formulation and micronized cabotegravir sodium and for providing technical input. The authors would also like to thank the PATH team members who contributed to the strategic direction of this work and/or provided review and support, including Ben Creelman, Abra Greene, Annie Rein-Weston, Jill Sherman-Konkle, Maggie Kilbourne-Brook, and Darin Zehrung.

Conflicts of Interest: Ryan F. Donnelly is the inventor of patents that have been licensed to companies developing microneedle-based products and a paid advisor to companies developing microneedle-based products. The potential conflict of interest has been disclosed and is being managed by Queen's University Belfast. The remaining authors declare no conflict of interest.

References

1. Girum, T.; Wasie, A.; Worku, A. Trend of HIV/AIDS for the Last 26 Years and Predicting Achievement of the 90–90-90 HIV Prevention Targets by 2020 in Ethiopia: A Time Series Analysis. *BMC Infect. Dis.* **2020**, *18*, 1–10. [CrossRef] [PubMed]
2. Barré-Sinoussi, F.; Chermann, J.C.; Rey, F.; Nugeyre, M.T.; Chamaret, S.; Gruest, J.; Dauguet, C.; Axler-Blin, C.; Vézinet-Brun, F.; Rouzioux, C.; et al. Isolation of a T-Lymphotropic Retrovirus from a Patient at Risk for Acquired Immune Deficiency Syndrome (AIDS). *Science* **1983**, *220*, 868–871. [CrossRef] [PubMed]
3. GBD 2017 HIV Collaborators. Global, Regional, and National Incidence, Prevalence, and Mortality of HIV, 1980–2017, and Forecasts to 2030, for 195 Countries and Territories: A Systematic Analysis for the Global Burden of Diseases, Injuries, and Risk Factors Study 2017. *Lancet HIV* **2019**, *6*, 831–859. [CrossRef]
4. Gottlieb, M.S.; Schroff, R.; Schanker, H.M.; Weisman, J.D.; Fan, P.T.; Wolf, R.A.; Saxon, A. Pneumocystis Carinii Pneumonia and Mucosal Candidiasis in Previously Healthy Homosexual Men. Evidence of a New Acquired Cellular Immunodeficiency. *J. Urol.* **1982**, *128*, 444. [CrossRef]
5. UNAIDS. *Danger: Unaids Global Aids Update*; UNAIDS: Geneva, Switzerland, 2022.
6. Sidebottom, D.; Ekström, A.M.; Strömdahl, S. A Systematic Review of Adherence to Oral Pre-Exposure Prophylaxis for HIV—How Can We Improve Uptake and Adherence? *BMC Infect. Dis.* **2018**, *18*, 581. [CrossRef]
7. Venkatesan, P. Long-Acting Injectable ART for HIV: A (Cautious) Step Forward. *Lancet Microbe* **2022**, *3*, e94. [CrossRef]
8. WHO. Recommends the Dapivirine Vaginal Ring as a New Choice for HIV Prevention for Women at Substantial Risk of HIV Infection. Available online: https://www.who.int/news/item/26-01-2021-who-recommends-the-dapivirine-vaginal-ring-as-a-new-choice-for-hiv-prevention-for-women-at-substantial-risk-of-hiv-infection (accessed on 3 November 2023).
9. Paik, J. Lenacapavir: First Approval. *Drugs* **2022**, *82*, 1499–1504. [CrossRef]
10. Pons-Faudoa, F.P.; Sizovs, A.; Di Trani, N.; Paez-Mayorga, J.; Bruno, G.; Rhudy, J.; Manohar, M.; Gwenden, K.; Martini, C.; Chua, C.Y.X.; et al. 2-Hydroxypropyl-β-cyclodextrin-enhanced pharmacokinetics of cabotegravir from a nanofluidic implant for HIV pre-exposure prophylaxis. *J. Control Release* **2019**, *306*, 89–96. [CrossRef]
11. Pons-Faudoa, F.P.; Di Trani, N.; Capuani, S.; Campa-Carranza, J.N.; Nehete, B.; Sharma, S.; Shelton, K.A.; Bushman, L.R.; Abdelmawla, F.; Williams, M.; et al. Long-acting refillable nanofluidic implant confers protection against SHIV infection in nonhuman primates. *Sci. Transl. Med.* **2023**, *15*, eadg2887. [CrossRef]
12. Massud, I.; Krovi, A.; Nishiura, K.; Ruone, S.; Li, L.; Holder, A.; Gary, J.; Mills, P.; Mitchell, J.; Khalil, G.; et al. Safety and efficacy of a biodegradable implant releasing tenofovir alafenamide for vaginal protection in a macaque model. *J. Antimicrob. Chemother.* **2022**, *77*, 2964–2971. [CrossRef]
13. Palanee-Phillips, T.; Baum, M.M.; Moss, J.A.; Clark, M.R.; Nuttall, J.; Romano, J.W. Drug-releasing vaginal rings for HIV/STI and pregnancy prevention: A review of recent advances and clinical applications. *Expert Opin. Drug Deliv.* **2022**, *19*, 47–58. [CrossRef] [PubMed]
14. Mgodi, N.M.; Murewanhema, G. Advancing the Use of Long-Acting Extended Delivery Formulations for HIV Prevention in Sub-Saharan Africa: Challenges, Opportunities, and Advancing the Use of Long-Acting Extended Delivery Formulations for HIV Prevention in Sub-Saharan Africa: Challeng. *J. Int. AIDS Soc.* **2023**, *26*, e26115. [CrossRef] [PubMed]
15. Tekko, I.A.; Vora, L.K.; Volpe-zanutto, F.; Moffatt, K.; Jarrahian, C.; Mccarthy, H.O.; Donnelly, R.F. Novel Bilayer Microarray Patch-Assisted Long-Acting Micro-Depot Cabotegravir Intradermal Delivery for HIV Pre-Exposure Prophylaxis. *Adv. Funct. Mater.* **2022**, *32*, 2106999. [CrossRef]
16. Moffatt, K.; Tekko, I.A.; Vora, L.; Volpe-Zanutto, F.; Hutton, A.R.J.; Mistilis, J.; Jarrahian, C.; Akhavein, N.; Weber, A.D.; McCarthy, H.O.; et al. Development and Evaluation of Dissolving Microarray Patches for Co-Administered and Repeated Intradermal Delivery of Long-Acting Rilpivirine and Cabotegravir Nanosuspensions for Paediatric HIV Antiretroviral Therapy. *Pharm. Res.* **2022**, *40*, 1673–1696. [CrossRef] [PubMed]
17. Zhang, C.; Vora, L.K.; Tekko, I.A.; Volpe-zanutto, F.; Peng, K.; Paredes, A.J.; Mccarthy, H.O.; Donnelly, R.F. Development of Dissolving Microneedles for Intradermal Delivery of the Long-Acting Antiretroviral Drug Bictegravir. *Int. J. Pharm.* **2023**, *642*, 123108. [CrossRef] [PubMed]
18. Paredes, A.J.; Ramöller, I.K.; Mckenna, P.E.; Abbate, M.T.A.; Volpe-zanutto, F.; Vora, L.K.; Kilbourne-brook, M.; Jarrahian, C.; Moffatt, K.; Zhang, C.; et al. Microarray Patches: Breaking down the Barriers to Contraceptive Care and HIV Prevention for Women across the Globe. *Adv. Drug Deliv. Rev.* **2021**, *173*, 331–348. [CrossRef] [PubMed]
19. Sabri, A.H.; Kim, Y.; Marlow, M.; Scurr, D.J.; Segal, J.; Banga, A.K.; Kagan, L.; Bong, J. Intradermal and Transdermal Drug Delivery Using Microneedles—Fabrication, Performance Evaluation and Application to Lymphatic Delivery. *Adv. Drug Deliv. Rev.* **2019**, *153*, 195–215. [CrossRef]
20. Sabri, A.; Ogilvie, J.; Abdulhamid, K.; Shpadaruk, V.; McKenna, J.; Segal, J.; Scurr, D.J.; Marlow, M. Expanding the Applications of Microneedles in Dermatology. *Eur. J. Pharm. Biopharm.* **2019**, *140*, 121–140. [CrossRef]
21. Ramöller, I.K.; Volpe-Zanutto, F.; Vora, L.K.; Abbate, M.T.A.; Hutton, A.R.J.; McKenna, P.E.; Peng, K.; Tekko, I.A.; Sabri, A.; McAlister, E.; et al. Intradermal delivery of the antiretroviral drugs cabotegravir and rilpivirine by dissolving microarray patches: Investigation of lymphatic uptake. *J. Control Release* **2024**, *366*, 548–566. [CrossRef]

22. Tekko, I.; Vora, L.; McCrudden, M.T.C.; Jarrahian, C.; Rein-Weston, A.; Zehrung, D.; Giffen, P.; McCarthy, H.O.; Donnelly, R.F. Novel Dissolving Bilayer Microarray Patches as a Minimally Invasive, Efficient Intradermal Delivery System for a Long-Acting Cabotegravir Nanosuspension. In Proceedings of the 2019 Controlled Release Society Annual Meeting & Exposition, Valencia, Spain, 21–24 July 2019.
23. Cordeiro, A.S.; Tekko, I.A.; Jomaa, M.H.; Vora, L.; McAlister, E.; Volpe-Zanutto, F.; Nethery, M.; Baine, P.T.; Mitchell, N.; McNeill, D.W.; et al. Two-Photon Polymerisation 3D Printing of Microneedle Array Templates with Versatile Designs: Application in the Development of Polymeric Drug Delivery Systems. *Pharm. Res.* **2020**, *37*, 174. [CrossRef]
24. Larrañeta, E.; Lutton, R.E.M.; Woolfson, A.D.; Donnelly, R.F. Microneedle Arrays as Transdermal and Intradermal Drug Delivery Systems: Materials Science, Manufacture and Commercial Development. *Mater. Sci. Eng. R Rep.* **2016**, *104*, 1–32. [CrossRef]
25. Tekko, I.A.; Permana, A.; Vora, L.; Hatahet, T.; Mccarthy, H.O.; Donnelly, R.F. Localised and Sustained Intradermal Delivery of Methotrexate Using Nanocrystal-Loaded Microneedle Arrays: Potential for Enhanced Treatment of Psoriasis. *Eur. J. Pharm. Sci.* **2020**, *152*, 105469. [CrossRef] [PubMed]
26. Volpe-Zanutto, F.; Vora, L.K.; Tekko, I.A.; McKenna, P.E.; Permana, A.D.; Sabri, A.H.; Anjani, Q.K.; McCarthy, H.O.; Paredes, A.J.; Donnelly, R.F. Hydrogel-Forming Microarray Patches with Cyclodextrin Drug Reservoirs for Long-Acting Delivery of Poorly Soluble Cabotegravir Sodium for HIV Pre-Exposure Prophylaxis. *J. Control Release* **2022**, *348*, 771–785. [CrossRef]
27. Zhang, Y.; Huo, M.; Zhou, J.; Xie, S. PKSolver: An Add-in Program for Pharmacokinetic and Pharmacodynamic Data Analysis in Microsoft Excel. *Comput. Methods Programs Biomed.* **2010**, *99*, 306–314. [CrossRef] [PubMed]
28. Herkenne, C.; Naik, A.; Kalia, Y.N.; Hadgraft, J.; Guy, R.H. Pig Ear Skin Ex Vivo as a Model for in Vivo Dermatopharmacokinetic Studies in Man. *Pharm. Res.* **2006**, *23*, 1850–1856. [CrossRef]
29. Makvandi, P.; Kirkby, M.; Hutton, A.R.J.; Shabani, M.; Yiu, C.K.Y.; Baghbantaraghdari, Z.; Jamaledin, R.; Carlotti, M.; Mazzolai, B.; Mattoli, V.; et al. *Engineering Microneedle Patches for Improved Penetration: Analysis, Skin Models and Factors Affecting Needle Insertion*; Springer: Singapore, 2021; Volume 13, ISBN 0123456789.
30. Anjani, Q.K.; Hidayat, A.; Sabri, B.; Moreno-castellanos, N.; Utomo, E.; Cárcamo-martínez, Á.; Domínguez-robles, J.; Ahmadi, L.; Wardoyo, H.; Donnelly, R.F. Soluplus®-Based Dissolving Microarray Patches Loaded with Colchicine: Towards a Minimally Invasive Treatment and Management of Gout. *Biomater. Sci.* **2022**, *10*, 5838–5855. [CrossRef] [PubMed]
31. Kurnia, Q.; Hidayat, A.; Sabri, B.; Domínguez-robles, J.; Donnelly, R.F. Metronidazole Nanosuspension Loaded Dissolving Microarray Patches: An Engineered Composite Pharmaceutical System for the Treatment of Skin and Soft Tissue Infection. *Biomater. Adv.* **2022**, *140*, 213073. [CrossRef]
32. Levintova, Y.; Plakogiannis, F.M.; Bellantone, R.A. An Improved in Vitro Method for Measuring Skin Permeability That Controls Excess Hydration of Skin Using Modified Franz Diffusion Cells. *Int. J. Pharm.* **2011**, *419*, 96–106. [CrossRef]
33. Nguyen, V.T.T.; Darville, N.; Vermeulen, A. Pharmacokinetics of Long-Acting Aqueous Nano-/Microsuspensions After Intramuscular Administration in Different Animal Species and Humans—A Review. *AAPS J.* **2023**, *25*, 4. [CrossRef]
34. Kovač, L.; Časar, Z.; Trdan Lušin, T.; Roškar, R. Development of an Analytical Method for Determination of Related Substances and Degradation Products of Cabotegravir Using Analytical Quality by Design Principles. *ACS Omega* **2022**, *7*, 8896–8905. [CrossRef]
35. Carter, N.W.; Rector, F.C.; Campion, D.S.; Seldin, D.W. Measurement of Intracellular PH of Skeletal Muscle with PH-Sensitive Glass Microelectrodes. *J. Clin. Investig.* **1967**, *46*, 920–933. [CrossRef]
36. Paredes, A.J.; Volpe-Zanutto, F.; Vora, L.K.; Tekko, I.A.; Permana, A.D.; Picco, C.J.; McCarthy, H.O.; Donnelly, R.F. Systemic Delivery of Tenofovir Alafenamide Using Dissolving and Implantable Microneedle Patches. *Mater. Today Bio* **2022**, *13*, 100217. [CrossRef] [PubMed]
37. Vora, L.K.; Donnelly, R.F.; Larrañeta, E.; González-Vázquez, P.; Thakur, R.R.S.; Vavia, P.R. Novel Bilayer Dissolving Microneedle Arrays with Concentrated PLGA Nano-Microparticles for Targeted Intradermal Delivery: Proof of Concept. *J. Control Release* **2017**, *265*, 93–101. [CrossRef] [PubMed]
38. Sabri, A.H.; Cater, Z.; Ogilvie, J.; Scurr, D.J.; Marlow, M.; Segal, J. Characterisation of Mechanical Insertion of Commercial Microneedles. *J. Drug Deliv. Sci. Technol.* **2020**, *58*, 101766. [CrossRef]
39. Donnelly, R.F.; Garland, M.J.; Morrow, D.I.J.; Migalska, K.; Raghu, T.; Singh, R.; Majithiya, R.; Woolfson, A.D. Optical Coherence Tomography Is a Valuable Tool in the Study of the Effects of Microneedle Geometry on Skin Penetration Characteristics and In-Skin Dissolution. *J. Control. Release* **2010**, *147*, 333–341. [CrossRef]
40. Gachigua, S.G.; Karuga, R.; Ngunjiri, A.; Jarrahian, C.; Coffey, P.S.; Kilbourne-Brook, M.; Otiso, L. Microarray Patch for HIV Prevention and as a Multipurpose Prevention Technology to Prevent HIV and Unplanned Pregnancy: An Assessment of Potential Acceptability, Usability, and Programmatic Fit in Kenya. *Front. Reprod. Health* **2023**, *5*, 1125159. [CrossRef]
41. Kinvig, H.; Rajoli, R.K.R.; Pertinez, H.; Vora, L.K.; Volpe-Zanutto, F.; Donnelly, R.F.; Rannard, S.; Flexner, C.; Siccardi, M.; Owen, A. Physiologically Based Pharmacokinetic Modelling of Cabotegravir Microarray Patches in Rats and Humans. *Pharmaceutics* **2023**, *15*, 2709. [CrossRef]

pharmaceutics

Article

Physiologically Based Pharmacokinetic Modelling of Cabotegravir Microarray Patches in Rats and Humans

Hannah Kinvig [1,2], Rajith K. R. Rajoli [1,2], Henry Pertinez [1,2], Lalitkumar K. Vora [3], Fabiana Volpe-Zanutto [3], Ryan F. Donnelly [3], Steve Rannard [2,4], Charles Flexner [5], Marco Siccardi [1] and Andrew Owen [1,2,*]

1 Department of Pharmacology and Therapeutics, Institute of Systems, Molecular and Integrative Biology, University of Liverpool, Liverpool L7 3NY, UK; hannah.kinvig@liverpool.ac.uk (H.K.)
2 Centre of Excellence in Long-Acting Therapeutics (CELT), University of Liverpool, Liverpool L7 3NY, UK
3 School of Pharmacy, Queen's University Belfast, Medical Biology Centre, 97 Lisburn Road, Belfast BT9 7BL, UK
4 Department of Chemistry, University of Liverpool, Liverpool L7 3NY, UK
5 Department of Pharmacology and Molecular Sciences, Johns Hopkins School of Medicine, Baltimore, MD 21205, USA; flex@jhmi.edu
* Correspondence: aowen@liverpool.ac.uk; Tel.: +44-(0)-151-794-8211

Abstract: Microarray patches (MAPs) are currently under investigation as a self-administered, pain-free alternative used to achieve long-acting (LA) drug delivery. Cabotegravir is a potent antiretroviral that has demonstrated superior results over current pre-exposure prophylaxis (PrEP) regimens. This study aimed to apply physiologically based pharmacokinetic (PBPK) modelling to describe the pharmacokinetics of the dissolving bilayer MAP platform and predict the optimal dosing strategies for a once-weekly cabotegravir MAP. A mathematical description of a MAP was implemented into a PBPK model, and empirical models were utilised for parameter estimation. The intradermal PBPK model was verified against previously published in vivo rat data for intramuscular (IM) and MAP administration, and in vivo human data for the IM administration of LA cabotegravir. The verified model was utilised for the prediction of 300 mg, 150 mg and 75 mg once-weekly MAP administration in humans. Cabotegravir plasma concentrations >4 × protein-adjusted 90% inhibitory concentration (PA-IC_{90}) (0.664 μg/mL) and >8 × PA-IC_{90} (1.33 μg/mL) were set as targets. The 75 mg, 150 mg and 300 mg once-weekly cabotegravir MAP regimens were predicted to sustain plasma concentrations >4 × PA-IC_{90}, while the 300 mg and 150 mg regimens achieved plasma concentrations >8 × PA-IC_{90}. These data demonstrate the potential for a once-weekly cabotegravir MAP using practical patch sizes for humans and inform the further development of cabotegravir MAPs for HIV PrEP.

Keywords: cabotegravir; HIV; PrEP; long acting; PBPK; MAP

1. Introduction

Microarray patches (MAPs) offer a simple drug delivery method that provides pain-free self-administration and has the potential to provide long-acting (LA) treatment strategies across a range of indications. MAPs are considered minimally invasive and consist of multiple, micron-scale needle-like structures that penetrate the stratum corneum, viable epidermis, and dermis skin layers, delivering drugs through intradermal depot formation [1–4]. There are a wide variety of MAPs currently under investigation due to their advantages over alternative LA administration routes, such as intramuscular or subcutaneous injections. Relevant MAPs centre around solid, coated, hollow, dissolvable, biodegradable, and hydrogel-forming technologies [5], and could play a significant role in low- and middle-income countries (LMICs) because of their minimal manufacturing, storage and transportation costs, as well as their discreetness and ease-of-use [1–4].

MAPs are of particular interest in the prevention and treatment of human immunodeficiency virus (HIV), with patient surveys highlighting the desire for long-acting therapeutic strategies [6]. Approximately 74% of the 37.7 million people living with HIV received antiretroviral therapy in 2020 [7]. Current antiretroviral regimens contain two or three drugs and are predominantly orally administered on a daily basis [8–11]. Additionally, antiretrovirals are utilised as pre-exposure prophylaxis (PrEP) and post-exposure prophylaxis in HIV-negative individuals at high risk of HIV exposure [12]. A distinct advantage of LA strategies vs. daily oral regimens is their ability to overcome adherence issues, which remains a key challenge in HIV [6,13]. Recently, the first long-acting intramuscular (IM) injectable (containing cabotegravir and rilpivirine) was approved by several global regulatory agencies for once-monthly administration. Recent studies have suggested that the concentrations may also support IM administration once every 2 months, but this remains unproven and contentious [14–16].

Cabotegravir is an integrase strand transfer inhibitor that prevents the integration of the host chromosome and viral DNA by inhibiting the integrase enzyme responsible for covalent bond formation [11,17,18]. LA cabotegravir has also demonstrated its efficacy in HIV PrEP and has proven superior to daily oral tenofovir disoproxil fumarate and emtricitabine, particularly in at-risk populations such as cisgender men who have sex with men and transgender women [19–24]. A recently published study investigated a novel, bilayer MAP design containing LA cabotegravir in rats, and demonstrated a promising MAP design with high cabotegravir loading per cm^2 [25]. A MAP intradermal PBPK model was previously reported, assessing the optimum characteristics for the MAP administration of cabotegravir and rilpivirine in HIV-negative humans, and described the movement of the drug from the MAP depot, through the layers of the skin and hair follicles, to its systemic circulation via the blood and lymphatic systems [26]. However, a specific empirical MAP design was not available at that time, and further modelling using published MAP characteristics is therefore warranted.

This study aimed to utilise in vivo rat data describing the cabotegravir pharmacokinetics resulting from dissolving bilayer MAPs for the qualification of a tailored MAP intradermal PBPK model. The qualified model was then used to predict cabotegravir pharmacokinetics in humans after once weekly MAP administration, identifying the optimal doses for sustaining the target plasma concentrations for HIV PrEP.

2. Materials and Methods

A PBPK model was developed for the prediction of cabotegravir pharmacokinetics in Simbiology v5.8, a product of Matlab 2018a (MathWorks, Natick, MA, USA). Excluding the intradermal MAP compartments, drug distribution in the PBPK model was described using blood-flow-limited, first-order kinetics with well-stirred compartments that assumed instant distribution of the drug. Drug distribution in the intradermal MAP compartments was described using permeability-limited, first-order kinetics. Physicochemical, pharmacokinetic, in vitro, and in vivo data for cabotegravir were sourced from the literature or, if unavailable, were estimated using an empirical pharmacokinetic model in the R programming environment v4.0.3 (The R Foundation, Vienna, Austria) and via curve-fitting to observed data in the PBPK model. Where applicable, concentration time profile data were extracted from graphs using the Plot Digitizer Tool v4.5 (WebPlotDigitizer, Pacifica, CA, USA). For PBPK model predictions in rats, virtual cohorts of 100 rats were simulated. For PBPK model predictions in humans, virtual cohorts consisting of 50 male and 50 female patients aged 18–60 years were simulated.

2.1. Human and Rat Physiological Parameters

Human weight, height, body mass index and body surface area for virtual male and female patients aged 18–60 years were described using data from the National Center for Health Statistics [27–29]. Anthropometric equations were used to calculate human organ weight and alongside organ density and organ volume [28,30]. Blood flow rates in humans were determined as fractions of the total cardiac output, with the exception of dermal blood flow. Rat weight was described using mean ± standard deviation data reported for each cohort of rats ($n = 6$) used in the cabotegravir IM and MAP regimens utilised for model verification [25]. Rat organ volume [31], blood flow rates [31] and tissue composition [32] were defined as previously reported. The systemic circulation of cabotegravir for both humans and rats was calculated using differential equations for eliminating and non-eliminating organs, as previously reported [33]. Similarly, for both humans and rats the volume of distribution was calculated as described previously, with the zwitterionic olive oil:buffer partition coefficient ($LogD^*_{vo:w}$) equation being applied for cabotegravir [32].

2.2. MAP Intradermal PBPK Model

The physicochemical and pharmacokinetic input parameters for the simulation of cabotegravir in the MAP intradermal PBPK model are shown in Table 1. Due to the lack of available data, quantitative structure–property relationship (QSPR) equations were used to calculate permeability and partition coefficients for each skin layer, as previously described [26,34]. In order to optimise model parameters describing the nanoparticle release rate (K_{NP}) in the microneedle depot and the rate of drug movement across the skin (K_{SKIN}), plasma concentration data for cabotegravir was initially fitted with an empirical pharmacokinetic model in the R programming environment v4.0.3 (The R Foundation, Vienna, Austria). This fitting made use of the Pracma library and lsqnonlin function for non-linear regression [35]. The empirical model described one-compartment pharmacokinetic disposition for IM administration and two-compartment pharmacokinetic disposition for MAP administration, with first-order input from a dosing depot compartment representing the IM injection or MAP. Descriptions of the one-compartment and two-compartment empirical models can be found in Figures S1 and S2 and Equations (S1) and (S2), in the Supplementary Material, respectively. Parameters describing K_{NP} and K_{SKIN} were then implemented in the PBPK model. MAP K_{NP} and K_{SKIN} values verified in rats were utilised for human MAP predictions. Additionally, the apparent clearance for cabotegravir IM ($CL/F_{IM,rat}$) and cabotegravir MAPs ($CL/F_{MAP,rat}$) were estimated via curve-fitting to available in vivo rat data in the PBPK model. The apparent clearance for cabotegravir MAPs in humans ($CL/F_{MAP,human}$) was estimated, allowing for a 20% bioavailability when compared to the apparent clearance of cabotegravir IM administration in humans ($CL/F_{IM,human}$), as previously reported in rats [25].

Table 1. Physicochemical and pharmacokinetic input parameters for cabotegravir.

Parameter	Cabotegravir	Reference
Molecular Weight (g/mol)	405.358	[36]
HBD	2	[36]
Log $P_{O:W}$	1.04	[36]
pK_a	10.04, −0.7	[36]
Protein Binding (%)	99.8	[37]
PSA (Å2)	99.18	[36]
R	0.5	[37]
Bioavailability (%)	20	[25]
$CL/F_{IM,human}$ (L/h)	0.197	[38]
$K_{np,IM,human}$ (h^{-1})	4.54×10^{-4}	[39]
$PC_{SC/W}$	0.304 *	[26]
$PC_{VE/W}$	0.106 **	[26]
$PC_{DE/W}$	0.106 **	[26,40]
$PC_{SC/VE}$	3.993 *	[26]
$PC_{VE/DE}$	1 **	[40]

HBD—hydrogen bond donor, Log $P_{O:W}$—partition coefficient between octanol and water, pK_a—logarithmic value of the dissociation constant, PSA—polar surface area, R—blood-to-plasma ratio, $CL/F_{IM,human}$—IM apparent clearance in humans, $K_{np,IM,human}$—IM nanoparticle release rate in humans, $PC_{sc/w}$—permeability coefficient between the stratum corneum and water, $PC_{VE/W}$—permeability coefficient between the viable epidermis and water, $PC_{DE/W}$—partition coefficient between dermis and water, $PC_{SC/VE}$—partition coefficient between the stratum corneum and viable epidermis, $PC_{VE/DE}$—partition coefficient between viable epidermis and dermis, *—calculated using QSPR equations as described previously [26], **—viable epidermis and dermis assumed to have similar compositions.

The MAP intradermal PBPK model was adapted from a previously reported model [26]. A schematic representation of the model is shown in Figure 1 and is described using differential Equations (1)–(9). Parameters utilised in Equations (1)–(9) are detailed in Table S1 in the Supplementary Material. The MAP drug dose was divided into the stratum corneum, viable epidermis and dermis layers of the skin based on the volume of the microneedle penetrating each layer. The thicknesses of the stratum corneum, viable epidermis and dermis implemented in the PBPK model were 18 μm, 32 μm and 2040 μm for rats and 17 μm, 47 μm and 2906 μm for humans, respectively [41]. As previously described in vivo, a dose loading of 5.86 mg/cm^2 was implemented, with each microneedle baseplate (0.49 cm^2) consisting of 16 × 16 right rectangular pyramid-shaped microneedles with a drug-free cuboidal base. The cuboidal base had a height of 250 μm with the right rectangular pyramid microneedle tip height equalling 600 μm. The width and length of each microneedle was 300 μm with a microneedle interspacing of 100 μm. A 97% insertion of the right rectangular pyramid microneedle tip was applied alongside no insertion of the cuboidal base [25]. Furthermore, the MAP intradermal PBPK model assumes that (1) only the drug released from the nanoparticle formulation can penetrate the layer of the skin and that this released drug can only travel in a unidirectional manner across different layers of the skin after it has left the microneedle depot; (2) the drug does not enter the hair follicles; (3) the drug enters and leaves the intradermal compartment via the bloodstream; and (4) K_{NP}, permeability coefficients, partition coefficients and K_{SKIN} are constant.

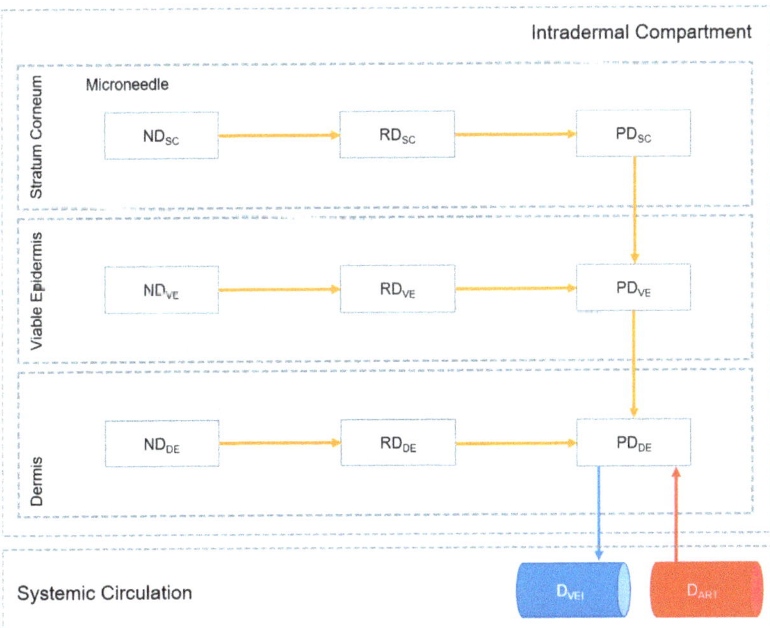

Figure 1. Schematic representation of the drug release pathway implemented in the MAP intradermal PBPK model. ND (nanoparticle drug), RD (released drug), PD (penetrating drug) and D (systemic drug) represent the drug in nanoparticle form in the microneedle, the drug released from its nanoparticle form in the microneedle, the released drug penetrating the layer of the skin and the drug in the systemic circulation, respectively. SC, VE, DE, VEI and ART represent the stratum corneum, viable epidermis, dermis, veins and arteries, respectively.

The drug in the stratum corneum:

$$\frac{dND_{SC}}{dt} = -K_{NP} \times ND_{SC} \quad (1)$$

$$\frac{dRD_{SC}}{dt} = K_{NP} \times ND_{SC} - (PC_{SC/W} \times SA_{MN,SC}) \times \frac{RD_{SC}}{V_{MN,SC}} \quad (2)$$

$$\frac{dPD_{SC}}{dt} = (PC_{SC/W} \times SA_{MN,SC}) \times \frac{RD_{SC}}{V_{MN,SC}} - \left(\frac{1}{PC_{SC/VE}} \times K_{SKIN}\right) \times \frac{PD_{SC}}{V_{SC}} \quad (3)$$

The drug in the viable epidermis:

$$\frac{dND_{VE}}{dt} = -K_{NP} \times ND_{VE} \quad (4)$$

$$\frac{dRD_{VE}}{dt} = K_{NP} \times ND_{VE} - (PC_{VE/W} \times SA_{MN,VE}) \times \frac{RD_{VE}}{V_{MN,VE}} \quad (5)$$

$$\frac{dPD_{VE}}{dt} = \left(\frac{1}{PC_{SC/VE}} \times K_{SKIN}\right) \times \frac{PD_{SC}}{V_{SC}} + (PC_{VE/W} \times SA_{MN,VE}) \times \frac{RD_{VE}}{V_{MN,VE}} - \left(\frac{1}{PC_{VE/DE}} \times K_{SKIN}\right) \times \frac{PD_{VE}}{V_{VE}} \quad (6)$$

The drug in the dermis:

$$\frac{dND_{DE}}{dt} = -K_{NP} \times ND_{DE} \quad (7)$$

$$\frac{dRD_{DE}}{dt} = K_{NP} \times ND_{DE} - (PC_{DE/W} \times SA_{MN,DE}) \times \frac{RD_{DE}}{V_{MN,DE}} \quad (8)$$

$$\frac{dPD_{VE}}{dt} = \left(\frac{1}{PC_{VE/DE}} \times K_{SKIN}\right) \times \frac{PD_{VE}}{V_{VE}} + (PC_{DE/W} \times SA_{MN,DE}) \times \frac{RD_{DE}}{V_{MN,DE}} + Q_{DE} \times \frac{D_{ART}}{V_{ART}} - Q_{DE} \times \frac{D_{VEI}}{V_{DE}/1000} \times \frac{R}{TP_{SKIN}} \quad (9)$$

where ND, RD, PD and D represent the drug in its nanoparticle form in the microneedle, the drug released from its nanoparticle form in the microneedle, the released drug penetrating the layer of the skin and the drug in the systemic circulation, respectively. SC, VE, DE, VEI, ART and MN represent the stratum corneum, viable epidermis, dermis, veins, arteries and microneedle, respectively. SA and V are the surface area and volume, respectively. $PC_{SC/W}$, $PC_{VE/W}$ and $PC_{DE/W}$ are the permeability coefficients between the stratum corneum and water, viable epidermis and water, and dermis and water, respectively. $PC_{SC/VE}$ and $PC_{VE/DE}$ are the partition coefficients between the stratum corneum and viable epidermis and between the viable epidermis and dermis. R and TP_{SKIN} are the blood-to-plasma ratio and tissue-to-plasma partition coefficient of the skin, respectively.

2.3. PBPK Model Verification

The MAP intradermal PBPK model was first verified against 2.5 mg/rat single-dose IM LA cabotegravir in vivo data in rats to validate the cabotegravir physicochemical parameters utilised in the model. The PBPK model was then verified against 11.72 mg/rat single-dose and once-weekly-dose LA cabotegravir MAP in vivo data in rats [25]. Following verification against rat data, human physiological characteristics were implemented into the MAP intradermal PBPK model and were verified against 800 mg single-dose IM LA cabotegravir in vivo data in HIV-negative humans [42]. The PBPK models were considered successfully verified if the ratio of predicted vs. observed pharmacokinetic values for the cabotegravir regimens in rats and humans were between 0.5–2, as per convention [26,43]. In addition, the absolute average fold error (AAFE) for the predicted vs. observed pharmacokinetic parameters and concentration time profiles were also calculated, as defined in Equation (10) [44]. AAFE values between 1 and 2 were considered successfully verified.

$$AAFE = 10^{\left|\frac{1}{N} \Sigma \log \frac{Predicted}{Observed}\right|} \quad (10)$$

2.4. Prediction of Cabotegravir MAP Pharmacokinetics in Humans

An initial MAP dosing regimen of once-weekly 300 mg cabotegravir with a dose loading of 5.86 mg/cm^2 and patch size of 51.2 cm^2 was simulated in humans based on previously reported in vivo studies in rats. MAP dosing regimens of 150 mg (25.6 cm^2) and 75 mg (12.8 cm^2) cabotegravir were also simulated to determine the smallest dose capable of sustaining effective plasma concentrations. K_{NP} and K_{SKIN} estimated in rats were applied in the human MAP intradermal PBPK model. A MAP bioavailability of 20% was applied in the prediction of cabotegravir pharmacokinetics in humans, as previously described [25]. A minimum cabotegravir plasma concentration of 4 × PA-IC_{90} (0.664 μg/mL) and 8 × PA-IC_{90} (1.33 μg/mL) of cabotegravir was targeted over a 6-month period [45].

3. Results

3.1. PBPK Model Verification

The MAP intradermal PBPK model was successfully verified according to the criteria by comparing the predicted pharmacokinetic parameters and concentration time profiles with the observed clinical data for IM and MAP administration. The AAFE and ratio verification results for the predicted cabotegravir pharmacokinetics from single-dose IM and MAP regimens in rats can be found in Tables 2 and 3, respectively. The concentration–time profiles for single-dose IM, single-dose MAP and once-weekly-dose MAP in rats are shown in Figures 2–4, respectively. The AAFE and ratio verification results for the predicted cabotegravir pharmacokinetics from single-dose IM administration in humans can be found in Table 4. The concentration–time profile for single-dose IM administration in humans is shown in Figure 5.

Table 2. Predicted vs. observed cabotegravir pharmacokinetics following single-dose IM administration of 2.5 mg cabotegravir in rats [25].

Parameter	Observed	Predicted	Ratio	AAFE
AUC_{0-28} (μg·day/mL)	1268.1 ± 290.3	1431.1 ± 67.9	1.13	1.13
C_{max} (μg/mL)	74.5 ± 11.6	61.1 ± 2.1	0.82	1.22
C_{28} (μg/mL)	14.5 ± 1.5	20 ± 1.3	1.38	1.38
Profile	-		1.10	-

Ratio data presented as predicted vs. observed. AAFE data were calculated as previously described. Observed clinical data presented as the mean ± standard deviation of six rats with a weight of 241 ± 16 mg [25]. Predicted data presented as the mean ± standard deviation of 100 simulated rats with a weight of 241 ± 16 mg. AUC_{0-28}—area under the curve over 0–28 days, C_{max}—maximum concentration, C_{28}—concentration at day 28, Profile—concentration–time profile, -—not applicable, AAFE—absolute average fold error.

Table 3. Predicted vs. observed cabotegravir pharmacokinetics following single-dose and once-weekly-dose MAP administration of 11.72 mg cabotegravir in rats [25].

Dose (mg)	Parameter	Observed	Predicted	Ratio	AAFE
11.72 (single dose)	AUC_{0-28} (μg·day/mL)	291.1 ± 23.4	304.6 ± 7.6	1.05	1.05
	C_{max} (μg/mL)	18.1 ± 2.3	20.7 ± 0.5	1.15	1.15
	C_{28} (μg/mL)	3.1 ± 1.1	3.3 ± 0.09	1.06	1.06
	Profile	-	-	-	1.12
11.72 (once-weekly dose)	C_{max-1} (μg/mL)	35.1 ± 4.6	28.7 ± 0.7	0.82	1.22
	C_{max-ss} (μg/mL)	60.3 ± 10.7	67.4 ± 1.8	1.12	1.12
	C_{42} (μg/mL)	11.2 ± 2.2	17.9 ± 0.6	1.60	1.60
	Profile	-	-	-	1.04

Ratio data presented as predicted vs. observed. AAFE data were calculated as previously described. Observed clinical data presented as the mean ± standard deviation of six rats with a weight of 284 ± 10 for the single-dose study and 198 ± 8 mg for the once-weekly-dose study [25]. Predicted data are presented as the mean ± standard deviation of 100 simulated rats with weights corresponding to each study. AUC_{0-28}—area under the curve over 0–28 days, C_{max}—maximum concentration, C_{max-1}—maximum concentration for the first dose, C_{max-ss}—maximum concentration at steady state, C_{28}—concentration at day 28, C_{42}—concentration at day 42, Profile—concentration–time profile, -—not applicable, AAFE—absolute average fold error.

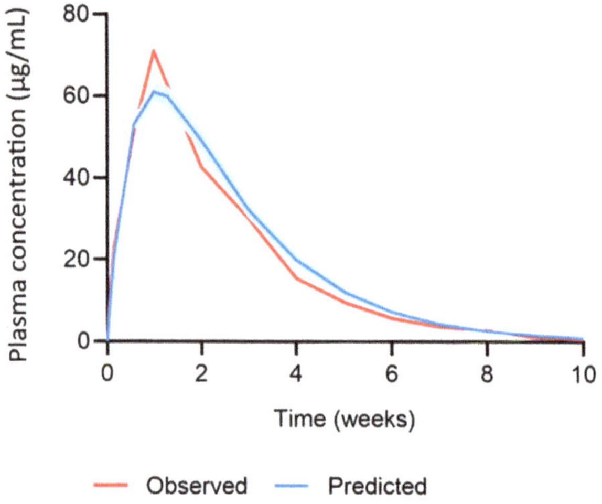

Figure 2. Observed vs. predicted concentration–time profile of cabotegravir after single-dose IM administration of 2.5 mg cabotegravir in rats. The red line represents the mean of the observed clinical data (µg/mL) [25]. The blue line and shaded area represent the mean plasma concentration ± standard deviation (µg/mL) of the predicted data.

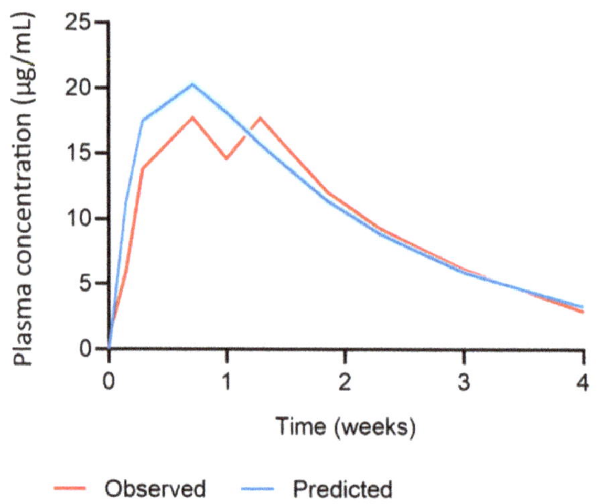

Figure 3. Observed vs. predicted concentration–time profile of cabotegravir after single-dose MAP administration of 11.72 mg cabotegravir in rats. The red line represents the mean of the observed clinical data (µg/mL) [25]. The blue line and shaded area represent the mean plasma concentration ± standard deviation (µg/mL) of the predicted data.

K_{NP} and K_{SKIN} were estimated and implemented in the PBPK models. The K_{NP} for IM and MAP administration in rats were 3×10^{-3} h^{-1} and 3.43×10^{-3} h^{-1}, respectively. The K_{SKIN} for MAP administration was 1.73×10^{-3} cm^3/h. $CL/F_{IM,rat}$ and $CL/F_{MAP,rat}$ were estimated as 0.6 L/h and 0.01 L/h, respectively. $CL/F_{MAP,human}$ was calculated as 0.985 L/h, as previously described [25]. A correction factor was applied to the equations describing the volume of distribution, as per convention. For cabotegravir IM and MAP administration in rats, correction factors of 5 and 0.05 were applied. For cabotegravir IM and MAP administration in humans, a correction factor of 0.01 was applied.

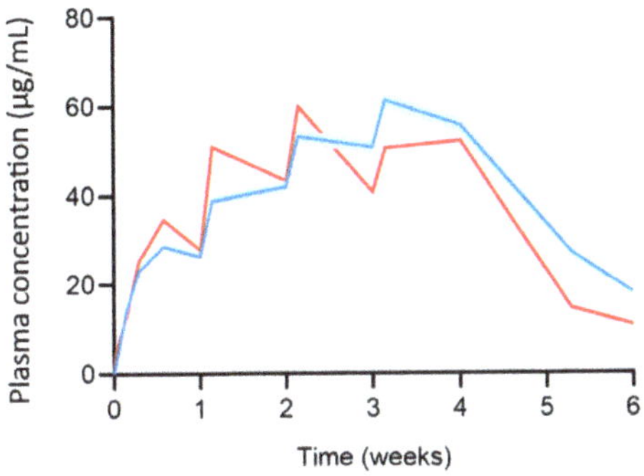

Figure 4. Observed vs. predicted concentration time profile of cabotegravir after once-weekly-dose MAP administration of 11.72 mg cabotegravir in rats. The red line represents the mean of the observed clinical data (µg/mL) [25]. The blue line and shaded area represent the mean plasma concentration ± standard deviation (µg/mL) of the predicted data.

Table 4. Predicted vs. observed cabotegravir pharmacokinetics following single-dose IM administration of 800 mg cabotegravir in humans [25].

Parameter	Observed	Predicted	Ratio	AAFE
$AUC_{0\text{–}w4}$ (µg·h/mL)	1497	1993.7 ± 48.3	1.33	1.33
$AUC_{0\text{–}w12}$ (µg·h/mL)	3851	4792.7 ± 65	1.24	1.24
C_{max} (µg/mL)	3.3	3.4 ± 0.06	1.02	1.02
C_{w4} (µg/mL)	2	2.8 ± 0.02	1.39	1.39
Profile	-	-	-	1.23

Ratio data presented as predicted vs. observed. AAFE data were calculated as previously described. Observed clinical data are presented as the mean of six HIV-negative persons [42]. Predicted data are presented as the mean ± standard deviation of 100 simulated healthy male (50%) and female (50%) patients aged 18–60 years. $AUC_{0\text{–}w4}$—area under the curve over 0–4 weeks, $AUC_{0\text{-}w12}$—area under the curve over 0–12 weeks, C_{max}—maximum concentration, C_{w4}—concentration at week 4, Profile—concentration–time profile, -—not applicable, AAFE—absolute average fold error.

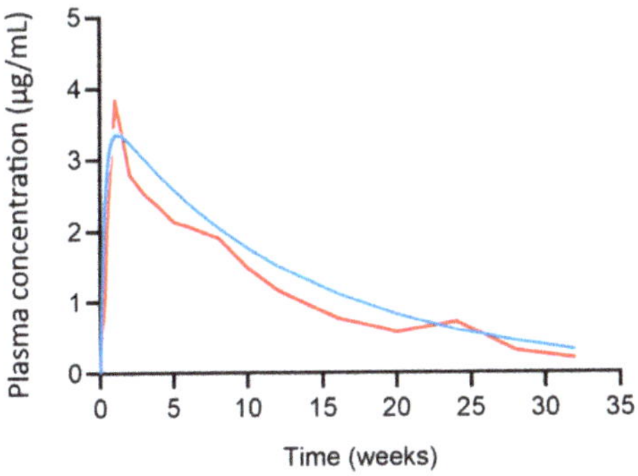

Figure 5. Observed vs. predicted concentration time profile of cabotegravir after single-dose IM administration of 800 mg cabotegravir in humans. The red line represents the mean of the observed clinical data (µg/mL) [42]. The blue line and shaded area represent the mean plasma concentration ± standard deviation (µg/mL) of the predicted data.

3.2. Predicted Cabotegravir MAP Pharmacokinetics in Humans

Human physiological characteristics were integrated into the MAP intradermal PBPK model to simulate cabotegravir pharmacokinetics resulting from the administration of MAPs in humans. A once-weekly cabotegravir MAP dose of 300 mg with a dose loading of 5.86 mg/cm^2 and patch size of 51.2 cm^2 was simulated with a nanoparticle release rate of 3.43×10^{-3} h^{-1} and MAP bioavailability of 20%, as previously described [25]. Additionally, once-weekly cabotegravir MAP doses of 150 mg (25.6 cm^2) and 75 mg (12.8 cm^2) were simulated. A minimum cabotegravir plasma concentration of 4 × PA-IC$_{90}$ (0.664 μg/mL) and 8 × PA-IC$_{90}$ (1.33 μg/mL) of cabotegravir were targeted [45]. The concentration–time profiles for all cabotegravir regimens can be found in Figure 6. The MAP intradermal PBPK model predicted that all cabotegravir MAP dosing regimens achieved cabotegravir concentrations greater than 4 × PA-IC$_{90}$, while only the 150 mg and 300 mg cabotegravir MAP dosing regimens achieved concentrations of cabotegravir greater than 8 × PA-IC$_{90}$. Specifically, cabotegravir MAP doses of 300 mg, 150 mg and 75 mg achieved a minimum concentration at steady state vs. 4 × PA-IC$_{90}$ of 4.33, 2.16 and 1.08, respectively. Cabotegravir MAP doses of 300 mg and 150 mg achieved a minimum concentration at steady state vs. 8 × PA-IC$_{90}$ of 2.16 and 1.08, respectively. However, while the 300 mg cabotegravir MAP sustained plasma concentrations above both minimum targets within 1 day of administration, the 150 mg and 75 mg cabotegravir MAPs did not. Plasma concentrations greater than 4 × PA-IC90 of cabotegravir were achieved 7 days and 28 days post MAP administration for the 150 mg and 75 mg doses, respectively, and plasma concentrations greater than 8 × PA-IC$_{90}$ of cabotegravir were achieved 28 days post MAP administration for the 150 mg weekly dose.

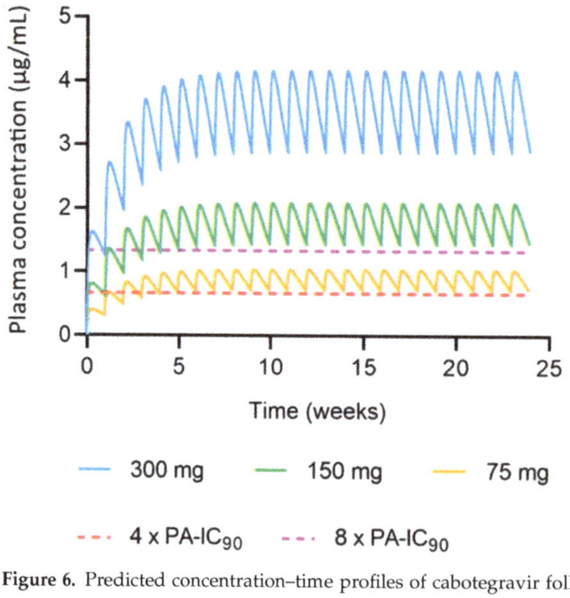

Figure 6. Predicted concentration–time profiles of cabotegravir following once-weekly MAP administration. The blue, green and yellow lines represent the mean plasma concentrations (μg/mL) of 300 mg (51.2 cm^2), 150 mg (25.6 cm^2) and 75 mg (12.8 cm^2) for cabotegravir MAP administration, respectively. The red and purple dashed lines represent the minimum cabotegravir target concentrations of 4 × PA-IC$_{90}$ (0.664 μg/mL) and 8 × PA-IC$_{90}$ (1.33 μg/mL), respectively.

4. Discussion

LA cabotegravir has recently been approved for long-acting IM administration in humans for HIV therapy (with LA rilpivirine) and PrEP (alone). Whilst current IM regimens offer once-monthly and bi-monthly administration [14–16], long-acting treatment options that are needle-free and self-administered are desirable. Several MAP technologies are currently under investigation and could offer a future alternative to oral and IM administration [1–5]. Unlike LA injectable formulations, MAPs are considered minimally invasive but are unlikely to provide an equivalent duration of exposure. Nonetheless, research has demonstrated an advantage to multiple treatment options, and MAPs may have a particular suitability for paediatric drug delivery, overcoming issues of tolerability and taste-masking in this oft under-studied patient group. In this study, in vivo rat data from a dissolving bilayer MAP device were utilised for the development and verification of a MAP intradermal PBPK model for the prediction of LA cabotegravir pharmacokinetics [25]. The PBPK model was used to predict the optimal MAP doses of cabotegravir for once-weekly administration in humans to evaluate cabotegravir MAPs as a potential candidate for PrEP.

The MAP intradermal PBPK model was verified against IM and MAP in vivo data in rats as well as IM clinical data in humans. Specifically, the simulated cabotegravir pharmacokinetics following IM administration in rats were within the conventional verification criteria. This dataset established the suitability of the parameters used to describe cabotegravir pharmacokinetics and rat physiology in the PBPK model. Generally, for IM administration in rats, the PBPK model tended to underpredict the C_{max} and overpredict the AUC, C_{28} and pharmacokinetic profile. The prediction of cabotegravir pharmacokinetics in rats following single-dose and multiple-dose MAP administration achieved the verification criteria, validating the mathematical description of the intradermal MAP in the PBPK model. When considering single-dose administration, all parameters were predicted well, with a tendency for overprediction. For multiple-dose MAP administration, the PBPK model underpredicted the C_{max} for the first dose, whilst the C_{max} at steady state was overpredicted. There was also a general trend in overpredicting the C_{42} and the pharmacokinetic profile. Interestingly, the C_{max} for the first dose was approximately 50% greater than the C_{max} found for the single-dose in vivo rat data, even though the MAP characteristics and dose were identical. This difference in pharmacokinetics creates challenges in PBPK model development and verification, as it is unclear as to whether the large variation is accurate or produced as a result of two differing experiments. Ideally, these studies would be replicated to confirm the current findings. Additionally, changes could be made to the in vivo protocol to verify further observations. For example, when analysing the predicted vs. observed pharmacokinetic profile in Figure 4, the observed C_{max} of the fourth dose is much lower than that of the second and third dose. This generates a discrepancy between the observed C_{max} and that of the predicted data, adversely affecting the ratio and AAFE verification results. The C_{max} observed for dose four may not have been accurately captured by the sampling protocol, whereby blood samples were taken prior to dose four, 24 h post dose and 7 days post dose. Considering the differences between doses, the C_{max} may have occurred in between these sampling timepoints.

Several parameters implemented in the PBPK model were estimated due to the lack of available data and must be taken into consideration when analysing the predicted PBPK model results. The K_{NP} for the IM and MAP administration of cabotegravir in rats as well as the K_{SKIN}, were estimated using an empirical model. The estimated K_{NP} values were similar between both methods of administration, which is justifiable as the same formulations were used in the in vivo studies. A previous MAP intradermal PBPK model found that K_{NP} values between 7×10^{-3} and 9×10^{-3} h^{-1} would provide optimal cabotegravir plasma concentrations for a once-weekly 60 mg dose [26]. The estimated K_{NP} values in this study were approximately two to three times faster, thus highlighting an area in which the current MAP platform could be improved. Furthermore, the K_{SKIN} value was estimated to be 1.73×10^{-3} cm^3/h^{-1}, although additional in vivo and in vitro studies are required to determine if this estimation is accurate. However, this would be particularly challenging as several parameters were estimated in the mathematical description of the drug's movement from the microneedle depot to systemic circulation, generating uncertainty in the reliability and accuracy of the predicted parameter values.

This holds true for the K_{NP} as well as the QSPR calculated partition and permeability coefficients [26,34].

In addition to K_{NP} and K_{SKIN}, $CL/F_{IM,rat}$ and $CL/F_{MAP,rat}$ were estimated in the PBPK model via curve-fitting to the observed clinical data. However, the CL/F estimated in the model for IM and MAP cabotegravir did not reflect the 20% bioavailability previously described [25]. This is the result of a nanoparticle dilution phenomenon reported in the in vivo rat studies. The 2.5 mg LA cabotegravir IM dose was simulated in the PBPK model due to its human-scaled dose being equivalent to the currently recommended 800 mg IM regimen. However, the 2.5 mg IM dose was diluted from the 10 mg dose stock during the in vivo study. Despite a difference in dose, the in vivo study reported similar C_{max} values for the 2.5 mg and 10 mg LA cabotegravir IM regimens. It is thought that this is due to an increase in absorption rate resulting from altered dissolution rates of the LA cabotegravir [25]. Due to a lack of data, the 20% bioavailability used to calculate $CL/F_{MAP,human}$ from $CL/F_{IM,human}$ remains justified, assuming that the LA cabotegravir simulated in humans is not produced via dilution. Lastly, a correction factor was applied to the volume of distribution to optimise model predictions for each regimen simulated in rats and humans, as per convention [46]. Nonetheless, the MAP intradermal PBPK model was verified against the criteria, demonstrating that the estimated parameters were capable of predicting the observed data.

The human physiological characteristics that were integrated into the MAP intradermal PBPK model and cabotegravir pharmacokinetics after IM administration in humans were predicted utilising available pharmacokinetic data. The PBPK model tended to over-predict the AUC, C_{max}, C_{W4} and the concentration–time profile, albeit within the criteria. The PBPK model was applied for the prediction of cabotegravir pharmacokinetics in humans after MAP administration. A previously reported in vivo study in rats suggested that, based on current MAP technologies, the once-weekly MAP administration of cabotegravir following an 800 mg IM cabotegravir lead-in could sustain effective plasma concentrations with an appropriate MAP size [25]. Considering a bioavailability of 20%, this equates to a cabotegravir dose of 300 mg with a MAP size of ~50 cm^2. Following these criteria, the PBPK model predicted that once-weekly MAP doses of cabotegravir between 300 mg and 75 mg (~13 cm^2) could achieve plasma concentrations above the minimum target criteria of 4 × PA-IC_{90} and 8 × PA-IC_{90} of cabotegravir without an IM loading dose. Lower doses were predicted to take longer to achieve these criteria, with doses ≤150 mg reaching 4 × PA-IC_{90} and 8 × PA-IC_{90} of cabotegravir 7 days (75 mg) and 28 days (150 mg) after MAP administration, respectively. Whilst this data present promising results for cabotegravir as a candidate for once-weekly MAP administration, cabotegravir lacks the potency for once-monthly administration based on the high dose required and the impractical MAP sizes resulting from current technologies. Specifically, limited MAP bioavailability and drug loading present significant barriers to the successful implementation of MAP administration across antiretroviral drugs [1–5]. As mentioned, MAP bioavailability has been estimated as 20% when compared to IM administration in rats [25]. Whilst intravenous cabotegravir data would provide a more accurate representation of MAP bioavailability, it is clear that a MAP bioavailability >20% could greatly improve plasma exposure, reducing the MAP's size and increasing MAP administration intervals across a range of antiretroviral drugs. Similarly, drug loading more than 3 mg/cm^2 could produce similar results.

As highlighted, the MAP intradermal PBPK model in this study was developed from a previously published model [26]. Overall, the model described herein represents a simplified description of intradermal MAP pharmacokinetics, limiting the parameter estimation and model assumptions. There are four major differences in comparison to the previously reported model. Firstly, it was assumed that no drug enters the hair follicles. A previous study reported that the inclusion of hair follicles had a minimal impact on PBPK model performance [34]. Additionally, hair follicles have been reported to cover 0.1–20% of total skin, although it is unclear how accurate these values are, especially considering their large variation [34,47,48]. With a negligible role of hair follicles reported for drug transport, it was deemed an unnecessary complication for inclusion at this stage of model development [34,47]. Secondly, the equations used to describe drugs entering the lymphatic system produced little difference in PBPK model predictions [49]. To reduce model complexity and based on current knowledge gaps, the lymphatic system was excluded from the MAP intradermal PBPK model. Thirdly, no data are currently available detailing the rate of movement of the drug from MAPs across the skin layers. Therefore, by assuming that only the free drug can move unidirectionally across skin layers to the systemic circulation, less parameters require estimation, generating a model that favours simplicity. Lastly, in contrast to the right circular cone-shaped microneedle defined in a previous PBPK model [26], a right rectangular pyramid-shaped microneedle with a drug-free cuboidal base was implemented in the PBPK model described herein. Additionally, while acknowledging the physiological differences in skin thickness between males and females, our current analysis utilised mean values for both rats and humans due to data availability constraints.

Research initiatives investigating the above-mentioned assumptions, as well as additional mechanisms such as dermis vascularisation, could help fill knowledge gaps and improve MAP intradermal PBPK models' prediction and reliability. However, future human pharmacokinetic data will be a prerequisite for identifying any inadequacies in the current model.

5. Conclusions

A MAP intradermal PBPK model was developed and verified utilising available rat and human pharmacokinetic data alongside empirical model-informed parameter estimation for the prediction of cabotegravir pharmacokinetics. Based on current MAP technologies, once-weekly cabotegravir doses of 75 mg, 150 mg and 300 mg were predicted to sustain effective plasma concentrations in humans with practical patch sizes between ~13 cm^2 and ~50 cm^2. Our study provides justification for further research into cabotegravir as a candidate for once-weekly MAP administration for HIV PrEP.

Supplementary Materials: The following supporting information can be downloaded at https://www.mdpi.com/article/10.3390/pharmaceutics15122709/s1, Figure S1: Schematic representation of the one-compartment empirical model describing the pharmacokinetic disposition of cabotegravir after intramuscular administration, Figure S2: Schematic representation of the two-compartment empirical model describing the pharmacokinetic disposition of cabotegravir after MAP administration, Table S1: MAP intradermal PBPK model parameters, Equation (S1): two-compartment empirical model, and Equation (S2): one-compartment empirical model. Reference [50] is cited in the Supplementary Materials.

Author Contributions: Conceptualization, A.O., H.K. and M.S.; methodology, H.K.; validation, H.K., H.P. and R.K.R.R.; formal analysis, H.K.; writing—original draft preparation, H.K.; writing—review and editing, H.K., R.K.R.R., H.P., L.K.V., F.V.-Z., R.F.D., S.R., C.F., M.S., A.O.; supervision, M.S. and A.O.; project administration, C.F., R.F.D. and A.O.; funding acquisition, A.O., C.F., M.S. and R.F.D. All authors have read and agreed to the published version of the manuscript.

Funding: This work was funded by the NIH (R24AI118397) and the EPSRC (EP/S028919/1). The authors also acknowledge funding from the NIH (R01AI134091), Unitaid (2020-38-LONGEVITY) and EPSRC (EP/R024804/1; EP/S012265/1).

Institutional Review Board Statement: Not applicable.

Informed Consent Statement: Not applicable.

Data Availability Statement: Data are contained within the article and Supplementary Material.

Conflicts of Interest: A.O. and S.R. are Directors of Tandem Nano Ltd. and co-inventors of patents relating to drug delivery. A.O. has been a co-investigator with funding received by the University of Liverpool from ViiV Healthcare and Gilead Sciences in the past 3 years. A.O. has received personal fees from Gilead and Assembly Biosciences in the past 3 years. S.R. has received research funding from ViiV and AstraZeneca and consultancy from Gilead not related to the current paper.

R.F.D. is an inventor of patents that have been licensed to companies developing microneedle-based products and is a paid advisor to companies developing microneedle-based products. The resulting potential conflicts of interest have been disclosed and are managed by Queen's University Belfast. The companies had no role in the design of the manuscript, in the collection, analyses, or interpretation of the various studies reviewed, in the writing of the manuscript or in the decision to publish. C.F. reports serving as a paid consultant for American Gene Technologies, Gilead Sciences, Janssen Pharmaceuticals, Merck Laboratories, Theratechnologies and ViiV Healthcare, and chairs the Scientific Advisory Board for the Navigen Corporation. He is also a co-inventor on three issued patents related to techniques for the long-acting delivery of antiretroviral drugs. No other conflicts are declared by the authors.

References

1. Li, W.; Li, S.; Fan, X.; Prausnitz, M.R. Microneedle patch designs to increase dose administered to human subjects. *J. Control. Release* **2021**, *339*, 350–360. [CrossRef] [PubMed]
2. Jung, J.H.; Jin, S.G. Microneedle for transdermal drug delivery: Current trends and fabrication. *J. Pharm. Investig.* **2021**, *51*, 503–517. [CrossRef] [PubMed]
3. Dugam, S.; Tade, R.; Dhole, R.; Nangare, S. Emerging era of microneedle array for pharmaceutical and biomedical applications: Recent advances and toxicological perspectives. *Future J. Pharm. Sci.* **2021**, *7*, 19. [CrossRef]
4. Vora, L.K.; Moffatt, K.; Tekko, I.A.; Paredes, A.J.; Volpe-Zanutto, F.; Mishra, D.; Peng, K.; Raj Singh Thakur, R.; Donnelly, R.F. Microneedle array systems for long-acting drug delivery. *Eur. J. Pharm. Biopharm.* **2021**, *159*, 44–76. [CrossRef] [PubMed]
5. Paredes, A.J.; Ramöller, I.K.; McKenna, P.E.; Abbate, M.T.A.; Volpe-Zanutto, F.; Vora, L.K.; Kilbourne-Brook, M.; Jarrahian, C.; Moffatt, K.; Zhang, C.; et al. Microarray patches: Breaking down the barriers to contraceptive care and HIV prevention for women across the globe. *Adv. Drug Deliv. Rev.* **2021**, *173*, 331–348. [CrossRef] [PubMed]
6. Scarsi, K.K.; Swindells, S. The Promise of Improved Adherence with Long-Acting Antiretroviral Therapy: What Are the Data? *J. Int. Assoc. Provid. AIDS Care* **2021**, *20*, 23259582211009011. [CrossRef] [PubMed]
7. UNAIDS. Global HIV & AIDS Statistics—2020 Fact Sheet. Available online: https://www.unaids.org/en/resources/fact-sheet (accessed on 17 February 2021).
8. NIH. FDA-Approved HIV Medicines. Available online: https://hivinfo.nih.gov/understanding-hiv/fact-sheets/fda-approved-hiv-medicines (accessed on 18 February 2021).
9. WHO. *Update of Recommendations on First—And Second-Line Antiretroviral Regimens*; World Health Organization: Geneva, Switzerland, 2019.
10. Menéndez-Arias, L.; Martín-Alonso, S.; Frutos-Beltrán, E. An Update on Antiretroviral Therapy. In *Antiviral Drug Discovery and Development*; Liu, X., Zhan, P., Menéndez-Arias, L., Poongavanam, V., Eds.; Springer: Singapore, 2021.
11. Maartens, G.; Celum, C.; Lewin, S.R. HIV infection: Epidemiology, pathogenesis, treatment, and prevention. *Lancet* **2014**, *384*, 258–271. [CrossRef]
12. CDC. Pre-Exposure Prophylaxis (PrEP) and Post-Exposure Prophylaxis (PEP). Available online: https://www.cdc.gov/hiv/risk/prep/index.html (accessed on 18 February 2021).
13. Cobb, D.A.; Smith, N.A.; Edagwa, B.J.; McMillan, J.M. Long-acting approaches for delivery of antiretroviral drugs for prevention and treatment of HIV: A review of recent research. *Expert Opin. Drug Deliv.* **2020**, *17*, 1227–1238. [CrossRef]
14. FDA. FDA Approves First Extended-Release, Injectable Drug Regimen for Adults Living with HIV. Available online: https://www.fda.gov/news-events/press-announcements/fda-approves-first-extended-release-injectable-drug-regimen-adults-living-hiv (accessed on 19 February 2021).
15. Overton, E.T.; Richmond, G.; Rizzardini, G.; Jaeger, H.; Orrell, C.; Nagimova, F.; Bredeek, F.; Garcia Deltoro, M.; Swindells, S.; Andrade-Villanueva, J.F.; et al. Long-acting cabotegravir and rilpivirine dosed every 2 months in adults with HIV-1 infection (ATLAS-2M), 48-week results: A randomised, multicentre, open-label, phase 3b, non-inferiority study. *Lancet* **2021**, *396*, 1994–2005. [CrossRef]
16. Jaeger, H.; Overton, E.T.; Richmond, G.; Rizzardini, G.; Andrade-Villanueva, J.F.; Mngqibisa, R.; Hermida, A.O.; Thalme, A.; Belonosova, E.; Ajana, F.; et al. Long-acting cabotegravir and rilpivirine dosed every 2 months in adults with HIV-1 infection (ATLAS-2M), 96-week results: A randomised, multicentre, open-label, phase 3b, non-inferiority study. *Lancet HIV* **2021**, *8*, e679–e689. [CrossRef]
17. Pau, A.K.; George, J.M. Antiretroviral therapy: Current drugs. *Infect. Dis. Clin. N. Am.* **2014**, *28*, 371–402. [CrossRef] [PubMed]
18. Wang, Y.; Gu, S.-X.; He, Q.; Fan, R. Advances in the development of HIV integrase strand transfer inhibitors. *Eur. J. Med. Chem.* **2021**, *225*, 113787. [CrossRef]
19. Hodge, D.; Back, D.J.; Gibbons, S.; Khoo, S.H.; Marzolini, C. Pharmacokinetics and Drug-Drug Interactions of Long-Acting Intramuscular Cabotegravir and Rilpivirine. *Clin. Pharmacokinet.* **2021**, *60*, 835–853. [CrossRef] [PubMed]
20. Landovitz, R.J.; Donnell, D.; Clement, M.E.; Hanscom, B.; Cottle, L.; Coelho, L.; Cabello, R.; Chariyalertsak, S.; Dunne, E.F.; Frank, I.; et al. Cabotegravir for HIV Prevention in Cisgender Men and Transgender Women. *N. Engl. J. Med.* **2021**, *385*, 595–608. [CrossRef] [PubMed]

21. Engelman, K.D.; Engelman, A.N. Long-Acting Cabotegravir for HIV/AIDS Prophylaxis. *Biochemistry* **2021**, *60*, 1731–1740. [CrossRef]
22. Lazarus, G.; Wangsaputra, V.K.; Christianto; Louisa, M.; Soetikno, V.; Hamers, R.L. Safety and Pharmacokinetic Profiles of Long-Acting Injectable Antiretroviral Drugs for HIV-1 Pre-Exposure Prophylaxis: A Systematic Review and Meta-analysis of Randomized Trials. *Front. Pharmacol.* **2021**, *12*, 664875. [CrossRef]
23. Bares, S.H.; Scarsi, K.K. A new paradigm for antiretroviral delivery: Long-acting cabotegravir and rilpivirine for the treatment and prevention of HIV. *Curr. Opin. HIV AIDS* **2022**, *17*, 22–31. [CrossRef]
24. Cambou, M.C.; Landovitz, R.J. Challenges and Opportunities for Preexposure Prophylaxis. *Top. Antivir. Med.* **2021**, *29*, 399–406.
25. Tekko, I.A.; Vora, L.K.; Volpe-Zanutto, F.; Moffatt, K.; Jarrahian, C.; McCarthy, H.O.; Donnelly, R.F. Novel Bilayer Microarray Patch-Assisted Long-Acting Micro-Depot Cabotegravir Intradermal Delivery for HIV Pre-Exposure Prophylaxis. *Adv. Funct. Mater.* **2022**, *32*, 2106999. [CrossRef]
26. Rajoli, R.K.R.; Flexner, C.; Chiong, J.; Owen, A.; Donnelly, R.F.; Larraneta, E.; Siccardi, M. Modelling the intradermal delivery of microneedle array patches for long-acting antiretrovirals using PBPK. *Eur. J. Pharm. Biopharm.* **2019**, *144*, 101–109. [CrossRef]
27. National Center for Health Statistics. *Anthropometric Reference Data for Children and Adults: United States, 2011–2014*; National Center for Health Statistics: Washington, DC, USA, 2016.
28. Bosgra, S.; van Eijkeren, J.; Bos, P.; Zeilmaker, M.; Slob, W. An improved model to predict physiologically based model parameters and their inter-individual variability from anthropometry. *Crit. Rev. Toxicol.* **2012**, *42*, 751–767. [CrossRef] [PubMed]
29. Shuter, B.; Aslani, A. Body surface area: Du bois and du bois revisited. *Eur. J. Appl. Physiol.* **2000**, *82*, 250–254. [CrossRef] [PubMed]
30. Brown, R.P.; Delp, M.D.; Lindstedt, S.L.; Rhomberg, L.R.; Beliles, R.P. Physiological parameter values for physiologically based pharmacokinetic models. *Toxicol. Ind. Health* **1997**, *13*, 407–484. [CrossRef] [PubMed]
31. Peters, S.A. *Physiologically-Based Pharmacokinetic (PBPK) Modeling and Simulations: Principles, Methods, and Applications in the Pharmaceutical Industry*, 1st ed.; John Wiley & Sons, Inc.: Hoboken, NJ, USA, 2012.
32. Poulin, P.; Theil, F.P. Prediction of pharmacokinetics prior to in vivo studies. 1. mechanism-based prediction of volume of distribution. *J. Pharm. Sci.* **2002**, *91*, 129–156. [CrossRef] [PubMed]
33. Peters, S.A. Evaluation of a generic physiologically based pharmacokinetic model for lineshape analysis. *Clin. Pharmacokinet.* **2008**, *47*, 261–275. [CrossRef]
34. Gajewska, M.; Worth, A.; Urani, C.; Briesen, H.; Schramm, K.W. Application of physiologically-based toxicokinetic modelling in oral-to-dermal extrapolation of threshold doses of cosmetic ingredients. *Toxicol. Lett.* **2014**, *227*, 189–202. [CrossRef] [PubMed]
35. Borchers, H.W. pracma: Practical Numerical Math Functions. R Package. Version 2.2.9. Available online: https://cran.r-project.org/web/packages/pracma/index.html (accessed on 17 February 2021).
36. DrugBank. Cabotegravir. Available online: https://go.drugbank.com/drugs/DB11751 (accessed on 20 December 2021).
37. ViiV-Healthcare. *Product Monograph Including Patient Medication Information: Vocabria & Cabenuva*; ViiV-Healthcare: Brentford, UK, 2020.
38. Markowitz, M.; Frank, I.; Grant, R.M.; Mayer, K.H.; Elion, R.; Goldstein, D.; Fisher, C.; Sobieszczyk, M.E.; Gallant, J.E.; Van Tieu, H.; et al. Safety and tolerability of long-acting cabotegravir injections in HIV-uninfected men (ECLAIR): A multicentre, double-blind, randomised, placebo-controlled, phase 2a trial. *Lancet HIV* **2017**, *4*, e331–e340. [CrossRef]
39. Rajoli, R.K.R.; Back, D.J.; Rannard, S.; Meyers, C.F.; Flexner, C.; Owen, A.; Siccardi, M. In Silico Dose Prediction for Long-Acting Rilpivirine and Cabotegravir Administration to Children and Adolescents. *Clin. Pharmacokinet.* **2018**, *57*, 255–266. [CrossRef]
40. Kattou, P.; Lian, G.; Glavin, S.; Sorrell, I.; Chen, T. Development of a Two-Dimensional Model for Predicting Transdermal Permeation with the Follicular Pathway: Demonstration with a Caffeine Study. *Pharm. Res.* **2017**, *34*, 2036–2048. [CrossRef]
41. Wei, J.C.J.; Edwards, G.A.; Martin, D.J.; Huang, H.; Crichton, M.L.; Kendall, M.A.F. Allometric scaling of skin thickness, elasticity, viscoelasticity to mass for micro-medical device translation: From mice, rats, rabbits, pigs to humans. *Sci. Rep.* **2017**, *7*, 15885. [CrossRef]
42. Spreen, W.; Ford, S.L.; Chen, S.; Wilfret, D.; Margolis, D.; Gould, E.; Piscitelli, S. GSK1265744 pharmacokinetics in plasma and tissue after single-dose long-acting injectable administration in healthy subjects. *J. Acquir. Immune Defic. Syndr.* **2014**, *67*, 481–486. [CrossRef] [PubMed]
43. Shebley, M.; Sandhu, P.; Emami Riedmaier, A.; Jamei, M.; Narayanan, R.; Patel, A.; Peters, S.A.; Reddy, V.P.; Zheng, M.; de Zwart, L.; et al. Physiologically based pharmacokinetic model qualification and reporting procedures for regulatory submissions: A consortium perspective. *Clin. Pharmacol. Ther.* **2018**, *104*, 88–110. [CrossRef] [PubMed]
44. Riley, R.J.; McGinnity, D.F.; Austin, R.P. A unified model for predicting human hepatic, metabolic clearance from in vitro intrinsic clearance data in hepatocytes and microsomes. *Drug Metab. Dispos.* **2005**, *33*, 1304–1311. [CrossRef] [PubMed]
45. Han, K.; Wannamaker, P.; Lu, H.; Zhu, B.; Wang, M.; Paff, M.; Spreen, W.R.; Ford, S.L. Safety, Tolerability, Pharmacokinetics, and Acceptability of Oral and Long-Acting Cabotegravir in HIV-Negative Chinese Men. *Antimicrob. Agents Chemother.* **2022**, *66*, e0205721. [CrossRef] [PubMed]
46. Kinvig, H. Integrated Experimental and Computational Approaches for the Prediction of Drug-Drug Interactions. Ph.D. Thesis, University of Liverpool, Liverpool, UK, 2021.
47. Exposure Assessment Group. *Dermal Exposure Assessment: Principles and Applications*; US Environmental Protection Agency, Office of Health and Environmental Assessment: Washington, DC, USA, 1992.

48. Otberg, N.; Patzelt, A.; Rasulev, U.; Hagemeister, T.; Linscheid, M.; Sinkgraven, R.; Sterry, W.; Lademann, J. The role of hair follicles in the percutaneous absorption of caffeine. *Br. J. Clin. Pharmacol.* **2008**, *65*, 488–492. [CrossRef]
49. Supersaxo, A.; Hein, W.R.; Steffen, H. Effect of molecular weight on the lymphatic absorption of water-soluble compounds following subcutaneous administration. *Pharm. Res.* **1990**, *7*, 167–169. [CrossRef]
50. Ibrahim, R.; Nitsche, J.M.; Kasting, G.B. Dermal clearance model for epidermal bioavailability calculations. *J. Pharm. Sci.* **2012**, *101*, 2094–2108. [CrossRef]

pharmaceutics

MDPI

Article

Dose-Ranging Plasma and Genital Tissue Pharmacokinetics and Biodegradation of Ultra-Long-Acting Cabotegravir In Situ Forming Implant

Isabella C. Young [1], Allison L. Thorson [2], Roopali Shrivastava [2], Craig Sykes [3], Amanda P. Schauer [3], Mackenzie L. Cottrell [3], Angela D. M. Kashuba [3] and Soumya Rahima Benhabbour [1,2,*]

1 Division of Pharmacoengineering and Molecular Pharmaceutics, UNC Eshelman School of Pharmacy, University of North Carolina at Chapel Hill, Chapel Hill, NC 27599, USA
2 Joint Department of Biomedical Engineering, North Carolina State University and The University of North Carolina at Chapel Hill, Chapel Hill, NC 27599, USA
3 Division of Pharmacotherapy and Experimental Therapeutics, UNC Eshelman School of Pharmacy, University of North Carolina at Chapel Hill, Chapel Hill, NC 27599, USA
* Correspondence: benhabs@email.unc.edu

Abstract: HIV continues to affect millions of men and women worldwide. The development of long-acting injectables for HIV prevention can overcome adherence challenges with daily oral prevention regimens by reducing dosing frequency and stigma. We previously developed an ultra-long-acting injectable, biodegradable, and removeable in situ forming implant (ISFI) with cabotegravir (CAB) that demonstrated protection after multiple rectal SHIV challenges in female macaques. Here, we sought to further characterize CAB ISFI pharmacokinetics (PK) in mice by assessing the effect of dose and number of injections on CAB PK, time to completion of CAB release and polymer degradation, long-term genital tissue PK, and CAB PK tail after implant removal. CAB concentrations in plasma were above the benchmark for protection for 11–12 months with proportionality between dose and drug exposure. CAB ISFI exhibited high concentrations in vaginal, cervical, and rectal tissues for up to 180 days. Furthermore, depots were easily retrievable up to 180 days post-administration with up to 34% residual CAB and near complete (85%) polymer degradation quantified in depots ex vivo. After depot removal, results demonstrated a median 11-fold decline in CAB plasma concentrations across all doses. Ultimately, this study provided critical PK information for the CAB ISFI formulation that could aid in its future translation to clinical studies.

Keywords: biodegradable; long acting; HIV; injectable; pharmacokinetics; polymer

1. Introduction

Globally, there are over 38 million people living with HIV with approximately 1.5 million new infections reported in 2021 [1]. Prior to late 2021, daily oral regimens containing tenofovir and emtricitabine were the only antiretroviral (ARV) prevention methods against HIV. However, effectiveness of daily oral regimens is strictly dependent on high patient adherence [2]. The average adherence to daily HIV pre-exposure prophylaxis (PrEP) regimens is only 70% and suboptimal adherence could lead to the development of drug-resistant virus and lack of protection [2,3]. The development of long-acting technologies for HIV prevention can overcome adherence challenges with daily oral PrEP due to reduced dosing frequency, which can increase patient adherence and reduce stigma [4,5].

In December 2021, the FDA approved the first long-acting injectable containing integrase inhibitor cabotegravir (CAB LA) for HIV PrEP given as bimonthly intramuscular injections (200 mg of CAB/mL, 3 mL). Although CAB LA represents a major progression in the HIV prevention landscape, there are a few limitations with the technology, such as large injection volumes (3 mL), injection site reactions [6,7] and need of physician assistance for administration. Furthermore, a long pharmacokinetic (PK) tail is present due to CAB LA's long half-life (40 days) [8] and the inability to remove the drug once administered to terminate treatment [9]. This PK tail elicits suboptimal but quantifiable levels of CAB in the plasma for over 1 year after the last CAB LA injection, which could cause drug resistant virus or breakthrough infections [9,10].

In situ forming implants (ISFIs) may overcome limitations associated with CAB LA due to their ability to achieve ultra-long-acting drug release, compatibility with many drugs including ARVs, biodegradable nature, and ability to be safely removed via a small incision to terminate treatment, if necessary [11–13]. ISFIs are comprised of a biodegradable and hydrophobic polymer (e.g., poly(lactic-*co*-glycolic acid) (PLGA)), water-miscible organic solvents (e.g., *N*-methyl-2-pyrrolidone (NMP) or dimethyl sulfoxide (DMSO)), and active pharmaceutical ingredients (APIs). When co-formulated, these components generate a syringeable and homogenous liquid suspension for injection into the subcutaneous or intramuscular space. Upon injection, a phase inversion occurs resulting in a solid depot with the API dispersed within the precipitated polymer matrix [14–16]. Drug release from the precipitated implant is driven by diffusion mechanisms or as the polymer degrades over time. PLGA degrades via ester hydrolysis and its degradation rate is largely dependent on molecular weight and its lactic acid to glycolic acid ratio [17]. In our previous ISFI reports, we utilized low molecular weight PLGA, such as 10 kDa or 27 kDa molecular weight, and a lactic acid to glycolic acid ratio of 50:50 to promote a degradation profile of approximately 3 months to ensure that complete polymer degradation occurs prior to subsequent ISFI injections to prevent polymer accumulation [11,13,17,18]. Furthermore, as previously mentioned, ISFIs involve the use of organic solvents within the formulation, such as NMP and DMSO [15,19], which can cause toxicity in large quantities. However, we previously developed an optimized ISFI formulation including NMP and DMSO that showed no local or chronic toxicity when delivered to mice or macaques and solvent quantities were below the median lethal dose (LD_{50}) and the no-observable-effect-level (NOEL) when administered to macaques [13,20–24].

Specifically, we previously developed a subcutaneously administered ultra-long-acting (>6 months) and well-tolerated PLGA-based ISFI with CAB (500 mg/mL) that demonstrated complete protection in female rhesus macaques after multiple rectal simian–human immunodeficiency virus (SHIV) challenges [13]. Furthermore, we showed that CAB ISFIs can be easily retrieved from macaques 90 days post-injection with a rapid decline in CAB plasma concentrations [13]. Reversibility and the ability to remove implants can aid in preventing a long PK tail, which can occur after discontinuation of treatment. The PK tail consists of low levels of drug remaining in the plasma after treatment termination, which could increase the risk for drug-resistance to the virus [9]. However, when CAB ISFIs were retrieved from macaques, complete CAB elimination in plasma was not achieved after depot removal. This was surprising as a previous study with a dolutegravir (DTG) ISFI (analog of CAB) demonstrated complete elimination of DTG in plasma within 7 days after ISFI removal in humanized mice [11]. Furthermore, due to the ultra-long-acting nature of CAB ISFI, the time to completion of CAB release and polymer degradation after injection and long-term vaginal and rectal tissue PK have yet to be established. Ultimately, detailed characterization of CAB ISFIs is needed for further translation to clinical studies.

Thus, we sought to perform a comprehensive dose-ranging PK study in mice with CAB ISFI (500 mg/mL) to investigate (1) CAB ISFI time to completion, (2) long-term (180 days) vaginal and rectal tissue PK, (3) CAB PK tail post-depot removal, (4) implant biodegradation, and (5) the effect of dose (50 μL (1215 mg/kg) vs. 100 μL (2430 mg/kg)) and number of injections (100 μL vs. 2× 50 μL) on CAB in vivo drug release (Figure 1). This study allowed us to further characterize the CAB ISFI formulation and obtain crucial PK information that can aid in establishing optimal formulation and administration parameters for clinical translation.

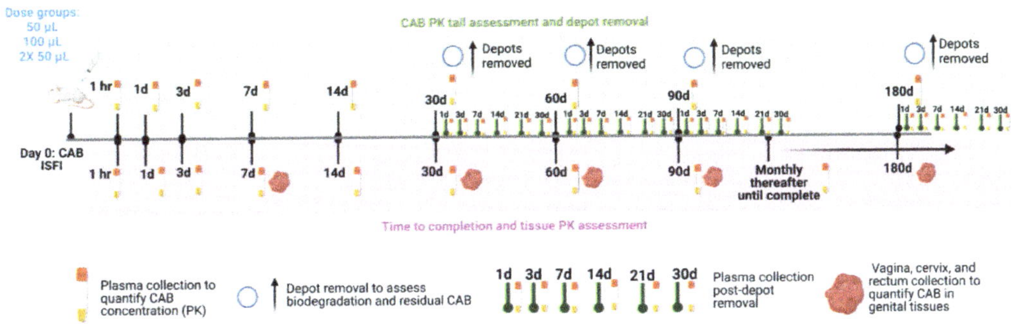

Figure 1. Study plan of dose-ranging CAB ISFI PK. Female BALB/c mice were subcutaneously injected with 50 µL (1215 mg/kg), 100 µL (2430 mg/kg), or 2× 50 µL (2430 mg/kg) of CAB ISFI to assess CAB PK tail and implant biodegradation (n = 6 mice/dose), CAB ISFI time to completion (n = 6 mice/dose), and long-term genital tissue (vagina, cervix, and rectum) PK (n = 3 mice/dose). Figure created with Biorender.com.

2. Materials and Methods

2.1. Materials

Poly(DL-lactide-co-glycolide, PLGA) containing 50:50 lactic acid:glycolic acid was purchased from LACTEL (Birmingham, AL, USA; Cat. No. B6017-1G, Lot# A15-108, molecular weight 10 kDa, inherent viscosity 0.15–0.25 Dl/g). N-methyl-2-pyrrolidone (NMP, USP) was received from ASHLAND (Wilmington, DE, USA; Product Code 851263, 100% NMP). Dimethyl sulfoxide (DMSO, ≥99.7%) was purchased from Sigma-Aldrich (St. Louis, MO, USA; Lot# RNBH5297). HPLC grade acetonitrile (ACN) and water were purchased from Sigma Aldrich (St. Louis, MO, USA). Tetrahydrofuran (THF) was purchased from Sigma Aldrich (St. Louis, Mo, USA; Cat No. SHBF9530V). High purity (≥99%) CAB was purchased from Selleckchem (Houston, TX, USA; Cat. No. S7766).

2.2. High-Performance Liquid Chromatography (HPLC)

A reverse-phase HPLC analysis was performed on an Agilent 1260 HPLC system (Agilent Technologies, Santa Clara, CA, USA) equipped with a Diode Array Detector and an LC pump with autosampler, as previously described [13]. In brief, an Inertsil ODS-3 column (5 µm, 4.6 mm × 150 mm 100 Å, (GL Sciences, Torrance, CA, USA)) maintained at 40 °C was used as the stationary phase, and the mobile phase consisted of 0.1% trifluoroacetic acid in water and ACN. We utilized a flow rate of 1.0 mL/min and a total run time of 25 min per sample. CAB concentration was analyzed at a wavelength of 254 nm with a retention time of 11.4 min. The calibration range was 0.48–250 µg/mL.

2.3. Preparation of ISFI Formulation

The preparation of CAB ISFIs has been previously described [13]. In brief, ISFI formulations for mouse studies were prepared under aseptic conditions in a biosafety cabinet. A 50:50 poly(DL-lactide-co-glycolide) (PLGA) (10 kDa molecular weight) was fully dissolved in the solvent (1:1 weight ratio (*w*/*w*) NMP:DMSO) at a ratio of 1:4 *w*/*w*. Solvents and placebo ISFI formulations were sterile filtered (Sterile Puradisc 13 mm Nylon Syringe Filter, 0.2 µm; Cat# 6786-1302) prior to incorporating the drug in the formulation. Cabotegravir (500 mg/mL) was added to the sterile placebo solution producing a stable ISFI suspension and mixed at 37 °C for 48 h. To ensure homogeneity of the drug suspension, sample aliquots were collected from the drug formulation and dissolved in ACN. Drug concentration was quantified by HPLC analysis.

2.4. Pharmacokinetic (PK) Studies in Female BALB/c Mice

All in vivo studies were conducted with female BALB/c mice (8–10 weeks, Jackson Laboratory). All experiments involving mice were carried out with an approved protocol by the University of North Carolina Animal Care and Use Committee. Housing conditions consisted of a 12 h/12 h light/dark cycle. The ambient temperature was 68–72 degrees Fahrenheit with 30–70% humidity. All injections were administered subcutaneously with a 19 G needle. All samples were stored at −80 °C until PK analysis.

2.4.1. Long-Term Plasma PK and Non-Compartmental Analysis

A 367-day in vivo study was conducted to assess long-term and time-to-completion PK of CAB ISFIs in female BALC/c mice. Mice were subcutaneously injected with 50 µL, 100 µL, or 2× 50 µL of CAB ISFI formulation (n = 6/dose). For this study, 2.5 mL of CAB ISFI formulation was prepared to ensure sufficient volume during injections, and an analytical CAB concentration of 499.1 ± 12.7 mg/mL was determined via HPLC analysis. At 1 h, 1 day, 3 days, 7 days, 30 days, and monthly thereafter, plasma samples were collected longitudinally from mice. It is important to note that CAB plasma concentrations up to 90 days post-injection for mice injected with 50 µL of CAB ISFI have been previously reported [13]. The data in this manuscript have been extended to 367 days post-injection. CAB was extracted from mouse plasma by protein precipitation with methanol containing the stable, isotopically labeled internal standard, 13C,2H2,15N-CAB before analysis by LC-MS/MS. Chromatographic separation was completed on a Waters Atlantis T3 (50 mm × 2.1 mm, 3 µm particle size) analytical column with detection on an AB Sciex API-5000 triple quadruple mass spectrometer. The calibration range was 25–75,000 ng/mL. Calibration standards and quality control samples were within 30% of nominal concentrations.

Area-under-the-curve (AUC) for all injection volumes were calculated using non-compartmental analysis (NCA) in Phoenix WinNonlin (version 8.3) (Certara, L.P., St. Louis, MO, USA). The log-linear trapezoidal method was used to calculate AUCs over the 367-day time course. PK data are represented as median with interquartile range.

2.4.2. Tissue PK

A 180-day study was conducted to assess the distribution of CAB in vaginal, cervical, and rectal tissue in female BALC/c mice. Mice were subcutaneously injected with 50 µL, 100 µL, or 2× 50 µL of CAB ISFI formulation (n = 3/dose per timepoint). For this study, 6.25 mL of CAB ISFI formulation was prepared to ensure sufficient volume during injections, and an analytical CAB concentration of 499.3 ± 8.6 mg/mL was determined via HPLC analysis. At 7, 30, 60, 90, and 180 days post-injection, vaginal, cervical, and rectal tissues were collected from each mouse, weighed, and stored at −80 °C until PK analysis. Weighed tissues were transferred into Precellys® hard tissue reinforced metal bead kit tubes (Cayman Chemical Company, Ann Arbor, MI, USA) containing 1 mL of 70:30 acetonitrile:water, homogenized, and then centrifuged. Following protein precipitation extraction with the isotopically labeled internal standard, 13C,2H2,15N-CAB, CAB was separated using reverse-phase chromatography via a Waters Atlantis T3 (50 mm × 2.1 mm, 3 µm particle size) analytical column. An AB Sciex API-5000 triple quadruple mass spectrometer was used to detect the analyte and internal standard under positive ion electrospray conditions. Precision and accuracy of the calibration standards and quality control samples were within 15% for the dynamic range of 1.00–4000 ng/mL. Final concentrations were normalized to tissue mass and reported in ng/g.

2.4.3. Depot Removal PK and Assessment of PK Tail

A 180-day in vivo study was conducted to assess plasma concentrations after CAB ISFI removal, residual CAB, and biodegradation of PLGA after ISFI removal in female BALB/c mice. Mice were subcutaneously injected with 50 µL, 100 µL, or 2× 50 µL of CAB ISFI formulation (n = 6/dose per depot removal timepoint). For this study, 9 mL of CAB ISFI formulation was prepared to ensure sufficient volume during injections and an analytical CAB concentration of 497.2 ± 12.6 mg/mL was determined via HPLC analysis. At 1 h, 1 day, 3 days, 7 days, 30 days, 60 days, 90 days, and 180 days (n = 6 mice/timepoint per dose), plasma samples were collected longitudinally from mice. At days 30, 60, 90, and 180 (n = 6 mice/timepoint per dose), CAB ISFI was carefully removed via a small skin incision at the injection site, flash frozen, and stored at −80 °C until further processing (Sections 2.5 and 2.6). Plasma samples (n = 6 mice/timepoint per dose) were collected at 1 day, 3 days, 7 days, 14 days, 21 days, and 30 days post-depot removal to assess CAB elimination (i.e., PK tail) over time. Plasma concentrations were analyzed as described in Section 2.4.1 above.

2.5. Residual Drug Quantification

CAB ISFIs were extracted to quantify the amount of CAB remaining in the implant after removal from mice at days 30, 60, 90, and 180 post-administration (n = 3/timepoint per dose). Each implant was massed to assess mass loss over time. To quantify residual drug concentration, implants were dissolved in ACN to extract residual CAB. Residual CAB from the implant was quantified by HPLC analysis. It is important to note that CAB residual drug measurements up to 90 days post-injection for mice injected with 50 µL of CAB ISFI have been previously reported [13]. The data in this manuscript have been extended to 180 days post-injection.

2.6. Gel Permeation Chromatography (GPC)

Gel permeation chromatography (GPC) was conducted as previously described [13]. In brief, GPC was conducted to determine the change in polymer molecular weight over time for ISFIs that were retrieved from mice at days 30, 60, 90, and 180 post-injection (n = 2–3/timepoint per dose). GPC analysis on neat polymer and CAB ISFIs was performed on a Tosoh Biosciences EcoSEC Elite HLC-8420 with a TSKgel GMH-M column (7.8 mm × 30 cm with a pore size of 5 microns). Tetrahydrofuran (THF) was selected as the mobile phase with a flow rate of 0.5 mL/min. Molecular weight was calculated relative to polystyrene standards and analyzed by refractive index detection. It is important to note that polymer molecular weight analysis up to 90 days post-administration for mice injected with 50 µL of CAB ISFI has been previously reported [13]. The data in this manuscript have been extended to 180 days post-injection.

2.7. Statistical Analysis

Statistical analyses were performed in GraphPad Prism version 9.4 (GraphPad Software, Inc., La Jolla, CA, USA). A Kruskal–Wallis one-way ANOVA followed by a Dunn's multiple comparisons test was performed to analyze differences in drug exposure (AUC) in plasma across doses and number of injections. To assess differences in CAB concentration in vaginal, cervical, and rectal tissues across doses and number of injections, a two-way ANOVA and Tukey's multiple comparisons test were conducted with respect to dose and timepoint. A p value of <0.05 (95% confidence level) was deemed significant.

3. Results and Discussion

3.1. Dose-Ranging Ultra-Long-Acting PK

Dose-ranging ultra-long-acting PK studies in female BALB/c mice were conducted over 367 days to assess the duration of complete CAB release across different injection volumes and doses (50 µL (1215 mg/kg), 100 µL (2430 mg/kg), and 2× 50 µL (2430 mg/kg)). Mice received a subcutaneous injection of 50 µL, 100 µL, or 2× 50 µL (n = 6 mice/dose) of CAB ISFI (500 mg/mL). CAB plasma concentrations for each dose were analyzed and plotted over 367 days (Figure 2). Mice elicited average plasma concentrations of CAB well above the 4× protein adjusted 90% inhibitory concentration (4× PA-IC_{90}, 664 ng/mL) for 11 months when injected with 50 µL of CAB ISFI and 12 months when injected with 100 µL or 2× 50 µL of CAB ISFI. Due to the longer than expected duration of the CAB ISFI, time to completion was not determined and this study was terminated at 367 days.

However, mice that were injected with 50 µL of CAB ISFI elicited a 19-fold decline in CAB plasma concentrations between day 330 and 367, exhibiting average concentrations near the 1× PA-IC_{90} (166 ng/mL). We could expect the 100 µL and 2× 50 µL of CAB ISFI groups to follow a similar trend if the study continued for an additional couple of months.

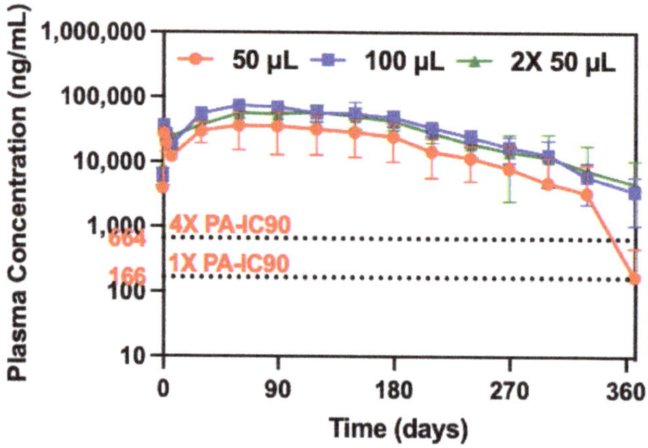

Figure 2. Ultra-long-acting CAB ISFI PK and overall drug exposure. Plasma concentrations of CAB after 50 µL, 100 µL, or 2× 50 µL injection of CAB ISFI in female BALB/c mice for 367 days. Data are presented as average ± standard deviation of n = 5–6 mice/timepoint per dose. Individual replicates are shown in Figure S1. CAB plasma concentration data for mice injected with 50 µL of CAB ISFI up to 90 days have been previously reported [13] (Reprinted with permission from Ref. [13]. 2023, Young, IC & Massud, I et al. Nature Communications).

Furthermore, the results demonstrated proportionality between dose (1215 mg/kg (50 µL) vs. 2430 mg/kg (100 µL)) and overall CAB exposure (AUC) (Table 1). Additionally, AUCs were not significantly different between the number of injections with the same dose (1 × 100 µL vs. 2× 50 µL) ($p = 0.3372$) (Table 1). Ultimately, these data suggest that (1) CAB ISFI can provide CAB plasma concentrations above the protective benchmark (4× PA-IC_{90}) for 11–12 months reducing dosing frequency compared with marketed products and (2) CAB ISFI dose is approximately proportional to overall drug exposure and there is no difference in drug exposure when administering the same dose with multiple injections. Knowing these characteristics of the formulation, if CAB ISFI were to fall below the protective benchmark in large animal or human studies, an increase in injection volume could provide the ability to reach protective plasma levels due to proportionality with drug exposure.

Table 1. CAB ISFI median AUCs over 367 days for all doses and injections.

Group	Dose (mg)	Median $AUC_{0\to367}$ (day × ng/mL)	25% Percentile	75% Percentile
50 µL	25	7,222,125	4,375,526	9,306,997
100 µL	50	14,731,084	13,075,480	15,226,388
2× 50 µL	50	12,634,860	11,474,641	12,906,369

3.2. Ultra-Long-Acting Genital Tissue PK

CAB concentrations were measured in vaginal, cervical, and rectal tissues in female BALB/c mice up to 180 days post-CAB ISFI injection. These genital tissues represent main entry points for HIV infection; thus, it is crucial to ensure CAB concentrations are high in these target tissues to provide protection against HIV infection. Figure 3A–C show

CAB concentrations in cervical, vaginal, and rectal tissues for 180 days, respectively. CAB concentrations were high in all tissues with slightly greater concentrations in cervical and vaginal tissues compared with rectal tissues. CAB concentrations increased across all tissues for 30 days and remained stable until 180 days. Furthermore, tissue-to-plasma ratios across doses remained stable over the study period (180 days) for vaginal, cervical, and rectal tissue (Table 2). Additionally, there was no significant difference between CAB tissue concentrations across doses ($p > 0.05$).

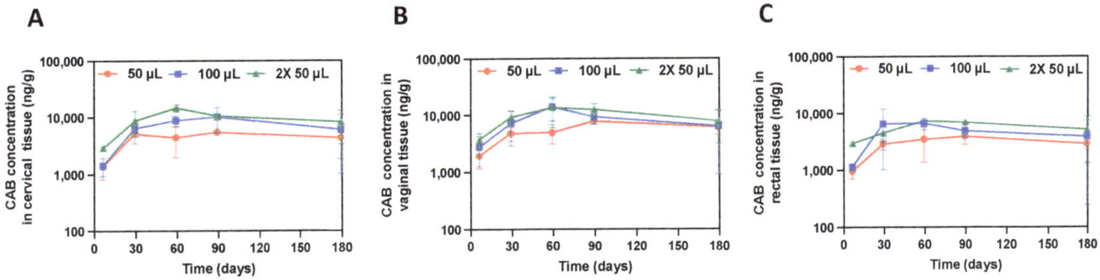

Figure 3. Genital tissue concentrations after CAB ISFI administration. (**A**) CAB concentration in cervical tissue (ng/g). (**B**) CAB concentration in vaginal tissue (ng/g). (**C**) CAB concentration in rectal tissue (ng/g). Data are presented as average ± standard deviation of n = 3 mice/timepoint per dose. Individual replicates are shown in Figure S2.

Table 2. Average tissue-to-plasma ratios after CAB ISFI injection in female BALB/c mice.

Group	Tissue	Day 7	Day 30	Day 60	Day 90	Day 180
50 µL	Cervical	0.114 ± 0.05	0.175 ± 0.06	0.123 ± 0.07	0.157 ± 0.02	0.177 ± 0.10
	Vaginal	0.160 ± 0.06	0.162 ± 0.04	0.137 ± 0.05	0.226 ± 0.03	0.247 ± 0.07
	Rectal	0.079 ± 0.02	0.097 ± 0.02	0.097 ± 0.06	0.111 ± 0.03	0.118 ± 0.02
100 µL	Cervical	0.078 ± 0.03	0.117 ± 0.02	0.120 ± 0.02	0.147 ± 0.07	0.124 ± 10
	Vaginal	0.152 ± 0.08	0.133 ± 0.08	0.192 ± 0.08	0.134 ± 0.03	0.132 ± 0.11
	Rectal	0.063 ± 0.01	0.119 ± 0.10	0.091 ± 0.02	0.071 ± 0.02	0.080 ± 0.07
2× 50 µL	Cervical	0.119 ± 0.01	0.240 ± 0.11	0.263 ± 0.04	0.195 ± 0.01	0.193 ± 0.11
	Vaginal	0.153 ± 0.04	0.255 ± 0.08	0.244 ± 0.14	0.232 ± 0.06	0.183 ± 0.11
	Rectal	0.121 ± 0.01	0.123 ± 0.03	0.129 ± 0.02	0.128 ± 0.01	0.120 ± 0.09

3.3. CAB ISFI Removal and Assessment of PK Tail

Currently available CAB LA for HIV prevention demonstrated a PK tail for at least 15 months due to CAB's long half-life (40 days) and inability to be removed to terminate treatment [9]. Investigating the PK tail is of upmost importance as suboptimal and detectable levels of CAB that remain in the plasma can cause drug-resistant virus or breakthrough infections. Since ISFIs are administered subcutaneously and form a depot under the skin, treatment can easily be terminated via a small skin incision at the injection site. However, it is important to note that CAB ISFIs are biodegradable and only need to be removed if necessary (i.e., allergic reaction, adverse events, desire to discontinue treatment, no longer at risk for HIV). In our previous work, we assessed the PK tail by removing CAB ISFIs from two rhesus macaques. It was found that 1 macaque demonstrated a 100-fold decline in CAB plasma concentration and fell below the limit of quantification within 2 weeks post-removal [13]. The other macaque only exhibited a 10-fold decline in CAB concentrations within 2 weeks after ISFI removal. Thus, we sought to further investigate this phenomenon by removing CAB ISFI from mice at days 30, 60, 90, and 180 post-administration and quantifying CAB plasma concentration 30 days post-ISFI removal (Figure 4).

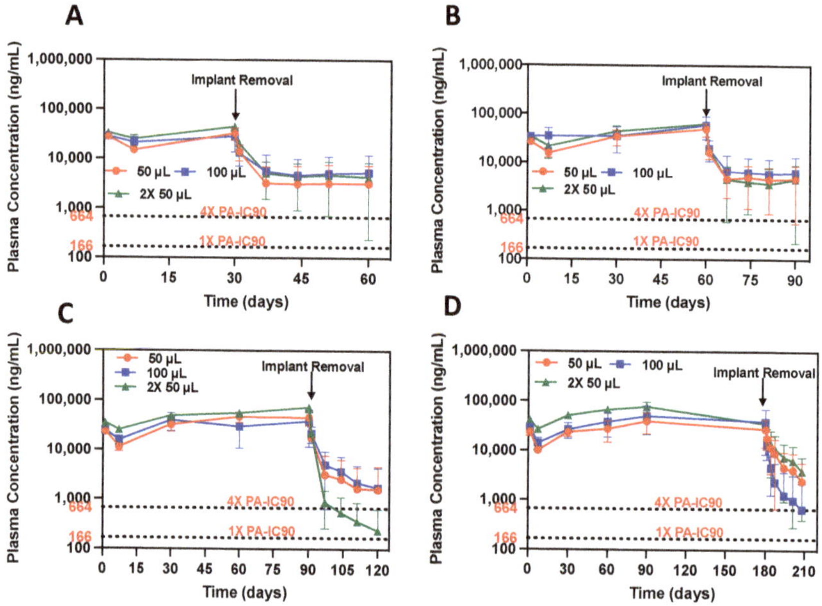

Figure 4. CAB ISFI removal and PK tail assessment. (**A–D**) CAB plasma concentration after ISFI removal at 30, 60, 90, and 180 days post-injection, respectively. Data are presented as average ± standard deviation for n = 6 mice/timepoint per dose. Individual replicates shown in Figure S3.

Results showed that plasma CAB levels exhibited a significant decrease after implant removal across doses (Figure 4A–D). Mean fold decline of CAB plasma concentrations across doses are illustrated in Table 3.

Table 3. Average fold decline in CAB plasma concentrations after 30 days post-ISFI removal.

Group	30-Day Implant Removal	60-Day Implant Removal	90-Day Implant Removal	180-Day Implant Removal
50 µL	10.11 ± 2.09	10.70 ± 3.55	28.68 ± 4.61	11.08 ± 4.81
100 µL	5.23 ± 2.16	9.61 ± 4.19	22.06 ± 10.44	58.37 ± 19.54
2× 50 µL	10.29 ± 1.87	13.92 ± 2.50	303.79 ± 36.41	10.81 ± 2.14

When CAB ISFIs were removed 30 and 60 days post-administration, mice exhibited a rapid decline in CAB plasma concentrations within 7 days post-removal, but CAB concentrations reached a plateau between day 7 and day 30 post-removal (Figure 4A,B). Interestingly, when CAB ISFIs were removed 90 and 180 days post-administration, there was a continuous decline in CAB plasma concentrations for 30 days post-removal (Figure 4C,D), however, complete CAB elimination was not achieved. From this, we hypothesize that there may be an accumulation of CAB in the tissues, particularly in the subcutaneous tissue at the injection site, due to the high dose of CAB administered (1215 mg/kg (50 µL) or 2430 mg/kg (100 µL or 2× 50 µL)). This potential reservoir may explain why CAB plasma levels were not completely eliminated after 30 days post-ISFI removal. Future studies will include assessing the complete biodistribution of the CAB ISFI after injection, particularly the CAB concentration at the subcutaneous tissue injection site to determine the presence of a drug reservoir. Additionally, we plan to determine the exact duration of the PK tail by quantifying CAB concentrations in plasma until they are no longer detectable. Ultimately, CAB ISFI removal resulted in a rapid and considerable decline in CAB concentration in plasma, but complete elimination was not achieved.

3.4. CAB ISFI Retrievability, Residual CAB, and Polymer Biodegradation

To assess CAB ISFI retrievability over time, we removed CAB ISFIs from mice via a small skin incision at the injection site at days 30, 60, 90, and 180 post-administration for each dose. Six mice per timepoint per dose were evaluated. However, 1 mouse that received a 100 μL injection for day 60 depot removal and 1 mouse that received a 2× 50 μL injection for day 180 depot removal were removed from the study prior to implant retrieval due to health concerns. ISFIs across doses were successfully removed from mice with no fibrotic tissue surrounding the depot (Figure 5A). For 1 of the mice, the depot retrieved at day 30 post-administration was significantly smaller (the mouse received a 100 μL injection), which was likely due to incomplete depot removal. There was a mass loss of 36.7 ± 5.2%, 44.5 ± 2.4%, 52.0 ± 6.5%, and 62.2 ± 11.6% relative to depot mass at day 0 across doses when CAB ISFIs were removed at days 30, 60, 90, and 180 post-administration, respectively (Figure 5B).

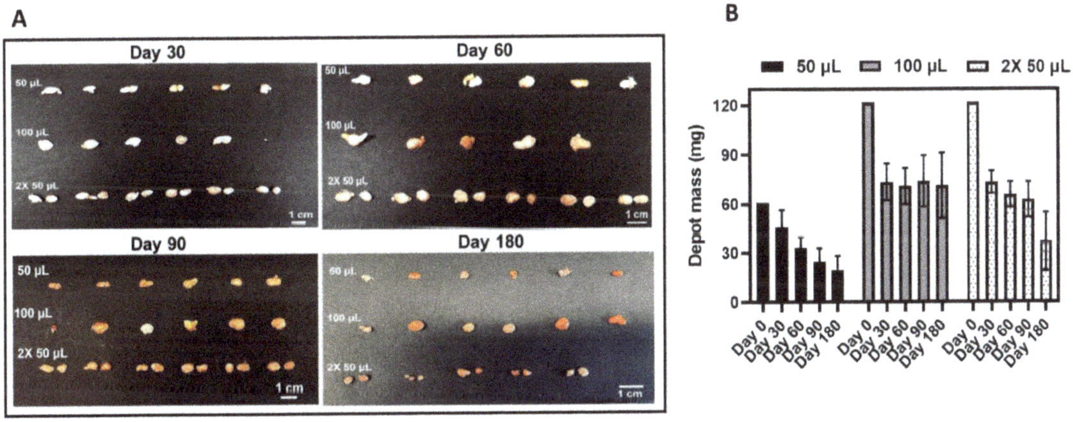

Figure 5. CAB ISFI removal and change in depot mass over 180 days. (**A**) Images of CAB ISFIs retrieved across doses after 30, 60, 90, and 180 days post administration. (**B**) CAB ISFI masses after retrieval at 30, 60, 90, and 180 days post-administration (data are represented as average ± standard deviation for n = 5 or 6 mice/timepoint) compared with initial ISFI mass (day 0) of 60.75 mg (50 μL injection volume) or 121.5 mg (100 μL total injection volume). Day 0 masses were determined based on the density of the formulation (1.215 g/mL). Depot images and masses of CAB ISFI for mice injected with 50 μL of CAB ISFI up to 90 days have been previously reported [13]. (Reprinted with permission from Ref. [13]. 2023, Young, IC & Massud, I et al. Nature Communications).

Depots were further processed to evaluate residual CAB and polymer degradation. After 180 days in mice, there was up to 34% CAB remaining (Figure 6A and Table 4) and approximately 15% polymer remaining (Figure 6B and Table 5). Notably, residual CAB and PLGA molecular weight over time were fairly consistent across doses, thus suggesting proportionality between dose and CAB release rate and polymer degradation. Furthermore, the depots were still intact and easily retrievable up to 180 days post-administration despite PLGA being almost completely degraded (85% PLGA molecular weight decrease). This is likely attributed to the high CAB content within the ISFI (41.2% CAB and 11.8% PLGA) which greatly contributes to the depot's structural integrity.

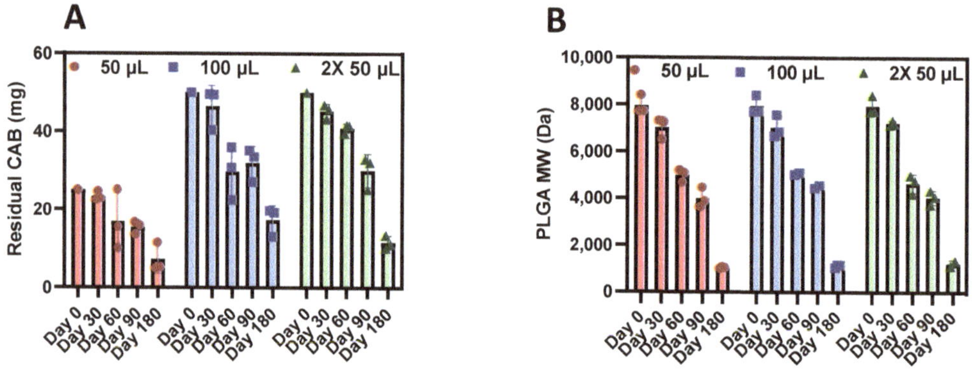

Figure 6. Residual CAB and PLGA biodegradation after CAB ISFI removal. (**A**) Residual CAB from ISFIs 30, 60, 90, and 180 days post-administration in BALB/c mice across injection volumes (average ± standard deviation of n = 3 samples per timepoint per injection volume). The initial CAB dose was 25 mg of CAB in 50 μL injection or 50 mg of CAB in 100 μL total injection. (**B**) PLGA molecular weight from CAB ISFIs 30, 60, 90, and 180 days post-administration in mice (average ± standard deviation of n = 2–3 samples per timepoint per injection volume) compared with neat PLGA (10 kDa). Residual CAB and polymer molecular weight data up to 90 days for mice injected with 50 μL of CAB ISFI have been previously reported [13]. (Reprinted with permission from Ref. [13]. 2023, Young, IC & Massud, I et al. Nature Communications).

Table 4. Summary of % CAB remaining after depot removal over time across injection volumes.

Injection Volume	% CAB Remaining			
	30 Days	60 Days	90 Days	180 Days
50 μL	93.2 ± 4.5%	67.6 ± 30.4%	61.7 ± 6.5%	23.2 ± 3.4%
100 μL	92.3 ± 10.5%	59.5 ± 13.6%	63.8 ± 8.5%	34.6 ± 7.4%
2× 50 μL	90.4 ± 3.6%	81.7 ± 2.3%	60.1 ± 8.5%	29.0 ± 14.8%

Table 5. Summary of % PLGA molecular weight (MW) decrease over time across injection volumes.

Injection Volume	% PLGA MW Decrease			
	30 Days	60 Days	90 Days	180 Days
50 μL	11.6 ± 5.6%	37.1 ± 3.4%	49.7 ± 5.4%	86.9 ± 0.43%
100 μL	11.6 ± 5.9%	36.3 ± 0.7%	43.3 ± 1.1%	85.8 ± 0.83%
2× 50 μL	9.1 ± 1.3%	41.5 ± 5.1%	49.1 ± 3.7%	84.5 ± 1.89%

Since only 23–34% of CAB remained after 180 days, it would be unlikely for CAB ISFI to produce protective concentrations for an additional 6 months. However, as previously seen in Figure 2, high concentrations of CAB in plasma were sustained for 330–367 days. This continues to suggest the possibility of a drug reservoir in tissues, particularly in the subcutaneous tissue at the injection site contributing to sustained plasma concentrations even if a negligible amount of CAB remains in the implant. Additionally, commercially available CAB LA is known to have a long half-life (40 days) [8], which could also attribute to drug accumulation and promote long-lasting CAB plasma concentrations. Therefore, it is critical to perform a complete biodistribution study in the future. Ultimately, these results further characterize the formulation defining CAB release from the implant and polymer biodegradation up to 180 days in vivo.

4. Conclusions

We performed a comprehensive dose-ranging PK study of CAB ISFIs in mice and assessed the effect of dose and number of injections on CAB ISFI PK, time to completion of CAB release and polymer degradation, long-term genital tissue PK, and CAB PK tail post-depot removal. After a single administration of CAB ISFI, plasma concentrations were maintained above the 4× PA-IC_{90} for 11–12 months, owing to its ultra-long-acting capability. Additionally, proportionality between CAB ISFI dose and overall drug exposure was observed. Furthermore, we demonstrated high CAB concentrations in vaginal, cervical, and rectal tissues for 180 days with no significant differences between doses or number of injections. When assessing the CAB PK tail, a significant decline in CAB plasma concentrations was observed within 30 days after depot removal, but complete CAB elimination was not achieved, which may be due to the presence of a CAB reservoir in the subcutaneous tissue. After 180 days post-injection, CAB ISFIs were easily retrievable with up to 34% CAB remaining and an 15% polymer remaining. These results confirm CAB ISFI's ultra-long-acting ability with high concentrations in plasma and target genital tissues, which are common sites of HIV infection. Ultimately, this study further characterized the PK of the CAB ISFI formulation and can assist with determining optimal administration parameters for clinical translation. In future, we plan to perform a complete biodistribution study in mice and similar studies should be performed in macaques to confirm results with clinically relevant injection volumes and doses.

5. Patents

SRB, ICY, and RS are inventors on a patent describing the Injectable Polymer-Based Drug Formulation for Ultra-Long-Acting Drug Delivery.

Supplementary Materials: The following supporting information can be downloaded at: https://www.mdpi.com/article/10.3390/pharmaceutics15051487/s1, Figure S1: Individual replicates of CAB plasma concentrations after 367 days post-injection. CAB plasma concentrations after 50 µL, 100 µL, or 2× 50 µL injection of CAB ISFI for 367 days. 4× PA-IC90 is 664 ng/mL and 1× PA-IC90 is 166 ng/mL. Each dose elicited n = 5–6 mice/timepoint; Figure S2: Individual replicates of CAB concentrations in tissues. (A) Individual replicates (n = 3/timepoint per dose) of CAB concentration in cervical tissue after CAB ISFI injection in female BALB/c mice. (B) Individual replicates (n = 3/timepoint per dose of CAB concentration in vaginal tissue after CAB ISFI injection in female BALB/c mice. (C) Individual replicates (n = 3/timepoint per dose) of CAB concentration in rectal tissue after CAB ISFI injection in female BALB/c mice; Figure S3: Individual replicates of CAB plasma concentrations after depot removal. (A–D) Individual replicates (n = 5–6/timepoint per dose) of CAB concentrations in plasma after ISFI removal at 30, 60, 90, and 180 days post-administration, respectively.

Author Contributions: Conceptualization, S.R.B. and I.C.Y.; methodology, S.R.B. and I.C.Y.; formal analysis, I.C.Y., C.S., A.P.S., M.L.C. and R.S.; resources, S.R.B., A.D.M.K. and M.L.C.; data curation, I.C.Y., A.L.T., C.S., A.P.S. and R.S.; writing—original draft preparation, I.C.Y., A.P.S., C.S. and S.R.B.; writing—review and editing, S.R.B. and I.C.Y.; supervision, S.R.B., A.D.M.K. and M.L.C.; funding acquisition, S.R.B. All authors have read and agreed to the published version of the manuscript.

Funding: This work was supported by the National Institute of Allergy and Infectious Diseases (grant numbers R01AI162246 and R01AI150358 to S.R.B.), the National Center for Advancing Translational Science (grant number KL2TR001109-04 to S.R.B.), University of North Carolina at Chapel Hill Center for AIDS Research (P30 AI050410), and the National Science Foundation Graduate Research Fellowship Program (grant number DGE-1650116 to I.C.Y.). The content is solely the responsibility of the authors and does not necessarily represent the official views of the National Institute of Allergy and Infectious Diseases. Any opinion, findings, and conclusions, or recommendations expressed in this material are those of the authors(s) and do not necessarily reflect the views of the National Science Foundation.

Institutional Review Board Statement: The animal study protocol was approved by the Institutional Animal Care and Use Committee at UNC Chapel Hill (protocol number/IACUC ID 20-276).

Informed Consent Statement: Not applicable.

Data Availability Statement: The data presented in this study are available upon request from the corresponding author.

Acknowledgments: We would like to thank the UNC Animal Studies Core (ASC), particularly Mark Ross, for assisting with the mouse ISFI injections and the UNC Clinical Pharmacology and Analytical Chemistry Core (CPAC), particularly Lauren Tompkins and Brian Van Horne.

Conflicts of Interest: S.R.B., I.C.Y., and R.S. are inventors on a patent describing the Injectable Polymer-Based Drug Formulation for Ultra-Long-Acting Drug Delivery. The other authors declare no conflict of interest.

References

1. UNAIDS. Global HIV and AIDS Statistics. Available online: https://www.unaids.org/en/resources/fact-sheet (accessed on 16 January 2023).
2. Cobb, D.A.; Smith, N.A.; Edagwa, B.J.; McMillan, J.M. Long-acting approaches for delivery of antiretroviral drugs for prevention and treatment of HIV: A review of recent research. *Expert Opin. Drug. Deliv.* **2020**, *17*, 1227–1238. [CrossRef] [PubMed]
3. Lima, V.D.; Harrigan, R.; Bangsberg, D.R.; Hogg, R.S.; Gross, R.; Yip, B.; Montaner, J.S. The combined effect of modern highly active antiretroviral therapy regimens and adherence on mortality over time. *J. Acquir. Immune. Defic. Syndr.* **2009**, *50*, 529–536. [CrossRef] [PubMed]
4. Tolley, E.E.; Li, S.; Zangeneh, S.Z.; Atujuna, M.; Musara, P.; Justman, J.; Pathak, S.; Bekker, L.G.; Swaminathan, S.; Stanton, J.; et al. Acceptability of a long-acting injectable HIV prevention product among US and African women: Findings from a phase 2 clinical Trial (HPTN 076). *J. Int. AIDS Soc.* **2019**, *22*, e25408. [CrossRef] [PubMed]
5. Minnis, A.M.; Montgomery, E.T.; Napierala, S.; Browne, E.N.; van der Straten, A. Insights for Implementation Science From 2 Multiphased Studies With End-Users of Potential Multipurpose Prevention Technology and HIV Prevention Products. *J. Acquir. Immune. Defic. Syndr.* **2019**, *82* (Suppl. 3), S222–S229. [CrossRef] [PubMed]
6. Landovitz, R.J.; Donnell, D.; Clement, M.E.; Hanscom, B.; Cottle, L.; Coelho, L.; Cabello, R.; Chariyalertsak, S.; Dunne, E.F.; Frank, I.; et al. Cabotegravir for HIV Prevention in Cisgender Men and Transgender Women. *N. Engl. J. Med.* **2021**, *385*, 595–608. [CrossRef] [PubMed]
7. Landovitz, R.J.; Li, S.; Grinsztejn, B.; Dawood, H.; Liu, A.Y.; Magnus, M.; Hosseinipour, M.C.; Panchia, R.; Cottle, L.; Chau, G.; et al. Safety, tolerability, and pharmacokinetics of long-acting injectable cabotegravir in low-risk HIV-uninfected individuals: HPTN 077, a phase 2a randomized controlled trial. *PLoS Med.* **2018**, *15*, e1002690. [CrossRef] [PubMed]
8. Trezza, C.; Ford, S.L.; Spreen, W.; Pan, R.; Piscitelli, S. Formulation and pharmacology of long-acting cabotegravir. *Curr. Opin. HIV AIDS* **2015**, *10*, 239–245. [CrossRef] [PubMed]
9. Landovitz, R.J.; Li, S.; Eron, J.J., Jr.; Grinsztejn, B.; Dawood, H.; Liu, A.Y.; Magnus, M.; Hosseinipour, M.C.; Panchia, R.; Cottle, L.; et al. Tail-phase safety, tolerability, and pharmacokinetics of long-acting injectable cabotegravir in HIV-uninfected adults: A secondary analysis of the HPTN 077 trial. *Lancet HIV* **2020**, *7*, e472–e481. [CrossRef] [PubMed]
10. Oliveira, M.; Ibanescu, R.I.; Anstett, K.; Mesplede, T.; Routy, J.P.; Robbins, M.A.; Brenner, B.G.; Montreal Primary, H.I.V.C.S.G. Selective resistance profiles emerging in patient-derived clinical isolates with cabotegravir, bictegravir, dolutegravir, and elvitegravir. *Retrovirology* **2018**, *15*, 56. [CrossRef] [PubMed]
11. Benhabbour, S.R.; Kovarova, M.; Jones, C.; Copeland, D.J.; Shrivastava, R.; Swanson, M.D.; Sykes, C.; Ho, P.T.; Cottrell, M.L.; Sridharan, A.; et al. Ultra-long-acting tunable biodegradable and removable controlled release implants for drug delivery. *Nat. Commun.* **2019**, *10*, 4324. [CrossRef] [PubMed]
12. Kovarova, M.; Benhabbour, S.R.; Massud, I.; Spagnuolo, R.A.; Skinner, B.; Baker, C.E.; Sykes, C.; Mollan, K.R.; Kashuba, A.D.M.; Garcia-Lerma, J.G.; et al. Ultra-long-acting removable drug delivery system for HIV treatment and prevention. *Nat. Commun.* **2018**, *9*, 4156. [CrossRef] [PubMed]
13. Young, I.C.; Massud, I.; Cottrell, M.L.; Shrivastava, R.; Maturavongsadit, P.; Prasher, A.; Wong-Sam, A.; Dinh, C.; Edwards, T.; Mrotz, V.; et al. Ultra-long-acting in-situ forming implants with cabotegravir protect female macaques against rectal SHIV infection. *Nat. Commun.* **2023**, *14*, 708. [CrossRef] [PubMed]
14. Eliaz, R.E.; Kost, J. Characterization of a polymeric PLGA-injectable implant delivery system for the controlled release of proteins. *J. Biomed. Mater. Res.* **2000**, *50*, 388–396. [CrossRef]
15. Parent, M.; Nouvel, C.; Koerber, M.; Sapin, A.; Maincent, P.; Boudier, A. PLGA in situ implants formed by phase inversion: Critical physicochemical parameters to modulate drug release. *J. Control. Release* **2013**, *172*, 292–304. [CrossRef] [PubMed]
16. Thakur, R.R.; McMillan, H.L.; Jones, D.S. Solvent induced phase inversion-based in situ forming controlled release drug delivery implants. *J. Control. Release* **2014**, *176*, 8–23. [CrossRef] [PubMed]
17. Makadia, H.K.; Siegel, S.J. Poly Lactic-co-Glycolic Acid (PLGA) as Biodegradable Controlled Drug Delivery Carrier. *Polymers* **2011**, *3*, 1377–1397. [CrossRef] [PubMed]

18. Joiner, J.B.; Prasher, A.; Young, I.C.; Kim, J.; Shrivastava, R.; Maturavongsadit, P.; Benhabbour, S.R. Effects of Drug Physicochemical Properties on In-Situ Forming Implant Polymer Degradation and Drug Release Kinetics. *Pharmaceutics* **2022**, *14*, 1188. [CrossRef] [PubMed]
19. Kempe, S.; Mader, K. In situ forming implants—An attractive formulation principle for parenteral depot formulations. *J. Control. Release* **2012**, *161*, 668–679. [CrossRef] [PubMed]
20. Gad, S.C.; Spainhour, C.B.; Shoemake, C.; Pallman, D.R.; Stricker-Krongrad, A.; Downing, P.A.; Seals, R.E.; Eagle, L.A.; Polhamus, K.; Daly, J. Tolerable Levels of Nonclinical Vehicles and Formulations Used in Studies by Multiple Routes in Multiple Species with Notes on Methods to Improve Utility. *Int. J. Toxicol.* **2016**, *35*, 95–178. [CrossRef] [PubMed]
21. WHO. N-methyl-2-pyrrolidone. 2001. Available online: http://apps.who.int/iris/bitstream/handle/10665/42404/9241530359.pdf;jsessionid=4F6D612136E539740D11CC34EB09A7EC?sequence=1 (accessed on 5 November 2022).
22. Becci, P.J.; Knickerbocker, M.J.; Reagan, E.L.; Parent, R.A.; Burnette, L.W. Teratogenicity study of N-methylpyrrolidone after dermal application to Sprague-Dawley rats. *Fundam. Appl. Toxicol.* **1982**, *2*, 73–76. [CrossRef] [PubMed]
23. EPA. Tolerance for N-methylpyrrolidone (NMP; CAS Reg. No. 872-50-4). 2006. Available online: https://www.epa.gov/sites/default/files/2015-04/documents/methyl.pdf (accessed on 5 November 2022).
24. Gaylord Chemical Company LLC. DMSO Health & Safety. 2021. Available online: https://www.gaylordchemical.com/environmental-health-safety/dmso-health-safety/ (accessed on 7 December 2021).

pharmaceutics

MDPI

Article

Pharmacokinetic Study of Islatravir and Etonogestrel Implants in Macaques

Michele B. Daly [1], Andres Wong-Sam [1], Linying Li [2], Archana Krovi [2], Gregory J. Gatto [2], Chasity Norton [2], Ellen H. Luecke [2], Victoria Mrotz [3], Catalina Forero [3], Mackenzie L. Cottrell [4], Amanda P. Schauer [4], Joy Gary [3], Josilene Nascimento-Seixas [3], James Mitchell [1], Ariane van der Straten [5,6], Walid Heneine [1], J. Gerardo García-Lerma [1,*], Charles W. Dobard [1] and Leah M. Johnson [2,*]

1 Division of HIV/AIDS Prevention, National Center for HIV/AIDS, Viral Hepatitis, STD, and TB Prevention, Centers for Disease Control and Prevention, Atlanta, GA 30329, USA
2 RTI International, Durham, NC 27709, USA
3 Division of Scientific Resources, National Center for Emerging and Zoonotic Infectious Diseases, Centers for Disease Control and Prevention, Atlanta, GA 30329, USA
4 Eshelman School of Pharmacy, University of North Carolina, Chapel Hill, NC 27599, USA
5 ASTRA Consulting, Kensington, CA 94708, USA
6 Center for AIDS Prevention Studies, Department of Medicine, University of California San Francisco, San Francisco, CA 94104, USA
* Correspondence: jng5@cdc.gov (J.G.G.-L.); leahjohnson@rti.org (L.M.J.)

Abstract: The prevention of HIV and unintended pregnancies is a public health priority. Multi-purpose prevention technologies capable of long-acting HIV and pregnancy prevention are desirable for women. Here, we utilized a preclinical macaque model to evaluate the pharmacokinetics of biodegradable ε-polycaprolactone implants delivering the antiretroviral islatravir (ISL) and the contraceptive etonogestrel (ENG). Three implants were tested: ISL-62 mg, ISL-98 mg, and ENG-33 mg. Animals received one or two ISL-eluting implants, with doses of 42, 66, or 108 μg of ISL/day with or without an additional ENG-33 mg implant (31 μg/day). Drug release increased linearly with dose with median [range] plasma ISL levels of 1.3 [1.0–2.5], 1.9 [1.2–6.3] and 2.8 [2.3–11.6], respectively. The ISL-62 and 98 mg implants demonstrated stable drug release over three months with ISL-triphosphate (ISL-TP) concentr54ations in PBMCs above levels predicted to be efficacious for PrEP. Similarly, ENG implants demonstrated sustained drug release with median [range] plasma ENG levels of 495 [229–1110] pg/mL, which suppressed progesterone within two weeks and showed no evidence of altering ISL pharmacokinetics. Two of the six ISL-98 mg implants broke during the study and induced implant-site reactions, whereas no reactions were observed with intact implants. We show that ISL and ENG biodegradable implants are safe and yield sufficient drug levels to achieve prevention targets. The evaluation of optimized implants with increased mechanical robustness is underway for improved durability and vaginal efficacy in a SHIV challenge model.

Keywords: HIV pre-exposure prophylaxis; islatravir; biodegradable implant; multipurpose prevention technologies; etonogestrel

1. Introduction

Globally, there is a growing recognition of HIV-associated gender disparities, with young women having increased rates of new HIV infections and higher rates of HIV-related mortality than men [1,2]. Pre-exposure prophylaxis (PrEP) is highly effective in preventing HIV; however, PrEP products must be desirable to the user, a quality that can vary by age, gender, and location, among others [3,4]. Women of reproductive age often seek medical care for pregnancy prevention, and offering PrEP in tandem with a contraceptive, as a multi-purpose prevention technology (MPT), has great potential to address reproductive health concerns for women while increasing PrEP use, thus mitigating HIV-related gender disparities [5,6].

Daily oral regimens are highly effective for both birth control and HIV PrEP if taken consistently. However, persistence to these oral medications is difficult for many users, and missed or late doses can reduce the contraceptive or HIV prevention effectiveness [7,8]. There are multiple long-acting contraceptive options, such as intrauterine devices, subdermal implants, intravaginal rings, and long-acting injectables, available to women who prefer these methods over daily oral medication or barrier methods [9]. Nexplanon®, a subdermal implant that releases the hormone etonogestrel (ENG) for three years, is the only hormonal contraceptive implant that is approved by the U.S. Food and Drug Administration (FDA) [10]. For long-acting PrEP, there are currently two options: Apretude®, a long-acting formulation of cabotegravir that is administered intramuscularly every 2 months [11], and the monthly intravaginal Dapivirine ring, approved in several African countries to prevent acquisition from vaginal sex [12]. Furthermore, there are many long-acting HIV PrEP products in the clinical pipeline, including subdermal implants, intravaginal rings, injectables, infusions, microarray patches, and oral regimens with reduced dosing frequency [13].

Implants are capable of ultra-long-acting drug release and, unlike injectables, can be removed if safety issues arise or user preferences change. A range of PrEP implants have been evaluated in early clinical or preclinical models including non-biodegradable implants (ethylene co-vinyl acetate [14–16], silicone [17,18], polyurethane [19], and titanium [20]) and implants constructed using biodegradable polymers (ε-polycaprolactone (PCL) [14,21,22] or polylactic acid [14]). Several of these implant technologies have been engineered to release the antiretroviral tenofovir alafenamide (TAF) and tested in preclinical models. Biodegradable PCL implants showed stable release of TAF (0.7 mg/day) and complete protection against vaginal SHIV infection in macaques; however, local inflammation and tissue necrosis developed at the implantation site after 8 weeks, a reaction that was not observed with the placebo PCL implants [23,24]. Implant-site necrosis was also observed with other TAF implant platforms [25–27], although some only induced a mild inflammatory response [26]. These TAF-related adverse reactions encouraged implant developers to pivot and investigate other antiretrovirals for long-term sustained release.

Islatravir (ISL) is a first-in-class nucleoside reverse transcriptase translocation inhibitor (NRTTI) that is phosphorylated within cells to ISL-triphosphate (ISL-TP), the active metabolite that inhibits viral replication [28]. The high potency and long half-life of ISL prompted the evaluation of multiple dosing strategies for HIV prevention or treatment in humans including monthly doses of ISL and sustained ISL release from nonerodable implants [15,16,29]. However, in 2021, the FDA placed clinical holds on all ISL trials after lymphopenia was observed across clinical studies [30]. Subsequent analyses indicated that ISL-induced lymphopenia was dose-exposure dependent and reversible after treatment discontinuation, but the exact mechanism of toxicity remains unknown [31]. ISL-TP concentrations and the corresponding changes in lymphocyte counts were used to model an optimized ISL dose that would be safe and effective [32]. Clinical trials are moving forward with an oral dose of 0.25 mg once daily (NCT05705349, NCT05631093, NCT05630755) or 2 mg once weekly (NCT05052996) for HIV treatment. The status of NCT05115838, which will evaluate ISL implants (47–57 mg) for HIV PrEP, is listed as ongoing but not yet recruiting. It is thus important to continue evaluating ISL doses capable of providing safe and effective PrEP.

In this pilot study, we evaluated the safety and pharmacokinetics of biodegradable PCL implants releasing ISL or ENG in a preclinical macaque model. The primary objectives were to monitor for adverse implant-site reactions and select implant candidates with desired pharmacokinetics for future PrEP efficacy studies. Three implants (ISL-62 mg, ISL-98 mg, and ENG-33 mg) were evaluated in normal cycling female pigtailed macaques. When co-implanted, ISL-62 mg and ISL-98 mg implants released ISL at levels that surpassed the predicted PrEP efficacy threshold for ISL-TP levels in peripheral blood mononuclear cells (PBMCs) (50 fmol/10^6 cells) [33]. ENG-33 mg implants inhibited progesterone production, indicating that the ENG levels were sufficient for suppressing ovulation in macaques. Two of the six ISL-98 mg implants broke during the study, resulting in swelling and redness at the implantation sites. However, intact ISL and ENG implants did not induce implant-site reactions, indicating that the reactions observed at the broken implant sites were likely due to sudden high drug exposure after rod breakage. Our data support the continued development of biodegradable ISL and ENG implants as an MPT for HIV PrEP and contraception.

2. Materials and Methods

2.1. Implant Fabrication and Sterilization

ISL, ENG, and pharmaceutical-grade sesame oil were purchased from WuXi AppTec (Shanghai, China), AdooQ® Bioscience (Irvine, CA, USA), and Croda (Snaith, UK), respectively. Medical-grade PCL (PC-17) pellets were purchased from Corbion (Amsterdam, The Netherlands). Hollow PCL tubes (2.5 mm outer diameter (O.D.) with a wall thickness of 100 μm or 300 μm to achieve targeted release rates) were manufactured via a hot-melt, single extrusion process using solid PC-17 pellets at GenX Medical (Chattanooga, TN, USA). The dimensions of the extruded tubes were verified with a 3-axis laser measurement system and light microscopy at GenX Medical.

PC-17 tubes were trimmed to the desired length (16 mm, 25 mm, or 40 mm) with 3 mm of headspace at each end for sealing and were enclosed at one end using custom-built sealing apparatus, as previously described [21,22]. ISL was mixed with sesame oil (1:1 *w/w* ratio) and loaded into the implant core using a 3 mL syringe fitted with a 14-gauge blunt-tip stainless steel needle. ENG was mixed with sesame oil (2:1 *w*/*w* ratio) and loaded into the implant core using a stainless-steel rod. Once the implants were loaded with a sufficient quantity of formulation that achieved the desired reservoir length, the other end of the implant was sealed. Implants were then subjected to gamma irradiation (18–24 kGy) on a continuous path for 8 h using a Cobalt-60 gamma ray source (Nordion Inc., Ottawa, ON, Canada) at STERIS Applied Sterilization Technologies (Libertyville, IL, USA).

2.2. Determining Release Rates (In Vitro) and Drug Purity (In Vitro and In Vivo)

To estimate the in vitro release rate, we followed previously validated methods [22]. Briefly, each implant (n = 3 per group) was submerged in an individual sterile polypropylene tube filled with 1× phosphate-buffered saline (PBS, pH 7.4) and incubated at 37 °C with gentle agitation (104 rpm). Implants were transferred (2×/week) to fresh buffer solutions to maintain sink conditions for the duration of the study. The concentrations of ISL or ENG in buffer release media were measured with high-performance liquid chromatography (HPLC) using an Agilent 1100/1200 HPLC-UV system equipped with an Agilent Zorbax Bonus-RP, 4.6 × 150 mm, 3.5 μm column. Samples were analyzed using an injection volume of 5 μL and a flow rate of 0.8 mL/min, with 0.01% TFA as Mobile Phase A and acetonitrile as Mobile Phase B. A run time of 30 min was used, along with a diode array detector (DAD) of 220 nm for ENG and a DAD of 260 nm for ISL.

At the terminal time points, the drug formulation was extracted from each implant (in vitro and in vivo) via sonication and analyzed for purity with HPLC using ISL and ENG standards purchased from WuXi AppTec (Shanghai, China) and AdooQ® Bioscience (Irvine, CA, USA), respectively. The excised implants from macaques were gamma sterilized prior to formulation extraction.

2.3. Animal Care Guidelines

This research was conducted under a Centers for Disease Control and Prevention (CDC) Institutional Animal Care and Use Committee (IACUC)-approved protocol in compliance with the Animal Welfare Act, PHS Policy, and other Federal statutes and regulations relating to the use of animals in research. Animals were housed in an AAALAC International accredited facility that adheres to principles stated in the 'Guide for the Care and Use of Laboratory Animals' [34]. All macaques were deemed healthy based on physical examination and baseline bloodwork. Every effort was made to minimize distress through enrichment opportunities and humane interactions. While assigned to the study, the macaques received food twice daily (LabDiet 5045 and 5049. LabDiet, St Louis, MO, USA) and water ad libitum via an automatic drinking valve. Wellness checks were conducted twice daily, and animals were given enrichment in the form of toys and food. Macaques were housed in ABSL2 rooms at 17.8–28.9 °C with a 30–70% relative humidity and 12:12 h light/dark cycle, and socially housed whenever possible. Prior to implantation, pair-housed and single-housed animals had cage dividers that, at minimum, allowed eye contact. After implantation, animals were placed in protected contact housing to minimize the potential for implant injury, where they were still able to see each other and have minimal contact.

All animal procedures were performed under anesthesia. The following sedation protocols were utilized throughout the study, chosen based on anesthetic needs and procedure time: ketamine (100 mg/mL; Dechra Veterinary Products, Overland Park, KS, USA) was administered at a dose of 10 mg/kg; dexmedetomidine (0.5 mg/mL Dexmedesed, Dechra Veterinary Products, Overland Park, KS, USA) was administered at a dose range of 0.01–0.013 mg/kg in combination with ketamine at a dose range of 3–5 mg/kg; atipamezole (Antisedan, 5.0 mg/mL, Zoetis, Parsippany, NJ, USA) was administered at a dose of 0.3 mg/kg to reverse the effects of dexmedetomidine; telazol (Tiletamine–zolazepam, Zoetis, Parsippany-Troy Hills, NJ, USA) at a dose range of 2–6 mg/kg. Drugs were administered intramuscularly (IM) using squeeze-back cages. Anesthetic events were conducted in ABSL2.

2.4. Implantation and Removal

Once anesthetized, the macaques were placed in dorsal/posterior recumbency with both arms extended. The implantation site (the medial aspect of the biceps brachii muscle) was clipped of hair and aseptically prepped with a triplicate application of Betadine solution (Povidone-Iodine 5%, Purdue Frederick Company, Stamford, CT, USA) and sterile saline. Clear plastic sterile drapes were placed over the site to maintain sterility during implantation. A small incision was made along the medial bicipital groove with a #10 scalpel. The implant was inserted subcutaneously using a 2.7 mm trocar kit (Shinva Ande Healthcare Apparatus Co., Ltd., Zibo City, China). Implants were removed in a similar manner, first clipping hair, aseptically preparing the site, and then making a small scalpel incision. The skin over the implant was grasped and the implant was pressed toward the incision site until visible. The tissue capsule was carefully snipped, allowing the implant to easily be pushed out. The incision was closed with GLUture® topical tissue adhesive (Formulated Cyanoacrylate, Abbott Animal Health, Abbott Park, IL, USA) after each procedure. After the implantation and removal procedures, Meloxicam SR (10 mg/mL, ZooPharm, Windsor, CO, USA) and Buprenorphine SR (10 mg/mL, ZooPharm, Windsor, CO, USA) were administered subcutaneously as a single dose of 0.6 mg/kg and 0.2 mg/kg, respectively.

2.5. Pharmacokinetic Study Design

The pharmacokinetic profile of implants releasing ISL with or without ENG was investigated in 6 female pigtailed macaques (mean age, 13 years [range 7–15]; mean weight 7.95 kg [range 7.10–11.45]). Using a staggered implantation design, three ISL doses were evaluated: low dose (one ISL-62 mg implant), mid dose (one ISL-98 mg implant), and high dose (one ISL-62 + one ISL-98 mg implant). ENG-33 mg implants were evaluated concurrently in the mid-ISL study group. Group 1 consisted of three SHIV+ macaques that were implanted with an ISL-62 mg implant in the right arm and monitored for 36 days to assess the low ISL dose. The same group of macaques received an ISL-98 mg implant in the left arm at day 36, receiving a high dose of ISL. Implanted animals were monitored for a total of 113–135 days prior to euthanasia. The second group of three macaques was implanted with an ISL-98 mg implant in the right arm (mid ISL dose), and 36 days later received an ENG-33 mg implant in their left arm to assess whether co-implantation altered the ISL pharmacokinetics. After 55 days of co-implantation, ISL implants were removed to determine the ISL PK tail. ENG implants were removed 198 days after implantation.

2.6. Blood, Tissue, and Swab Processing

Blood was collected in BD Vacutainer® CPT™ Cell Preparation Tubes and PBMCs were isolated following the manufacturer's protocol. Plasma aliquots of 500 µL were immediately frozen at −70 °C for ISL, ENG and progesterone analyses. PBMCs were treated with BioLegend red blood cell (RBC) lysis buffer (BioLegend, San Diego, CA, USA) and then counted using trypan blue exclusion on the Countess II FL (Invitrogen, Carlsbad, CA, USA). Vaginal and rectal mucosal fluids and tissue biopsies were collected once every two weeks. Swabs (Fisherbrand™ Synthetic-tipped Applicator, Waltham, MA, USA) were inserted in the vagina and rectum for five minutes to ensure uptake of mucosal secretions. The volume of fluid for each sample was determined by weighing the swabs pre- and post-collection. Tissue samples were taken with medical forceps (Boston Scientific Radial Jaw 4 Forceps, 2.2 mm jaw, 160 length), and three pinches were attempted at each site. Extraneous liquid was carefully removed, and tissue was weighed on an analytical balance.

2.7. Measurement of Intracellular ISL-TP and dATP from PBMCs and Tissue

Following an established protocol, 5 million live PBMCs were resuspended in 500 µL ice-cold 80% methanol, vortexed for 1 min and immediately frozen at −70 °C [35]. ISL-TP and dATP were quantified using previously published LC-MS/MS methods [36]. The calibrated ranges were 0.05–200 and 0.005–100 ng/mL for ISL-TP and dATP, respectively. Accuracy and precision were within 20%. Concentrations were normalized to live cells and reported in fmol/million cells. Tissue biopsies were transferred to 1.5 mL tubes prefilled with ceramic beads (Omni International, Kennesaw, GA, USA), and 800 µL 80% MeOH was added. Tissues underwent bead mill homogenization using the Omni Bead Ruptor 24 and were then frozen at −70 °C. ISL, ISL-TP and dATP were quantified using LC-MS/MS. Briefly, analytes were extracted using protein precipitation with isotopically labeled internal standards ($^{13}C_{15}$, N_3-EFdA (Wuxi AppTec, Shanghai, China); $^{13}C_{10}$, $^{15}N_5$-dGTP (Sigma-Aldrich, St. Louis, MO, USA), and $^{13}C_{10}$; $^{15}N_5$-dATP (Sigma-Aldrich, St. Louis, MO, USA) for ISL; and ISL-TP and dATP, respectively) then dried under nitrogen gas and reconstituted with 125 µL of 1 mM ammonium phosphate. Analytes were separated on an XTERRA MS C18 (40 × 2.1 mm 3.5 µm) column and then detected on an AB Sciex API-5000 triple quadrupole mass spectrometer under positive ionization mode. The calibrated range was 0.025–50 ng/mL. Accuracy and precision were within 20%. Concentrations were normalized to tissue mass and reported in ng/g (ISL) or fmol/g (ISL-TP and dATP) units.

2.8. Measurement of ISL and ENG in Plasma and Mucosal Fluids

Plasma ISL and ENG were quantified by adapting previously published LC-MS/MS rat plasma methods to NHP plasma [22]. The calibrated ranges were 0.1–100 and 0.2–200 ng/mL for ISL and ENG, respectively. Accuracy and precision were within 15% with 20% at the lower limit of quantification (LLOQ).

Mucosal fluids were eluted off of the swabs in 1 mL of 80:20 methanol:water and a 500 µL aliquot of the elution fluid was sent for ISL quantification by LC-MS/MS. Briefly, ISL was extracted using protein precipitation with isotopically labeled internal standard ($^{13}C_{15}$, N_3-EFdA (Wuxi AppTec, Shanghai, China)) or ENG (Toronto Research Chemicals, North York, ON, Canada) and then dried under nitrogen gas and reconstituted in 75 µL of purified water. ISL was separated on a Waters Atlantis T3 (50 × 2.1 mm 3 µm) column and then detected on an AB Sciex API-5000 triple quadrupole mass spectrometer under positive ionization mode. The calibrated range was 0.02–100 ng/mL. Accuracy and precision were within 30%.

2.9. Measurement of Progesterone in Plasma

Plasma progesterone was measured by Wisconsin National Primate Center Assay Services. Briefly, plasma, standards and quality control samples were diluted in 500 µL of ultrapurified water (Fisher Scientific, Waltham, MA, USA). Methyl tert butyl ether (Fisher Scientific) was then added, vortexed vigorously, and incubated at room temperature for 5 min. The organic phase containing steroids was transferred into a new tube, evaporated to dryness using an air stream and heated water bath (60 °C), and then resuspended in 50 µL of 20% acetonitrile in water and analyzed using LC-MS/MS. The calibration curve ranged from 0.195 to 25 ng/mL and the linearity was $r > 0.9990$. The intraassay CV was 4.24% and the interassay CV was 6.47%.

2.10. Implant-Site Reactions

Implantation sites were monitored weekly for adverse reactions including but not limited to erythema, edema, and necrosis. A modified Draize scale was used to grade erythema and edema of implant-site reactions (0, no reaction–4, severe), either in person or with high-resolution digital photography [23,37]. Draize scores from two veterinarians were averaged. After implant removal, two skin biopsies (4 mm, Integra, Princeton, NJ, USA) were collected at the medial or distal end of the implant site, H&E stained, and scored semiquantitatively by two pathologists using a lightly modified version of the international standard for biological evaluation of medical devices (ISO-10993-6 [38], severity scores for each parameter ranging from 0 (not present) to 5 (severe)).

2.11. Statistical Analysis

Simple linear regression was performed, and all figures were created using GraphPad Prism version 9.1.0 for Windows, GraphPad Software, San Diego, CA, USA.

3. Results

3.1. Implant Characteristics and Macaque Study Design

We evaluated two ISL implants (ISL 62 mg and ISL-98 mg) and one ENG implant (ENG-33 mg), which are described in Table 1. The ISL and ENG implants had wall thicknesses of 100 µm and 300 µm, respectively. The two ISL implants differed in the amount of drug loaded, length, and in vitro release rates. Figure 1 shows the in vitro drug release profile of the ISL and ENG implants.

Table 1. Biodegradable ε-Polycaprolactone Implant Characteristics.

Implant	Length (mm)	Wall Thickness (µm)	In Vitro Release Rate (µg/Day)	In Vitro Purity (%)	In Vivo Purity (%)
ISL-62 mg [1]	25	100	41.9 ± 5.5	97.1 ± 0.04	97.1 ± 0.09
ISL-98 mg [1]	40	100	65.6 ± 8.2	97.1 ± 0.02	97.1 ± 0.12
ENG-33 mg [2]	16	300	31.1 ± 9.6	99.3 ± 0.01	98.9 ± 0.01

[1] ISL implants were formulated 1:1 with a sesame oil excipient. Average release rate ± standard deviation was determined from day 0–135. [2] ENG-33 mg implants were formulated 2:1 with a sesame oil excipient. Average release rate ± standard deviation was determined from day 0–215.

For the in vivo macaque study, we used a staggered implantation design to achieve three fixed doses of ISL based on in vitro release rates: low dose (one ISL-62 mg implant, in vitro release rate of 41.9 µg/day), mid dose (one ISL-98 mg implant, in vitro release rate of 65.6 µg/day), and high dose (one ISL-62 and one ISL-98 mg implants for a total cumulative in vitro release rate of 107.5 µg/day). For the low-dose group, 3 macaques had an ISL-62 mg implant inserted in their right arm and approximately one month later received an ISL-98 mg implant in their left arm, thus transitioning to the high-dose group (Figure 2A). Both implants were removed at euthanasia, approximately 3–4 months following the insertion of the ISL-98 and ISL-62 implants, respectively. A second group of three animals received an ISL-98 mg implant in the right arm and then received an

ENG-33 mg implant in the left arm approximately one month later. The ISL-98 implant was removed ~3 months after implantation and the ENG-33 mg implant was removed ~6 months after implantation (Figure 2B).

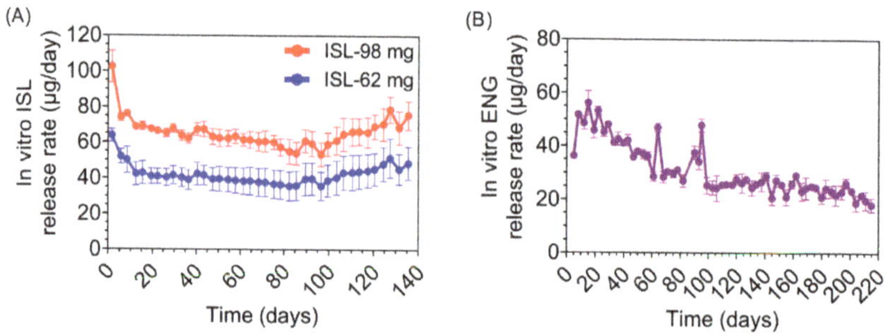

Figure 1. In vitro drug release from PCL implants. (**A**) Daily in vitro release profiles of ISL from implants run in parallel with the in vivo study in macaques with average release rates of 41.9 ± 5.5 μg/day (ISL-62 mg) and 65.6 ± 8.2 μg/day (ISL-98 mg). (**B**) Daily in vitro release profile of ENG from implants run in parallel with the in vivo study with an average release rate of 31.1 ± 9.6 μg/day.

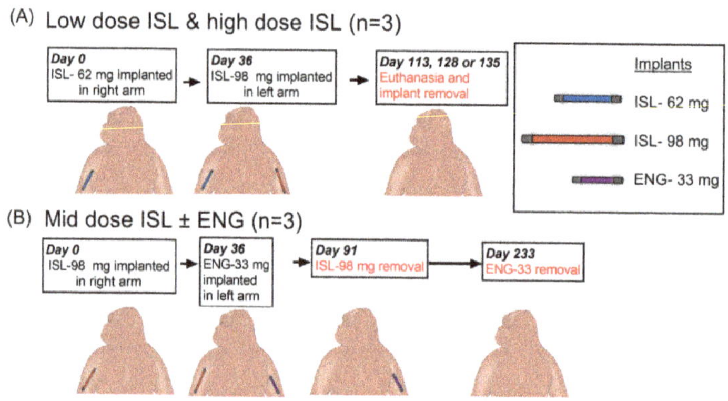

Figure 2. Study Design. (**A**) The first group of 3 macaques received ISL-62 mg implant (right arm) to assess the low ISL dose and a month later received an ISL-98 mg implant in their left arm to evaluate the high ISL dose. (**B**) The second group of 3 macaques received ISL-98 mg implant (right arm) to assess the mid ISL dose and received an ENG-33 mg implant in their left arm a month later. The ISL-98 mg implant was removed at 3 months. The ENG implant was removed after 6 months.

3.2. *Pharmacokinetic Profile of ISL and ISL-TP in Blood*

The concentrations of ISL in plasma and ISL-TP in PBMCs were measured as a function of time post-implantation (Figure 3A,B). Both analytes showed stable and sustained drug release, with the high-dose ISL group (ISL-62 mg and ISL-98 mg) achieving the PrEP target of 50 fmol ISL-TP/10^6 PBMCs (Figure 3B and Table 2). The levels of plasma ISL and PBMC ISL-TP exhibited dose linearity and correlated with the in vitro release rate (Figure 3C). To better visualize the PK differences between the three dose levels, Figure 3A,B show all dosage groups on the same axis. For the low-dose (ISL-62 mg; blue) and mid-dose (ISL-98 mg; red) groups, day zero started at the time of implantation. However, for the high-dose (ISL-62 and ISL-98 mg; green) group, day zero started the day of the second implantation of the ISL-98 mg implant.

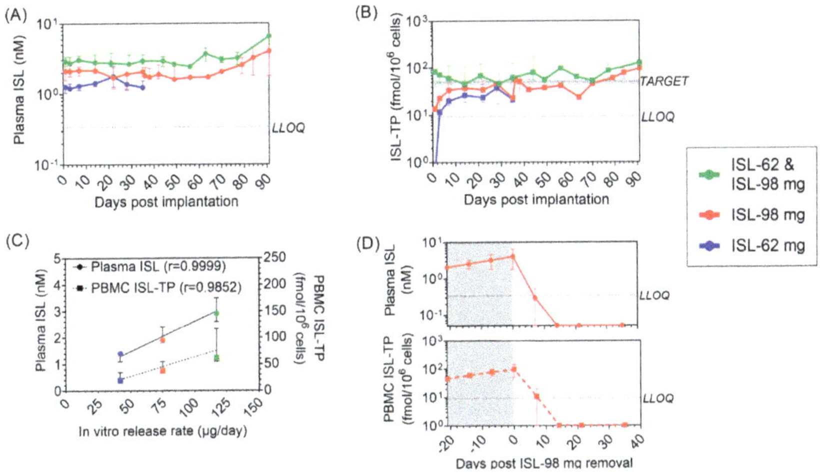

Figure 3. Pharmacokinetics of ISL and ISL-TP. (**A**) Median [range] plasma ISL and (**B**) PBMC ISL-TP levels for the low-dose ISL-62 mg (blue), mid-dose ISL-98 mg (red), and high-dose ISL-62 and ISL-98 mg (green). Day zero for ISL-62 mg and ISL-98 mg animals was the day of the first implantation. Day zero for the ISL-62 & ISL-98 mg was the day of the second implantation. The pharmacokinetic benchmark of 50 fmol ISL-TP/10^6 PBMCs (Target) and lower limit of quantification (LLOQ) are shown with dashed lines. (**C**) Dose linearity between the in vitro release rate and plasma ISL (left axis, circles) and PBMC ISL-TP (right axis, squares). (**D**) Pharmacokinetic tail of ISL plasma ISL (top) and PBMC ISL-TP (bottom) after removal of ISL-98 mg from the mid-dose group.

Table 2. Pharmacokinetics of ISL and ENG implants.

	ISL-62 mg (*n* = 3)	**ISL-98 mg ± ENG-33 (*n* = 3)**	**ISL-62 mg + ISL-98 mg (*n* = 3)**
Plasma			
Islatravir (nM)	1.3 [1.0–2.5]	1.9 [1.2–6.3]	2.8 [2.3–11.6]
ISL C_{max} (nM)	1.9 [1.5–2.5]	3.7 [2.2–6.3]	6.1 [2.9–11.6]
ISL T_{max} (days)	26 [22–35]	37 [7–91]	43 [7–91]
Etonogestrel (pg/mL)	n/a	495 [229–1100]	n/a
ENG C_{max} (nM)	n/a	875 [660–1110]	n/a
ENG T_{max} (days)	n/a	12 [7–22]	n/a
Peripheral blood mononuclear cells			
Islatravir-triphosphate (fmol/10^6 cells)	20.5 [7.8–45.4]	37.7 [13.3–143.0]	63.6 [40.3 -229.0]
ISL-TP C_{max} (fmol/10^6 cells)	36.4 [26.3–45.4]	96.8 [59.7–143.0]	164.2 [89.5–229.0]
ISL-TP T_{max} (days)	21 [14–28]	69 [36–91]	67 [44–100]

Table 2. *Cont.*

	ISL-62 mg (*n* = 3)	ISL-98 mg ± ENG-33 (*n* = 3)	ISL-62 mg + ISL-98 mg (*n* = 3)
Mucosal tissues			
Vaginal ISL-TP (pmol/g)	9.6 [a] [8.5–11.6]	13.5 [a] [8.4–21.6]	30.5 [b] [10.7–64.1]
Rectal ISL-TP (pmol/g)	18.5 [a] [8.6–27.3]	20.5 [a] [12.5–34.0]	36.9 [b] [26.0–45.4]

Median [range] is shown for ISL and ENG in plasma and ISL-TP in PBMCs. Averages [range] are shown for C_{max}, T_{max}, and mucosal tissues. Data from broken implants was excluded. [a] Tissue concentrations 21 days after the first implant, either ISL-62 mg or ISL-98 mg, was inserted. [b] Tissue concentrations 21 days after the second implant, ISL-98 mg, was inserted.

Two ISL-98 mg implants, one from the mid-dose ISL group and one from the high-dose ISL group, were deemed broken during the implant-site assessment on day 49 and 56 post-implantation, respectively (Supplemental Figure S1A). This observation coincided with a rapid increase in plasma ISL and PBMC ISL-TP. Due to the breakage, the amount of ISL being released was unknown; therefore, these data were excluded from the comprehensive pharmacokinetic analyses but shown individually (Supplemental Figure S1B,C). Overall, intact implants demonstrated stable release of ISL over the 3–4 months of study.

To evaluate the pharmacokinetic tail of ISL after implant removal, the ISL-98 mg implant was surgically removed from the mid-dose group at day 91. Plasma ISL and PBMC ISL-TP were undetectable within 7–14 days after removal (Figure 3D).

3.3. *ISL Penetration in Rectal and Vaginal Tissues*

ISL-TP was detected in tissue homogenates from both vaginal and rectal compartments at all dose levels (Supplemental Figure S2). In vaginal tissues, average ISL-TP concentrations at day 21 achieved with the low, mid, and high ISL doses were 9.6 (range, 8.5–11.6), 13.5 (range, 8.4–21.6), and 30.5 (range, 10.7–64.1) pmol/g of tissue, respectively (Table 2). Average ISL-TP levels in rectal tissues at day 21 were higher than in vaginal tissues; 18.5 (range, 8.6–27.3) pmol/g for the low dose, 20.5 (range, 12.5–34.0) pmol/g for the mid dose, and 36.9 (range, 26.0–45.4) pmol/g of tissue for the high dose (Table 2). ISL was also consistently detected in rectal and vaginal fluids (Supplemental Figure S2).

3.4. *Suppression of Progesterone Production with ENG Implants*

The mid-dose ISL macaques (*n* = 3) received one ENG-33 mg implant and were monitored for progesterone production and ENG levels in plasma (Figure 4). Plasma ENG concentrations reached C_{max} within 1–3 weeks (875 pg/mL [range 660–1110]) post-implantation and remained above 229 pg/mL for 198 days (Table 2). Prior to ENG-33 mg implantation, all animals had fluctuating progesterone concentrations indicative of the luteal and follicular phase of the menstrual cycle. After ENG-33 mg implantation, progesterone levels were suppressed in all animals below 0.3 ng/mL within 2 weeks. Plasma progesterone was suppressed during the entire ENG-33 mg implantation period, demonstrating the ability of ENG-eluting implants to effectively inhibit ovulation over several months. After ENG-33 mg implant removal, ENG was undetectable in plasma within 5 days and plasma progesterone levels began to rebound within 20 days (Figure 4).

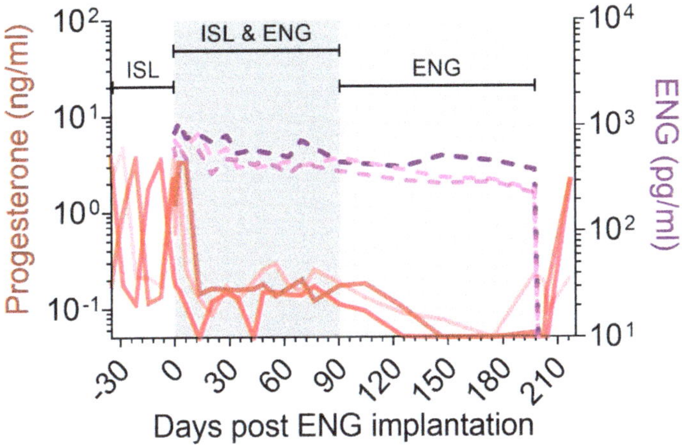

Figure 4. Plasma progesterone (left axis, red) and ENG (right axis, purple) in macaques that received ISL-98 mg and ENG-33 mg implants.

3.5. Effect of ISL and ENG Co-Implantation on ISL Pharmacokinetics

To assess how co-implantation of ISL and ENG may alter pharmacokinetics, we compared ISL levels in blood and tissues from ISL-98 mg implanted macaques prior to and after ENG-33 mg implantations. The median plasma ISL levels were 2.1 nM (range, 1.2–2.8) and 1.8 nM (range, 1.6–6.3) prior to and after ENG-33 mg insertion, respectively (Figure 5A). The median ISL-TP levels in PBMCs two weeks after ISL-98 mg implantation were 33.6 fmol/10^6 PBMCs (range, 21.2–61.2) and, after ENG-33 mg implantation, they were 48.8 (range, 22.5–143.0) (Figure 5B). In vaginal tissues, the ISL-TP levels were similar with or without ENG. In rectal tissues, ISL-TP was lower in the presence of ENG; however, this data set was limited as ISL-TP was detectable in rectal tissues at only one timepoint prior to ENG implantation (Figure 5C and Supplemental S2C).

ISL-TP competes with the natural nucleotide dATP for incorporation by reverse transcriptase, and some studies suggest that hormonal contraceptives can alter nucleotide metabolism [39–41]. To determine whether ENG affected nucleotide metabolism and possibly competition between ISL-TP and dATP, we compared ratios between ISL-TP and dATP among ISL-98 mg-implanted animals before and after ENG-33 mg implantation. A higher ratio indicates that more ISL-TP is present, increasing the likelihood of incorporation by reverse transcriptase and successful viral inhibition. Figure 5D shows that the ISL-TP:dATP ratios were unaffected in PBMCs and rectal tissues but increased in vaginal tissues in the presence of ENG. These data indicate that ENG does not reduce the active pharmaceutic agent, ISL-TP, in blood or tissues, and will likely have no adverse impact on viral inhibition as a result of enhanced dATP levels.

3.6. Implant Site Reactions

Implant sites were scored for erythema and edema using a modified Draize scale: 0 (no reaction) to 4 (severe) (Figure 6A). The median [range] Draize score was 0.0 [0.0–0.8] for the ENG-33 mg, 0.0 [0.0–1.5] for the ISL-62 mg and 0.3 [0.0–1.3] for the ISL-98 mg implants, excluding scores after implant breakage (Figure 6B). The two ISL-98 implants that broke on day 49 and 56 after implantation resulted in mild to well-defined redness and swelling, with a median Draize score (range) of 1.25 (range 0.8–2.0). There was no obvious cause of the break, e.g., scratching or picking at the implant site. Overall, intact ISL implants (n = 7/9) and ENG implants (n = 3/3) were safe and well-tolerated in macaques.

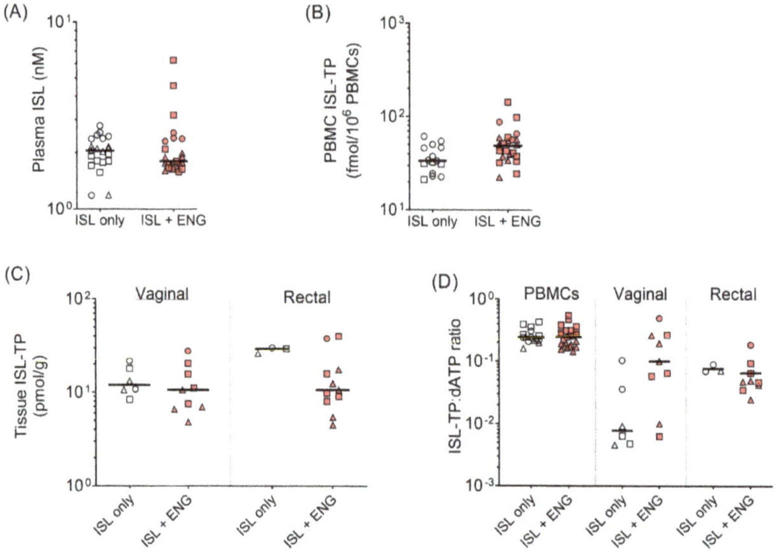

Figure 5. The effect of ISL and ENG co-implantation on ISL pharmacokinetics was determined by evaluating ISL analytes. (**A**) Plasma ISL, (**B**) PBMC ISL-TP, and (**C**) mucosal tissue ISL-TP when animals had a single ISL-98 mg implant (open shapes) compared to after co-implantation with ENG (red shapes). (**D**) Ratios of ISL-TP to the natural nucleotide competitor, dATP, were determined for each tissue. Each individual animal (n = 3) is shown with a different shape.

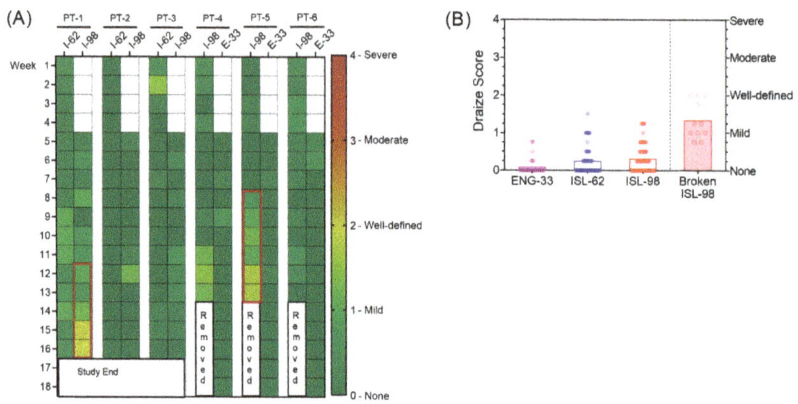

Figure 6. Implant site reactions were scored weekly for erythema and edema using a modified Draize scale. (**A**) Heat map of weekly Draize scoring of implant site for each animal (columns PT1-PT6) and each implant (ISL-62, ISL-98 or ENG-33 mg). Colors indicate severity of implant site reaction (0 = none/green, 1 = mid, 2 = well-defined, 3 = moderate, 4 = severe/red). Red boxes indicate when ISL-98 mg implants were deemed broken. (**B**) Median Draize scores for ENG-33 (purple), ISL-62 (blue), intact ISL-98 mg implants (red, unfilled box), and broken ISL-98 mg implants (red, filled box).

Skin biopsies were collected immediately after ISL-98 mg implant removal at two sites, one central to the implant (medial) and one at the end (distal). Biopsies were H&E stained and semi-quantitatively scored using a slightly modified version of the international standard for biomedical device evaluation (ISO-10993-6 [38], severity scores for each parameter ranging from 0 (not present) to 5 (severe)). H&E staining showed that intact implants were pathologically unremarkable with a median score of 1 (range 0–55, adding the severity scores across all evaluated histologic parameters) (Table 3). A biopsy taken directly at the breakage site of the ISL-98 mg implant showed edema, early fibrosis, and mild necrosis/inflammation, receiving a score of 27 out of 55 (PT-5, medial). Interestingly, the biopsy taken distal to the breakage site was pathologically unremarkable with a score of 1, suggesting that the pathology was limited to the point of the breakage.

Table 3. Implant site pathology.

Macaque ID	PT-4		PT-5		PT-6	
Biopsy Site	Medial	Distal	Medial (at Break)	Distal	Medial	Distal
Lymphocytes	1	0	3	1	0	0
Plasma cells	2	0	3	0	0	0
Histiocytes/Macrophages	1	0	3	0	0	0
Multinucleated giant cells	0	0	1	0	0	1
Polymorphonuclear cells	1	0	3	0	0	0
Fibrosis	0	0	3	0	0	0
Reactive fibroblasts	1	0	3	0	0	0
Hemorrhage	0	0	2	0	0	0
Neovascularization	0	0	2	0	0	0
Necrosis + type	0	0	2	0	0	0
Edema	0	0	3	0	0	0
Total Score	6	0	27	1	0	1

Total possible score: 55. Scoring criteria 0 = none; 1 = rare/minimal; 2 = mild/scattered, 3 = moderate; 4 = moderate to marked; 5 = abundant, extensive, severe. Scores recorded for lymphocytes, plasma cells, histiocytes/macrophages, polymorphonuclear cells, fibrosis, proliferating fibroblasts, and acute hemorrhage in the subcutaneous tissue (0 to 5 scale). Extravascular red blood cells only scored as hemorrhage if they were present deeper within tissue (not just at margins). Superficially adhered red blood cells consistent with acute hemorrhage at the time of surgical excision were not scored.

3.7. *Purity of Drug and Predicted Duration of Release*

After implant removal, the residual drug was assessed for purity (Table 1), showing >97% drug purity for ISL-62 mg and ISL-98 mg and >99% drug purity for ENG-33. Considering the in vitro release rates, the ISL-62, ISL-98, and ENG-33 implants have an expected duration of 4.1, 4.1, and 2.9 years, respectively. The estimated durations assume zero-order release kinetics and do not account for potential changes in dosing over time or tailing of the release profiles.

4. Discussion

MPTs for HIV PrEP and contraception can offer women combined protection in a single product [42]. Subdermal implants capable of long-term drug release and preventative efficacy are intriguing as MPTs as they can reduce the burden of daily oral medications and lessen clinic visits required for monthly ring replacement or injectable contraception or PrEP. Moreover, implants comprised of biodegradable polymers can advantageously bypass the need for removal, thus eliminating a minor surgical procedure to retrieve spent implants. A subdermal biodegradable PCL implant releasing tenofovir alafenamide (TAF) recently showed extended drug release and protective efficacy in a macaque SHIV challenge model, although further clinical advancement was halted due to drug-related implant-site reactions [23,24,27]. Here, we report on the preclinical assessment of next-generation PCL implants releasing the antiretroviral ISL, or the hormonal contraceptive ENG. We document prolonged release of ISL from two implant configurations that varied in length, drug load, and release rate, and demonstrated sustained release of ENG at clinically relevant levels. We also observed little to no implant site reactions when implants kept their integrity. Additional studies are needed to address the release kinetics of ISL and ENG co-formulated in a single implant.

Pharmacokinetic analysis in macaques of ISL implants showed that ISL analytes increased linearly with dose as estimated by the in vitro release rate. The data also revealed a strong correlation between ISL levels in plasma and ISL-TP concentrations in PBMCs, suggesting that plasma ISL may serve as a good surrogate for estimating ISL-TP levels in HIV target cells. Macaques that received two ISL implants, ISL-62 and ISL-98, had ISL-TP concentrations above the predicted protective PrEP benchmark of 50 fmols/10^6 PBMCs [33]. In addition, ISL-TP was also detected in vaginal and rectal tissue, both important sites for sexual HIV transmission. Notably, the tissue ISL-TP levels in macaques (10–30 pmols/g) were slightly higher than those observed in the small number of human tissue samples [16,43]. The ISL levels in plasma were between 1.3 and 2.8 nM, which are within the range predicted to be safe and moving forward in clinical trials [32,44]. Compared with human studies, including the nonerodable ISL implant, we observed lower ISL-TP concentrations at equivalent plasma ISL values [15,16,44,45]. However, we calculated the ISL-TP:ISL ratios in other macaque studies and found a similar trend of lower ISL-TP in macaques compared to humans [14,36,46,47]. This could be attributed to differential rates of phosphorylation between species or possibly the sample processing method [36].

Recently, a refillable ISL implant was evaluated as PrEP in both vaginal and rectal SHIV challenge models. Although complete protection was achieved, the ISL levels were well above the limit deemed to be safe in humans [46]. Next-generation PCL implants releasing ISL at levels that are predicted to be safe were assessed as PrEP in the vaginal SHIV challenge model, and the threshold of protection is very close to the human PrEP benchmark of 50 fmol/10^6 PBMCs [48]. Defining the threshold of protection for various routes of infection and confirming safe and sustained release are priorities for ISL implant modalities.

Although we document the sustained release of ISL from PCL implants, two out of six ISL-98 mg implants broke approximately 2 months after insertion. Implant breakage was associated with a concurrent spike in plasma ISL and ISL-TP in PBMCs, which led to redness and swelling at the implant site, albeit not as severe as intact TAF implants [24]. Increased levels of ISL and ISL-TP were maintained for over a month before the implants were removed, indicating that the drug remained in the implant despite the PCL wall being compromised. The cause of the implant breakages cannot be determined with certainty but may be related to several factors including the wall thickness, length, or a combination of the wall thickness and length of the implant. It is important to note that the ISL-62 mg implants, which had the same wall thickness (100 μm) as the ISL-98 implants but were shorter in length, all remained intact. Additional studies are underway to increase the mechanical robustness of PCL implants to ensure implants are capable of long-term safe and effective drug release [48].

Biodegradable PCL implants releasing ENG were also assessed to determine the compatibility of ISL and ENG as an MPT. ENG-33 mg implants placed in macaques achieved similar plasma ENG pharmacokinetics to those observed in humans implanted with Nexplanon implants [10]. ENG prevents pregnancy by inhibiting ovulation, which precludes the development of the corpus luteum, in turn limiting the production of endogenous progesterone [49]. To assess the efficacy of ENG, we used pigtailed macaques, which, like humans, have a lunar menstrual cycle and associated monthly hormone fluctuations. We measured progesterone as a surrogate of ovulation and found that progesterone production was suppressed until ENG implant removal, after which the progesterone levels rebounded, indicating a return to fertility [50].

Based on the different metabolic pathways of ISL and ENG, we did not expect significant drug interactions. Islatravir does not induce or inhibit major drug-metabolizing enzymes, including cytochrome50 (CYP), which is likely involved in ENG metabolism and has previously shown no DDIs with another progestogen contraceptive, levonorgestrel [51,52]. Consistent with this expectation, we found that co-administration of ISL and ENG resulted in no significant change to the PK profile of either drug. We found that ENG released in plasma had no effect on the pharmacokinetics of ISL in plasma or ISL-TP in PBMCs. However, we noted an unexpected decrease in dATP in vaginal tissues in the presence of ENG, which increased ISL-TP:dATP ratios. As dATP and ISL-TP compete for the active site of reverse transcriptase, a higher ratio of ISL-TP:dATP indicates a higher likelihood of ISL-TP incorporation and viral inhibition. Other studies have noted compartment-specific effects of hormonal treatments on nucleotides, both endogenous and antiretrovirals [39–41]. Further studies are needed to define the vaginal PrEP efficacy of ISL and possible modulations of activity in the presence of ENG or other hormonal contraceptives.

This study assessed two ISL implants at three dose levels. However, the study had several limitations including a small sample size (n = 3 animals per dose) and a relatively short duration of 3 months, both of which hindered our ability to precisely determine the in vivo release rates and predict the duration of release. The accuracy of estimating in vivo release rates relies on achieving a zero-order release profile over the duration of the study. Additional investigations are required to further characterize the release profile, especially during the depletion phase. As the implant nears depletion, physiochemical differences could lead to altered release rates. Pharmacologic models, such as deconvolution, can estimate release rates based on the PK profile of ISL and that of the controlled-release formulation. Thus, these modeling techniques are not reliant on achieving a zero-order profile over the duration of the study to accurately estimate the release rate. A larger sample size and an extended study period will be important to determine drug pharmacokinetics through implant depletion, as well as inform on long-term safety, particularly regarding ISL exposure and potential effects on lymphocytes.

5. Conclusions

In summary, we describe a PCL implant with viable MPT attributes that can release ISL above predicted PrEP protection benchmarks and deliver ENG at levels that are sufficient to suppress ovulation for several months. The evaluation of optimized ISL-eluting implants with increased wall thickness is underway to assess long-term vaginal PrEP efficacy in a validated SHIV challenge model.

Supplementary Materials: The following supporting information can be downloaded at: https://www.mdpi.com/article/10.3390/pharmaceutics15122676/s1, Figure S1: Adverse reactions and pharmacokinetics of broken ISL-98 mg implants; Figure S2: Islatravir from (A) rectal and (B) vaginal secretions. ISL-TP from (C) rectal, and (D) vaginal tissue biopsies.

Author Contributions: Conceptualization, M.B.D., A.v.d.S., W.H., J.G.G.-L., C.W.D. and L.M.J.; data curation, M.B.D., L.L. and A.K.; formal analysis, M.B.D., L.L. and A.K.; funding acquisition, A.v.d.S. and L.M.J.; investigation, M.B.D., A.W.-S., L.L., A.K., V.M., C.F., M.L.C., A.P.S., J.G., J.N.-S. and J.M.; methodology, M.B.D., L.L., A.K., G.J.G., A.v.d.S., W.H., J.G.G.-L., C.W.D. and L.M.J.; project administration, M.B.D., C.N., E.H.L., W.H., J.G.G.-L., C.W.D. and L.M.J.; resources, L.L., A.K., J.M., J.G.G.-L. and L.M.J.; supervision, C.N., E.H.L., A.v.d.S., J.G.G.-L., C.W.D. and L.M.J.; validation, M.B.D., L.L., A.K., G.J.G., W.H., J.G.G.-L., C.W.D. and L.M.J.; visualization, M.B.D.; writing—original draft, M.B.D.; writing—review and editing, M.B.D., L.L., A.K., G.J.G., C.N., E.H.L., A.v.d.S., W.H., J.G.G.-L., C.W.D., A.W.-S., V.M., C.F., M.L.C., A.P.S., J.G., J.N.-S., J.M. and L.M.J. All authors have read and agreed to the published version of the manuscript.

Funding: This research was supported by CDC intramural funds and by the generous support of the American people through the U.S. President's Emergency Plan for AIDS Relief (AID-OAA-A-17-00011, AID-OAA-A-14-00012) through the United States Agency for International Development (USAID). The contents are the responsibility of the authors and do not necessarily reflect the views of RTI, USAID, PEPFAR, or the United States Government. This research was also supported by the University of North Carolina at Chapel Hill Center for AIDS Research (CFAR), an NIH-funded program P30AI050410. WNPRC Assay services performed the plasma progesterone analysis which was supported in part by the Office of the Director, National Institute of Health under Award Number P51OD011106 to the Wisconsin National Primate Research Center, University of Wisconsin-Madison.

Institutional Review Board Statement: The animal study protocol was approved by the Institutional Animal Care and Use Committee (IACUC) of the Centers for Disease Control and Prevention (DOBMONC3002, approved on: 01 October 2019).

Informed Consent Statement: Not applicable.

Data Availability Statement: The data presented in this study are available upon request from the corresponding author.

Acknowledgments: We thank Daniel Kim, Kenji Nishiura, and Mara Sterling for lab assistance; Marie Brake, Rachel Wier, Jillian Condrey, and Rex Howard for veterinary expertise; and Frank Deyounks and Ryan Johnson for serving as animal technicians.

Conflicts of Interest: The authors (L.M.J., A.v.d.S., L.L., A.K. and E.H.L.) are inventors on pending patent applications filed by RTI. Author Ariane van der Straten is employed by the company ASTRA Consulting. The remaining authors declare that the research was conducted in the absence of any commercial or financial relationships that could be construed as a potential conflict of interest.

References

1. GBD 2015 HIV Collaborators. Estimates of global, regional, and national incidence, prevalence, and mortality of HIV, 1980-2015: The Global Burden of Disease Study 2015. *Lancet HIV* **2016**, *3*, e361–e387. [CrossRef]
2. UNAIDS. Gender Every Minute. Available online: https://www.unaids.org/sites/default/files/20120607_infographic_gender-every-minute_en.pdf (accessed on 20 June 2023).
3. Wulandari, L.P.L.; He, S.Y.; Fairley, C.K.; Bavinton, B.R.; Marie-Schmidt, H.; Wiseman, V.; Guy, R.; Tang, W.; Zhang, L.; Ong, J.J. Preferences for pre-exposure prophylaxis for HIV: A systematic review of discrete choice experiments. *EClinicalMedicine* **2022**, *51*, 101507. [CrossRef]
4. Quaife, M.; Eakle, R.; Cabrera Escobar, M.A.; Vickerman, P.; Kilbourne-Brook, M.; Mvundura, M.; Delany-Moretlwe, S.; Terris-Prestholt, F. Divergent Preferences for HIV Prevention: A Discrete Choice Experiment for Multipurpose HIV Prevention Products in South Africa. *Med. Decis. Mak.* **2018**, *38*, 120–133. [CrossRef]
5. Browne, E.N.; Montgomery, E.T.; Mansfield, C.; Boeri, M.; Mange, B.; Beksinska, M.; Schwartz, J.L.; Clark, M.R.; Doncel, G.F.; Smit, J.; et al. Efficacy is Not Everything: Eliciting Women's Preferences for a Vaginal HIV Prevention Product Using a Discrete-Choice Experiment. *AIDS Behav.* **2020**, *24*, 1443–1451. [CrossRef]
6. Shapley-Quinn, M.K.; Manenzhe, K.N.; Agot, K.; Minnis, A.M.; van der Straten, A. "We are not the same": African women's view of multipurpose prevention products in the TRIO clinical study. *Int. J. Womens Health* **2019**, *11*, 97–107. [CrossRef]
7. Dimitrov, D.T.; Mâsse, B.R.; Donnell, D. PrEP Adherence Patterns Strongly Affect Individual HIV Risk and Observed Efficacy in Randomized Clinical Trials. *J. Acquir. Immune Defic. Syndr.* **2016**, *72*, 444–451. [CrossRef]
8. Trussell, J. Contraceptive failure in the United States. *Contraception* **2011**, *83*, 397–404. [CrossRef]
9. Trussell, J.A.A.; Micks, E.; Guthrie, K.A. Efficacy, safety, and personal considerations. In *Contraceptive Technology*, 21st ed.; Ayer Company Publishers, Inc.: New York, NY, USA, 2018.
10. *Nexplanon (Etonegestrel) FDA Insert, Reference ID 3808594*; Application No. 021529/S11; Merck & Co, Inc.: Rahway, NJ, USA, 2001.
11. *Apretude (Cabotegravir) FDA Insert, Reference ID 4908014*; ViiV Healthcare: London, UK, 2021.
12. PrEPWatch an initiative of AVAC. The Dapivirine Vaginal Ring. Available online: https://www.prepwatch.org/products/the-dapivirine-vaginal-ring/ (accessed on 6 September 2023).
13. Agrahari, V.; Anderson, S.M.; Peet, M.M.; Wong, A.P.; Singh, O.N.; Doncel, G.F.; Clark, M.R. Long-acting HIV pre-exposure prophylaxis (PrEP) approaches: Recent advances, emerging technologies, and development challenges. *Expert. Opin. Drug Deliv.* **2022**, *19*, 1365–1380. [CrossRef]
14. Barrett, S.E.; Teller, R.S.; Forster, S.P.; Li, L.; Mackey, M.A.; Skomski, D.; Yang, Z.; Fillgrove, K.L.; Doto, G.J.; Wood, S.L.; et al. Extended-duration MK-8591-eluting implant as a candidate for HIV treatment and prevention. *Antimicrob. Agents Chemother.* **2018**, *62*, 10–1128. [CrossRef]
15. Matthews, R.P.; Patel, M.; Barrett, S.E.; Haspeslagh, L.; Reynders, T.; Zhang, S.; Rottey, S.; Goodey, A.; Vargo, R.C.; Grobler, J.A.; et al. Safety and pharmacokinetics of islatravir subdermal implant for HIV-1 pre-exposure prophylaxis: A randomized, placebo-controlled phase 1 trial. *Nat. Med.* **2021**, *27*, 1712–1717. [CrossRef]
16. Matthews, R.P.; Zang, X.; Barrett, S.E.; Koynov, A.; Goodey, A.; Heimbach, T.; Weissler, V.L.; Leyssens, C.; Reynders, T.; Xu, Z.; et al. A Randomized, Double-Blind, Placebo-Controlled, Phase 1 Trial of Radiopaque Islatravir-Eluting Subdermal Implants for Pre-exposure Prophylaxis Against HIV-1 Infection. *JAIDS J. Acquir. Immune Defic. Syndr.* **2023**, *92*, 310–316. [CrossRef]
17. Gunawardana, M.; Remedios-Chan, M.; Miller, C.S.; Fanter, R.; Yang, F.; Marzinke, M.A.; Hendrix, C.W.; Beliveau, M.; Moss, J.A.; Smith, T.J.; et al. Pharmacokinetics of long-acting tenofovir alafenamide (GS-7340) subdermal im-plant for HIV prophylaxis. *Antimicrob. Agents Chemother.* **2015**, *59*, 3913–3919. [CrossRef]
18. Gengiah, T.N.; Karim, Q.A.; Harkoo, I.; Mansoor, L.; Zuma, N.Y.; Radebe, P.; Samsunder, N.; Baxter, C.; Maharaj, B.; Baum, M.M.; et al. CAPRISA 018: A phase I/II clinical trial study protocol to assess the safety, acceptability, tolerability and pharmacokinetics of a sustained-release tenofovir alafenamide subdermal implant for HIV prevention in women. *BMJ Open* **2022**, *12*, e052880. [CrossRef]

19. Simpson, S.M.; Widanapathirana, L.; Su, J.T.; Sung, S.; Watrous, D.; Qiu, J.; Pearson, E.; Evanoff, A.; Karunakaran, D.; Chacon, J.E.; et al. Design of a Drug-Eluting Subcutaneous Implant of the Antiretroviral Tenofovir Alafenamide Fumarate. *Pharm. Res.* **2020**, *37*, 83. [CrossRef]
20. Chua, C.Y.X.; Jain, P.; Ballerini, A.; Bruno, G.; Hood, R.L.; Gupte, M.; Gao, S.; Di Trani, N.; Susnjar, A.; Shelton, K.; et al. Transcutaneously refillable nanofluidic implant achieves sustained level of tenofovir diphosphate for HIV pre-exposure prophylaxis. *J. Control. Release* **2018**, *286*, 315–325. [CrossRef]
21. Johnson, L.M.; Krovi, S.A.; Li, L.; Girouard, N.; Demkovich, Z.R.; Myers, D.; Creelman, B.; van der Straten, A. Characterization of a Reservoir-Style Implant for Sustained Release of Tenofovir Alafenamide (TAF) for HIV Pre-Exposure Prophylaxis (PrEP). *Pharmaceutics* **2019**, *11*, 315. [CrossRef]
22. Li, L.; Gatto, G.J.; Brand, R.M.; Krovi, S.A.; Cottrell, M.L.; Norton, C.; van der Straten, A.; Johnson, L.M. Long-acting biodegradable implant for sustained delivery of antiretrovirals (ARVs) and hormones. *J. Control. Release* **2021**, *340*, 188–199. [CrossRef]
23. Massud, I.; Krovi, A.; Nishiura, K.; Ruone, S.; Li, L.; Holder, A.; Gary, J.; Mills, P.; Mitchell, J.; Khalil, G.; et al. Safety and efficacy of a biodegradable implant releasing tenofovir alafenamide for vaginal protection in a macaque model. *J. Antimicrob. Chemother.* **2022**, *77*, 2964–2971. [CrossRef]
24. Gatto, G.J.; Krovi, A.; Li, L.; Massud, I.; Holder, A.; Gary, J.; Mills, P.; Mitchell, J.; Luecke, E., Demkovich, Z.R.; et al. Comparative Pharmacokinetics and Local Tolerance of Tenofovir Alafenamide (TAF) From Subcutaneous Implant in Rabbits, Dogs, and Macaques. *Front. Pharmacol.* **2022**, *13*, 923954. [CrossRef]
25. Su, J.T.; Simpson, S.M.; Sung, S.; Tfaily, E.B.; Veazey, R.; Marzinke, M.; Qiu, J.; Watrous, D.; Widanapathirana, L.; Pearson, E.; et al. A Subcutaneous Implant of Tenofovir Alafenamide Fumarate Causes Local Inflammation and Tissue Necrosis in Rabbits and Macaques. *Antimicrob. Agents Chemother.* **2020**, *64*. [CrossRef]
26. Zane, D.; Roller, S.; Shelton, J.; Singh, R.; Jain, R.; Wang, Y.; Yang, B.; Felx, M.; Alessi, T.; Feldman, P.L. A 28-Day Toxicity Study of Tenofovir Alafenamide Hemifumarate by Subcutaneous Infusion in Rats and Dogs. *Microbiol. Spectr.* **2021**, *9*, e0033921. [CrossRef]
27. Romano, J.W.; Baum, M.M.; Demkovich, Z.R.; Diana, F.; Dobard, C.; Feldman, P.L.; Garcia-Lerma, J.G.; Grattoni, A.; Gunawardana, M.; Ho, D.K.; et al. Tenofovir Alafenamide for HIV Prevention: Review of the Proceedings from the Gates Foundation Long-Acting TAF Product Development Meeting. *AIDS Res. Hum. Retroviruses* **2021**, *37*, 409–420. [CrossRef]
28. Markowitz, M.; Grobler, J.A. Islatravir for the treatment and prevention of infection with the human immunodeficiency virus type 1. *Curr. Opin. HIV AIDS.* **2020**, *15*, 27–32. [CrossRef]
29. Hillier, S.; Bekker, L.-G.; Riddler, S.A.; Hendrix, C.W.; Badal-Faesen, S.; Rasmussen, S.; Schwartz, H.; Macdonald, P.; Lombaard, J.; Caraco, Y.; et al. Safety and Pharmacokinetics of Oral Islatravir Once Monthly for HIV Pre-exposure Prophylaxis (PrEP): Week 24 Analysis of a Phase 2a Trial. In Proceedings of the 11th IAS Con-ference on HIV Science, Virtual, 18–22 July 2021.
30. Merck & Co., Inc. Merck Announces Clinical Holds on Studies Evaluating Islatravir for the Treatment and Pre-Vention of HIV-1 Infection. Available online: https://www.merck.com/news/merck-announces-clinical-holds-on-studies-evaluating-islatravir-for-the-treatment-and-prevention-of-hiv-1-infection/ (accessed on 20 June 2023).
31. Squires, K.E.; Correll, T.A.; Robertson, M.N.; Klopfer, S.O.; Hwang, P.M.T.; Zhou, Y.-P.; Martin, E.; Rhee, E.G. Effect of ISL on Total Lymphocytes and Lymphocytes Subset Counts. In Proceedings of the Conference on Retroviruses and Opportunistic Infections, Seattle, WA, USA, 19–22 February 2023.
32. Vargo, R.C.; Robey, S.; Zang, X.; Du, L.; Kandala, B.; Roberts, J.; Klopfer, S.; Squires, K. Modeling to optimize islatravir QW dose in HIV virologically supressed PWH. In Proceedings of the Conference on Retroviruses and Opportunistic Infections, Seattle, WA, USA, 19–22 February 2023.
33. Patel, M.; Zang, X.; Cao, Y.; Matthews, R.P.; Plank, R.; Sklar, P.; Grobler, J.; Robertson, M.; Vargo, R. Islatravir PK Threshold & Dose Selection for Monthly Oral PrEP. In Proceedings of the Conference on Retroviruses and Opportunistic Infections, Virtual, 6–11 March 2021.
34. Institute of Laboratory Animal Resources (US); Committee on Care and Use of Laboratory Animals. *Guide for the Care and Use of Laboratory Animals*, 8th ed.; National Academies Press: Washington, DC, USA, 2011.
35. Kuklenyik, Z.; Martin, A.; Pau, C.P.; Holder, A.; Youngpairoj, A.S.; Zheng, Q.; Cong, M.E.; Garcia-Lerma, J.G.; Heneine, W.; Pirkle, J.L.; et al. On-line coupling of anion exchange and ion pair chromatography for measurement of intracel-lular triphosphate metabolites of reverse transcriptase inhibitors. *J. Chromatogr. B* **2009**, *877*, 3659–3666. [CrossRef]
36. Sykes, C.; Van Horne, B.; Jones, J.; Kashuba, A.D.M.; Gatto, G.; Van Der Straten, A.; Johnson, L.; Cottrell, M.L. Intracellular islatravir pharmacology differs between species in an in vitro model: Implications for preclinical study design. *J. Antimicrob. Chemother.* **2022**, *77*, 1000–1004. [CrossRef]
37. Farage, M.A.; Maibach, H.I.; Andersen, K.E.; Lachapelle, J.M.; Kern, P.; Ryan, C.; Ely, J.; Kanti, A. Historical perspective on the use of visual grading scales in evaluating skin irritation and sensitization. *Contact Dermatitis* **2011**, *65*, 65–75. [CrossRef]
38. *ISO 10993-6:2016*; Biological Evaluation of Medical Devices. Part 6: Tests for Local Effects after Implantation. ISO: Geneva, Switzerland, 2016.
39. Shen, Z.; Rodriguez-Garcia, M.; Patel, M.V.; Bodwell, J.; Kashuba, A.D.M.; Wira, C.R. Hormonal Contraceptives Differentially Suppress TFV and TAF Inhibition of HIV Infection and TFV-DP in Blood and Genital Tract CD4+ T cells. *Sci. Rep.* **2017**, *7*, 17697. [CrossRef]

40. Cottrell, M.L.; Prince, H.M.A.; Schauer, A.P.; Sykes, C.; Maffuid, K.; Poliseno, A.; Chun, T.W.; Huiting, E.; Stanczyk, F.Z.; Peery, A.F.; et al. Decreased Tenofovir Diphosphate Concentrations in a Transgender Female Cohort: Implications for Human Immunodeficiency Virus Preexposure Prophylaxis. *Clin. Infect. Dis.* **2019**, *69*, 2201–2204. [CrossRef]
41. Daly, M.B.; Sterling, M.; Holder, A.; Dinh, C.; Nishiura, K.; Khalil, G.; García-Lerma, J.G.; Dobard, C. The effect of depot medroxyprogesterone acetate on tenofovir alafenamide in rhesus macaques. *Antivir. Res.* **2021**, *186*, 105001. [CrossRef]
42. Bershteyn, A.; Resar, D.; Kim, H.Y.; Platais, I.; Mullick, S. Optimizing the pipeline of multipurpose prevention technologies: Opportunities across women's reproductive lifespans. *Front. Reprod. Health* **2023**, *5*, 1169110. [CrossRef]
43. Hendrix, C.W.; Hillier, S.L.; Bekker, L.-G.; Badal-Faesen, S.; Riddler, S.A.; Fuchs, E.J.; Macdonald, P.J.; Edick, S.; Plank, R.M.; Chavez-Eng, C.M. Islatravir distribution in mucosal tissues, PBMC & plasma after monthly oral dosing. In Proceedings of the Conference on Retroviruses and Opportunistic Infections, Virtual, 12–16 February 2022.
44. Matthews, R.P.; Jackson Rudd, D.; Zhang, S.; Fillgrove, K.L.; Sterling, L.M.; Grobler, J.A.; Vargo, R.C.; Stoch, S.A.; Iwamoto, M. Safety and Pharmacokinetics of Once-Daily Multiple-Dose Administration of Islatravir in Adults Without HIV. *J. Acquir. Immune Defic. Syndr.* **2021**, *88*, 314–321. [CrossRef]
45. Matthews, R.P.; Ankrom, W.; Friedman, E.; Jackson Rudd, D.; Liu, Y.; Mogg, R.; Panebianco, D.; De Lepeleire, I.; Petkova, M.; Grobler, J.A.; et al. Safety, tolerability, and pharmacokinetics of single- and multiple-dose administration of islatravir (MK-8591) in adults without HIV. *Clin. Transl. Sci.* **2021**, *14*, 1935–1944. [CrossRef]
46. Pons-Faudoa, F.P.; Di Trani, N.; Capuani, S.; Campa-Carranza, J.N.; Nehete, B.; Sharma, S.; Shelton, K.A.; Bushman, L.R.; Abdelmawla, F.; Williams, M.; et al. Long-acting refillable nanofluidic implant confers protection against SHIV infection in nonhuman primates. *Sci. Transl. Med.* **2023**, *15*, eadg2887. [CrossRef]
47. Markowitz, M.; Gettie, A.; St Bernard, L.; Andrews, C.D.; Mohri, H.; Horowitz, A.; Grasperge, B.F.; Blanchard, J.L.; Niu, T.; Sun, L.; et al. Once-Weekly Oral Dosing of MK-8591 Protects Male Rhesus Macaques from Intrarectal Challenge with SHIV109CP3. *J. Infect. Dis.* **2020**, *221*, 1398–1406. [CrossRef] [PubMed]
48. Daly, M.B.; Kim, D.; Li, L.; Krovi, A.; Norton, C.; Forero, C.; Brake, M.; Dinh, C.; Edwards, T.I.; Mitchell, J.; et al. Vagi-nal PrEP efficacy of biodegradable islatravir implants in macaques. In Proceedings of the Conference on Retro-viruses and Opportunistic Infections, Seattle, WA, USA, 19–22 February 2023.
49. Bennink, H.J. The pharmacokinetics and pharmacodynamics of Implanon, a single-rod etonogestrel contraceptive implant. *Eur. J. Contracept. Reprod. Health Care* **2000**, *5* (Suppl. S2), 12–20. [CrossRef] [PubMed]
50. Slayden, O.D.; Luo, F.; Bishop, C.V. Physiological Action of Progesterone in the Nonhuman Primate Oviduct. *Cells* **2022**, *11*, 1534. [CrossRef] [PubMed]
51. Bleasby, K.; Houle, R.; Hafey, M.; Lin, M.; Guo, J.; Lu, B.; Sanchez, R.I.; Fillgrove, K.L. Islatravir Is Not Expected to Be a Victim or Perpetrator of Drug-Drug Interactions via Major Drug-Metabolizing Enzymes or Transporters. *Viruses* **2021**, *13*, 1566. [CrossRef]
52. Ankrom, W.; Jackson Rudd, D.; Zhang, S.; Fillgrove, K.L.; Gravesande, K.N.; Matthews, R.P.; Brimhall, D.; Stoch, S.A.; Iwamoto, M.N. A phase 1, open-label study to evaluate the drug interaction between islatravir (MK-8591) and the oral contraceptive levonorgestrel/ethinyl estradiol in healthy adult females. *J. Int. AIDS Soc.* **2021**, *24*, e25858. [CrossRef]

pharmaceutics

Article

Pharmacokinetic Modeling to Guide Preclinical Development of an Islatravir-Eluting Reservoir-Style Biodegradable Implant for Long-Acting HIV PrEP

Talisa S. Kinsale [1], Mackenzie L. Cottrell [1,*], Linying Li [2], Rhonda Brand [3], Greg Gatto [4], Ellen Luecke [4], Chasity Norton [2], Archana Krovi [2], Julie B. Dumond [1], Gauri Rao [1], Shekhar Yeshwante [1], Brian Van Horne [1], Ariane Van Der Straten [5,6], Angela D. M. Kashuba [1] and Leah M. Johnson [2]

1 Division of Pharmacotherapy and Experimental Therapeutics, Eshelman School of Pharmacy, University of North Carolina, Chapel Hill, NC 27599, USA; tkinsale@email.unc.edu (T.S.K.); jdumond@unc.edu (J.B.D.); gaurirao@live.unc.edu (G.R.); shekhary@email.unc.edu (S.Y.); bvanhorn@email.unc.edu (B.V.H.); akashuba@unc.edu (A.D.M.K.)
2 Biomedical Technologies RTI International, 3040 E. Cornwallis Road, Research Triangle Park, NC 27709, USA; ali@rti.org (L.L.); cnorton@rti.org (C.N.); akrovi@rti.org (A.K.)
3 Magee-Womens Research Institute, University of Pittsburgh, Pittsburgh, PA 15213, USA; rmb91@pitt.edu
4 Global Public Health Impact Center, RTI International, Research Triangle Park, NC 27709, USA; ggatto@rti.org (G.G.); eluecke@rti.org (E.L.)
5 Center for AIDS Prevention Studies (CAPS), Department of Medicine, University of California San Francisco, San Francisco, CA 94104, USA; ariane.vanderstraten@ucsf.edu
6 ASTRA Consulting, Kensington, CA 94708, USA
* Correspondence: mlcottre@email.unc.edu; Tel.: +1 919-843-0321

Abstract: Long-acting injectable cabotegravir is more effective than daily oral PrEP at preventing HIV transmission due to improved adherence, but requires bi-monthly large-volume intramuscular injections. Subcutaneous (SC) contraceptive implants can be formulated with antiretrovirals for extended-duration HIV PrEP. Islatravir (ISL) is a first-in-class, investigational antiretroviral with pharmacologic properties well-suited for implant delivery. We performed preclinical studies for the development of a reservoir-style, poly(ε-caprolactone) ISL-eluting implant by conducting a single-dose SC ISL dose-ranging pharmacokinetic (PK) study of 0.1, 0.3, and 1 mg/kg in adult Wistar rats. Non-compartmental analysis was conducted, and dose proportionality assessed for ISL plasma and intracellular islatravir-triphosphate (ISL-tp). Population PK models estimated ISL's unit impulse response to deconvolve ISL-implant in vivo absorption rate (mg/day) and cumulative mass (mg) from published rat plasma PK (n = 10). Drug release was interpreted using four kinetic models. Dose proportionality was affirmed for ISL and ISL-tp. A first-order, two-compartment model fitted the SC ISL bolus data. Mean (SD) absorption rate from 0 to 154 days was 0.072 ± 0.024 mg/day, and cumulative mass at 154 days was 8.67 ± 3.22 mg. ISL absorption was well-described by zero-order (r^2 = 0.95) and Ritger–Peppas (r^2 = 0.98). Our zero-order ISL-release poly(ε-caprolactone) implant is projected to achieve clinical PK above ISL-tp's PrEP efficacy threshold. Continued development for HIV PrEP applications is warranted.

Keywords: HIV prevention; preexposure prophylaxis; antiretroviral agents; islatravir; extended-duration; long-acting implant

1. Introduction

Fixed-dose combinations of daily oral antiretrovirals (emtricitabine 200 mg with tenofovir disoproxil fumarate 300 mg or tenofovir alafenamide 25 mg) have been shown to reduce the risk of HIV acquisition in multiple randomized placebo-controlled trials (RCT) and their open label extensions [1,2]. However, two large oral preexposure prophylaxis (PrEP) trials enrolling exclusively cisgender female populations were stopped early due to futility—later attributed to incomplete adherence to the study treatment [3,4]. Recently, a long-acting (LA) injectable formulation of the ARV, cabotegravir, demonstrated superiority over oral PrEP in preventing HIV acquisition in both male and female populations [5,6]. While injectable LA PrEP is more effective than oral PrEP due to improved adherence, it requires a large-volume (3 mL) intramuscular injection every other month with injection site reactions reported by 50% of patients [7]. Additionally, this injectable exhibits a sustained depletion phase (termed the PK tail), where the ARV (e.g., cabotegravir) is detectable at subtherapeutic levels [8]. This tail increases the risk of breakthrough infection with ARV resistance in patients who choose to discontinue LA PrEP or fall out of care.

Subcutaneous implants, such as those used for hormonal contraceptives, can offer therapeutic advantages over currently approved drug delivery approaches for LA PrEP [9]. Advantages include the potential for extended therapeutic intervals, retrievability in the event of adverse drug reactions or alternative reasons to discontinue PrEP, and improved adherence. The design of an implant comprising biodegradable polymers, such as poly(ε-caprolactone) (PCL), offers an additional advantage of bypassing the need for device removal, which could prove desirable for health care providers and end-users in HIV PrEP applications [10,11].

Islatravir (4′-ethynyl-2-fluoro-2′-deoxyadenosine [ISL]) is a prodrug, first-in-class nucleoside reverse transcriptase translocation inhibitor (NRTTI) deoxyadenosine analog that suppresses wild type HIV-1 virus at subnanomolar concentrations [12]. Its activity in preclinical and clinical studies demonstrated its antiviral potency as an investigational PrEP agent [13], and the active islatravir triphosphate (ISL-tp) metabolite exhibited an extended intracellular half-life (120–177 h) following oral dosing [14,15]. Pharmacokinetic/pharmacodynamic (PK/PD) modeling of clinical and preclinical data estimated an inhibitory quotient (IQ = 5) for ISL-tp, which defined a therapeutic threshold of 0.05 pmol/million peripheral blood mononuclear cells (PBMCs) for prevention of HIV-1 acquisition [16].

PrEP trials of an ISL-eluting prototype polymer implant (54 mg and 62 mg) [17] and downstream next-generation radiopaque [18] implant maintained ISL-tp PBMC concentrations above the PrEP efficacy threshold for 12 weeks. Secondary safety endpoints were similar to those observed in earlier oral studies, with no discontinuations due to adverse events [19]. In 2021, clinical trials evaluating ISL were placed on hold following findings of decreased $CD4^+$ T cell and total lymphocyte counts in participants taking oral ISL that ranged from 0.75 mg daily to 120 mg monthly [20]. Subsequent meta-analyses and modeling demonstrated a dose-dependent relationship for this adverse event and established 0.25 mg daily as a safe upper dosing limit for the oral formulation [21]. Importantly, clinical trials evaluating the ISL-eluting implants did not report decreased $CD4^+$ T cells likely due to the tested release rates in these studies falling below this upper dosing limit [17,18].

A biodegradable reservoir-style implant comprised of PCL has been previously formulated to co-deliver contraceptive hormone and ISL for 6 months in rats [22]. Dose optimization efforts for this type of controlled-release formulation can be enhanced through characterization of the mass transport mechanisms dictating drug dissolution. These mechanisms have been previously described by well-defined empirical mathematical equations such as zero-order, first-order, Higuchi and Ritger–Peppas.

Here, we conducted a preclinical SC bolus study in rats for deconvolution of ISL's in vivo absorption PK to optimize implant release rates for LA PrEP. We leveraged PK modeling to fully characterize the in vivo absorption and distribution kinetics as a guide to interspecies dose-translation efforts during development of a long-acting controlled-release implant for HIV PrEP.

2. Materials and Methods

2.1. Drug Preparation

ISL was provided by Pharmaron (Beijing, China). Drug purity was determined on a 5-point calibration curve by linear regression analysis to be 97.3% and titrated as 1 mg/mL stock solutions in DMSO at 37 °C and prepared 24 h prior to administration.

2.2. Preclinical Islatravir PK Study

A SC bolus animal study was conducted to match the species and route of administration for the published SC ISL-eluting implant study under the University of Pittsburgh Institutional Animal Care and Use Committee (IACUC) (Protocol 20056840) at Magee-Womens Research Institute and Foundation (Pittsburgh, PA, USA), in collaboration with Biomedical Technologies (RTI International, Research Triangle Park, NC, USA). All procedures met the principles of the NIH Office of Laboratory Animal Welfare, Guide for the Care and Use of Laboratory Animals [23]. Seventy-two adult female Wistar rats with mean (±standard deviation, SD) body weight of 0.307 ± 0.02 kg, were included in this dose-ranging study designed to characterize the 24 h plasma and PBMC PK profile for three ISL doses, 0.1 mg/kg (low), 0.3 mg/kg (mid), and 1 mg/kg (high), selected to encompass the projected daily dose from the published optimized implant design [22].

2.3. Dose Administration and Sparse Sampling Schema

Animals were observed for general health daily and weighed prior to each dose. Sparse timepoints collected were 0.17, 0.5, 1, 2, 4, 6, and 24 h for each dose level. Predose samples were collected before ISL bolus administration, and all were below limits of quantification. For a single timepoint, three rats were administered a SC bolus under the neck skin and sampled at each pre-determined timepoint, contributing one measure per animal. The dosing schedule within each dose level was staggered to reduce blood loss so that each subset (n = 3) of animals contributed a single measure per timepoint within the 24 h period. Blood was drawn from the lateral tail vein at each set timepoint (1.2–1.7 mL), and samples were processed to plasma or PBMCs by centrifugation for the determination of ISL and ISL-tp using previously published procedures [22].

2.4. Islatravir Quantification

ISL was quantified in plasma and ISL-tp in PBMC cell lysate using previously published liquid-chromatography tandem mass spectrometry (LC-MS/MS) methods [22,24]. In short, ISL and ISL-tp were prepared in 100 µL of 70:30 methanol/water and extracted with isotopically labeled internal standards (IS-$^{3}C^{15}N_3$-ISL and $^{3}C_{10}$,$^{15}N_5$-dGTP, respectively) and detected on an AB Sciex API-5000 triple quadrupole mass spectrometer under positive ionization mode (SCIEX, Framingham, MA, USA). A dynamic calibration range of 0.1–100 ng/mL was used for ISL with precision and accuracy acceptance criteria within ±15%. ISL-tp was analyzed using the same mass spectrometer calibrated for 0.05–125 ng/mL (ISL-tp). ISL-tp concentrations were normalized to representative live cell counts determined by Trypan blue staining and a hemocytometer and reported as pmol/million cells. Concentrations below the limit of quantification (LLOQ) for each were censored according to the Beal M2 method [25].

2.5. Noncompartmental Analysis

SC bolus plasma ISL (ng/mL) and intracellular ISL-tp (pmol/million cells) concentration data were visually inspected on a concentration–time plot. A sparse, extravascular, weight (kg) normalized, non-compartmental analysis (NCA) was used to estimate pharmacokinetic endpoints for maximal concentration (C_{max}), time to C_{max} (T_{max}) and area under the curve to the last measurable timepoint (AUC_{last}) by linear up-log down trapezoidal method. Half-life ($t_{1/2}$) was estimated from 4 to 24 h for both ISL and ISL-tp profiles that had 3 quantifiable datapoints, except the low dose (0.1 mg/kg) plasma ISL. The average concentration (C_{avg}) of plasma ISL and PBMC ISL-tp were estimated by NCA estimated AUC divided by 24 h. The ratio of ISL-tp:ISL was calculated assuming an approximate PBMC volume of 0.2 pL per cell determined for 1 million live cells in each PBMC sample [17]. Dose proportionality was assessed for ISL and ISL-tp by unweighted linear regression of the natural log (ln) transformed parameters, C_{max} and AUC_{last}, versus ln transformed absolute dose (mg) of each dose level. The slope was reported, and linearity was declared for values approximating 1 and coefficient of r^2 > 0.8.

2.6. Pharmacokinetic Analysis

We tested 1- and 2-compartment, macro parameterized population PK (PopPK) models with first-order linear kinetics fit to the SC bolus plasma ISL concentration data for individual dose levels (0.1, 0.3, 1 mg/kg) and a collapsed simultaneous model of all dose levels. Each population model was conducted with a naive pooled algorithm for multiplicative, additive, and log-additive error models. They were each evaluated by: (1) goodness-of-fit plots (residual, predicted and quantile-quantile), (2) model diagnostics (log-likelihood and akaike/Bayesian criterion), and (3) reduction in parameter/model variability (%CV < 30) for the final model. To enhance robustness, each model eta (η) diagnostics were tested for shrinkage (>0.3) and nested models were additionally prioritized by minimum objective function value. The final popPK model estimates for macro-constants (A, B ng/mL) were divided by the average dose (145,803 ng) to attain the unit impulse response (UIR) for plasma ISL in rats. Each dose profile was assessed for outliers by Grubb's test which were excluded, and PK analyses were performed in Phoenix NLME™ v6.3 (Certara, Inc., Princeton, NJ, USA), with estimates reported as mean ± SD per dose level.

2.7. Deconvolution of In Vivo Absorption

To perform deconvolution, we utilized previously published plasma PK profiles from 10 adult Wistar rats with a SC ISL-eluting implant (40 mm long, 100 μm wall-thickness) comprised of extruded PCL from PURASORB PC-17 pellets (106 kDa) (Amsterdam, The Netherlands) formulated with a 1:1 mass ratio of ISL: sesame oil [22]. In this previously published study animals were treated with either ISL-only, which included one ISL-eluting implant and one excipient-only implant; or ISL-plus with one ISL-eluting implant and one etonogestrel (ENG)-eluting implant (hormone data not discussed here). While PK samples were collected over 197 days of implantation in this previous published study, increasing inter-individual variability (%CV > 90) in plasma concentrations was observed beyond 100 days, which may have been associated with decreased mechanical integrity of the implants. As such, the deconvolution analyses for in vivo drug absorption were constrained to 154 days (%CV ≤ 60). Two animals were identified as outliers contributing >30% variability to the mean PK profiles and were excluded from downstream analyses. The 2-compartmental model-derived UIR was used to deconvolve the individual in vivo animal absorption profiles to estimate: (1) rate of absorption (mg/day), (2) cumulative mass absorbed (mg), and (3) fraction of mass released from the implant loaded dose. Welch's *t*-test was utilized to compare the rate of absorption for ISL-only and ISL-plus groups at 90 and 154 days. Drug recovery was calculated as mean cumulative mass absorbed (mg) at day 154 subtracted from mean loaded dose (98.6 ± 3.31 mg) in the reservoir-style PCL implant. Analyses were performed in Phoenix WinNonlin™ Deconvolution Toolkit v6.3 (Certara, Inc., Princeton, NJ, USA) and reported as mean ± SD for each profile.

2.8. In Vivo Drug Absorption Models

Plasma ISL in vivo absorption kinetics were characterized by fitting the individual animal profiles (n = 10) of deconvolved cumulative mass (mg) absorbed over time (days) to 4 kinetic models describing drug distribution from the published ISL-eluting PCL implant [22]. Cumulative mass (mg) data were fit to (pseudo)linear equations of each kinetic model using SAS® Studio v9.04 (Enterprise 3.81, SAS Institute, Cary, NC, USA) to fit regression models to their release profiles using the "proc model" function (Figure S1). The individual r-squared (r^2) correlation coefficient was determined for each animal profile (n = 10) and prioritized by highest mean r^2 with a minimum threshold of $r^2 > 0.75$. For the Ritger–Peppas model the drug release exponent, n, was characterized to describe mechanisms of non-Fickian release. The mathematical equations describing each model for: Zero-order (constant release); First-order (concentration-dependent release with respect to time); Higuchi (square root of a time-dependent release from planar matrix systems); and Ritger–Peppas (power law exponential diffusion from polymeric systems) were:

$$M_t = M_0 + K_Z t, \text{ Zero-order} \quad (1)$$

$$M_t = M_0 + K_F t, \text{ First-order} \quad (2)$$

$$M_t = M_0 + K_H t^{0.5}, \text{ Higuchi} \quad (3)$$

$$M_t / M_{inf} = K_R t^n, \text{ Ritger-Peppas} \quad (4)$$

where M_t is the cumulative mass absorbed from the implant at time t, M_0 is the mass of drug absorbed at $t = 0$ (attributable to instantaneous burst release), M_{inf} is the total dug loaded in the implant, n is the power exponential (Ritger–Peppas only), and K is the absorption rate constant for each respective model estimated by regression analysis. We assumed that M_0 was negligible for this reservoir-style implant system. Each mathematical model was leveraged as reported by Bruschi et al. and Costa et al., to characterize in vivo drug absorption kinetics of ISL from the PCL implant [26,27].

3. Results

3.1. Noncompartmental PK Analysis

SC bolus plasma concentrations were plotted against time for $n = 57$ quantifiable ISL values. Visual inspection of each concentration–time profile elucidated a multi-phasic drug distribution and clearance (Figure 1a); 0–5 h for the first phase, and 5 h onwards for the second, excluding the lowest dose level which was not quantifiable at 24 h. Six values were below the assay limit of quantification (BLQ), and one outlier (confirmed by Grubb's test) was censored. The mean T_{max} was at 10 min except for the mid dose level (0.3 mg/kg) which occurred at 30 min. Estimates for C_{max}, C_{avg} and AUC_{last} were: 22.6, 44.7, 201.3 ng/mL, 0.675; 2.283, 9.566 ng/mL; and 16.2, 54.8, 229.6 h*ng/mL for dose levels 0.1, 0.3, 1.0 mg/kg, respectively. Plasma ISL C_{max} and AUC_{last} exhibited dose proportionality with slopes of 0.95 ($r^2 = 0.961$) and 1.15 ($r^2 = 0.995$) for ln[Dose] vs. ln[C_{max}] and ln[AUC_{last}], respectively.

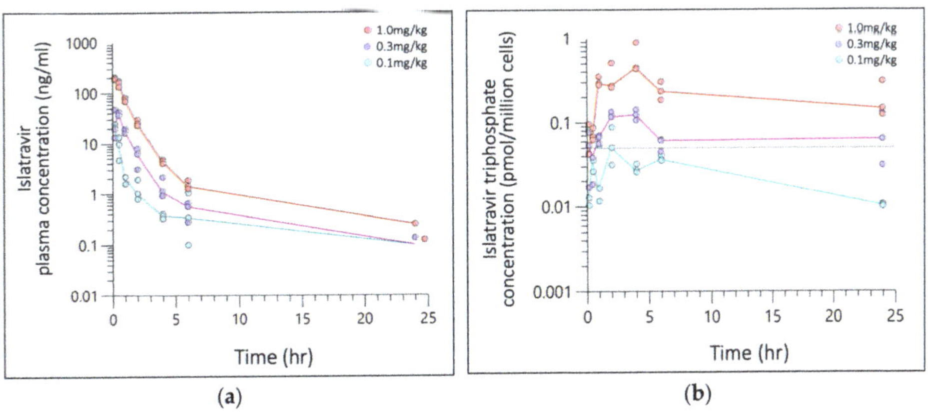

Figure 1. Semi-log concentration–time plot of (**a**) Islatravir (ISL) plasma concentration (ng/mL) and (**b**) ISL-triphosphate (ISL-tp) concentration (pmol/million cells) following a single SC administration at 0.1, 0.3, or 1.0 mg/kg. Plasma and PBMCs was collected from 3 rats per time point, with 24 rats per dosing level. Markers are individual plasma concentration with median (line) categorized by dose (color). The grey reference line (dash) is clinical threshold of 0.05 pmol/million cells.

SC bolus ISL-tp PBMC concentrations were plotted against time for $n = 43$ quantifiable ISL-tp values (Figure 1b) and 20 BLQ values. Time to maximal concentration ranged from

2–4 h and $t_{1/2}$ was 11.8, 27.1, 17.8 hr^{-1}, respectively. Estimates for C_{max}, C_{avg} and AUC_{last} were: 0.057, 0.125, 0.593 pmol/million cells; 0.025, 0.073, 0.254 pmol/million cells; and 0.609, 1.746, 6.086 h*pmol/million cells for dose levels 0.1, 0.3, 1.0 mg/kg, respectively. PBMC ISL-tp C_{max} and AUC_{last} exhibited dose proportionality with slopes of 1.02 (r^2 = 0.974) and 0.99 (r^2 = 0.998) for ln[Dose] vs. ln[Cmax] and ln[AUC_{last}], respectively. The ratio of C_{avg} for ISL-tp:ISL for 0.1, 0.3 and 1.0 mg/kg doses were 100, 85 and 70, respectively.

3.2. Pharmacokinetic Modeling

The 24 h plasma-concentration PK profile for all SC ISL bolus doses were best described by a simultaneously modeled two-compartment, multiplicative error, naïve pooled PopPK model, with 76% residual error (Figure 2) for all dose levels. Extravascular absorption rate (Ka) was >10 h^{-1} for individual dose level models and estimated at 12 h^{-1} for the final simultaneously modeled extravascular two-comp PopPK macro-model. The popPK macro-parameters were fit to each compartment describing the parameter estimates for determining the UIR of ISL in rats after subcutaneous administration (Figure 2, inset table).

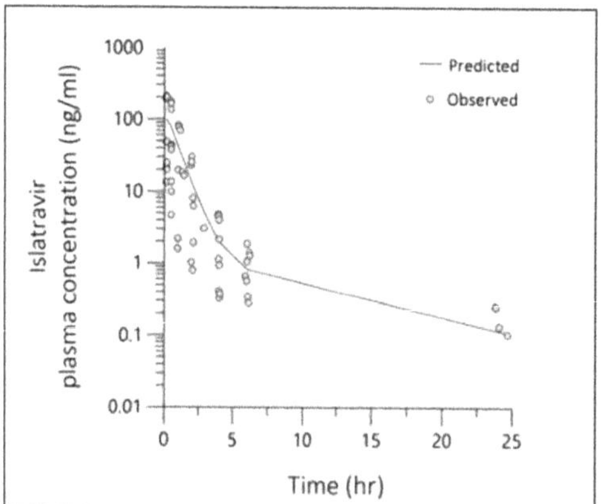

Parameter	Ka (1/h)	A (ng/ml)	alpha (1/h)	B (ng/ml)	beta (1/h)
Estimate(CV%)	12 (17)	142 (26)	1.2 (5)	1.4 (24)	0.103 (7)

Figure 2. Two-compartment population pharmacokinetic (PopPK) model with multiplicative error was fit to ISL plasma concentrations observed in rats following single SC administration at 0.1, 0.3 and 1 mg/kg doses. Macro-parameter constants (A, B) and exponential rates (α, β) are described in the inset table.

No preclusive deviations from normality were identified in the final model diagnostic plots which were evaluated for; (Figure 3a) individual unweighted residual distribution, (Figure 3b) observed–predicted unweighted distribution about the line of unity, and (Figure 3c) the quantile–quantile plot describing the normality of the model residuals. Lower concentrations in the 0.1 mg/kg dose level approaching the limit of quantification had overall negligible bias in the final model. Tailing of the quantiles indicated some deviations from normality in the extremities. Overall, the diagnostics show adequate fit for fit-for-purpose deconvolution, with best prediction for doses 0.3 mg/kg and higher. The mean UIR estimates for ISL in rats were $A_1 = 9.8 \times 10^{-4}$ ng/mL/ng, $A_2 = 9.6 \times 10^{-6}$ ng/mL/ng, $\alpha_1 = 1.2\ h^{-1}$ and $\alpha_2 = 0.103\ h^{-1}$.

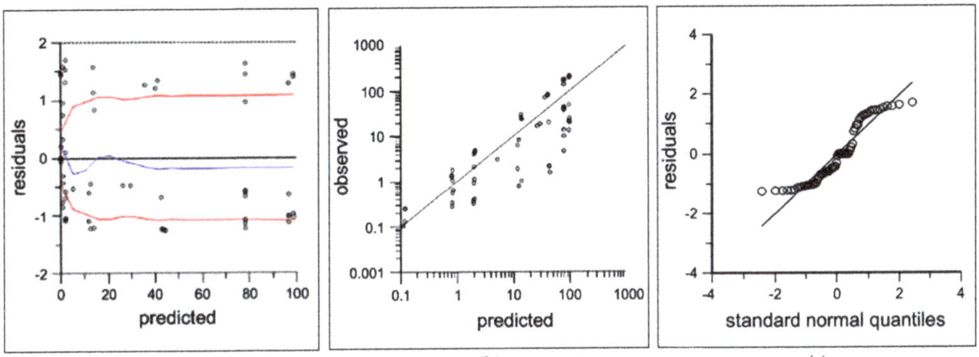

(a) (b) (c)

Figure 3. Two-compartment population PK diagnostics showing observed (open circles) and predicted (black lines) for (**a**) individual predicted concentrations vs. residuals with loess fit (blue) and absolute trend lines (red), (**b**) individual predicted vs. observed log-concentrations, and (**c**) quantile–quantile normal plot.

3.3. In Vivo Absorption

The mean absorption rate from 0 to 154 days in the ISL-only and -plus groups were 0.084 ± 0.025 and 0.061 ± 0.018 mg/day with an overall mean of 0.072 ± 0.024 mg/day (Figure 4a). The mean cumulative mass absorbed from 0–154 days for the ISL-only and ISL-plus groups was 8.82 ± 4.07 and 8.53 ± 2.60 mg, respectively with an overall mean of 8.67 ± 3.22 mg (Figure 4b).

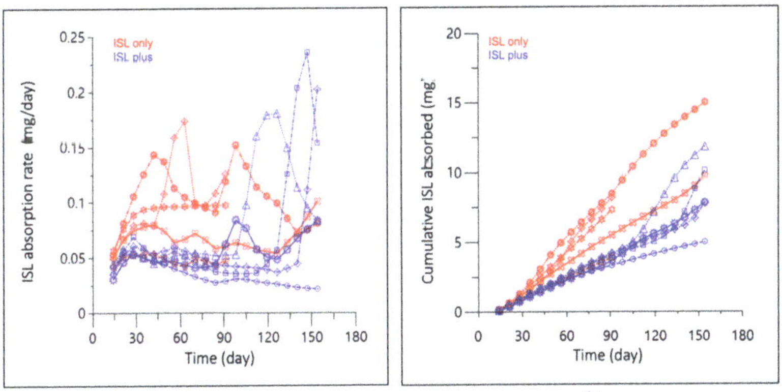

(a) (b)

Figure 4. (**a**) Absorption rate, and (**b**) cumulative mass absorbed of subcutaneous ISL-eluting implants in rats over 154 days. Red represents animals receiving ISL only, while blue received ISL alongside a hormone implant (ISL plus). Lines and markers (shape) represent each individual animal sample collected at approximately 15-day increments. Three implants were removed after 100 days in the ISL only group.

The best-fitting pseudo (linear) mathematical models describing in vivo absorption from the ISL-eluting implant were Zero-order and Ritger–Peppas (Table 1) with a mean (SD) r^2 of 0.946 (0.04) and 0.985 (0.02), respectively. The mean (SD) power exponent estimated for Ritger–Peppas was 1.3 (0.29), with a non-reportable (N.R.) estimate for one animal due to non-convergence. First-order and Higuchi models did not meet the predetermined threshold of mean $r^2 > 0.75$ with, mean r^2 of −4.757 (1.08) and 0.699 (0.08), respectively. For the Higuchi model, only 3 of 10 animals had individual fits of $r^2 > 0.75$.

Table 1. Regression coefficient (r^2) kinetic models of drug absorption from the SC ISL eluting implant.

Group	Zero-Order r^2	First-Order r^2	Higuchi r^2	Ritger–Peppas [b] r^2
ISL_only	0.9680	−6.6068	0.7054	N.R. [a]
ISL_only	0.8933	−4.0118	0.5972	0.9909
ISL_only	0.9409	−4.5061	0.6568	**0.9944**
ISL_only	0.9776	−3.0757	0.7388	0.9908
ISL_only	**0.9888**	−4.3850	0.7672	0.9942
ISL_plus	0.9805	−6.5246	0.8384	0.9855
ISL_plus	0.8667	−4.2326	0.5859	0.9844
ISL_plus	0.8971	−4.7319	0.6481	0.9399
ISL_plus	**0.9825**	−4.8870	0.7619	0.9878
ISL_plus	0.9615	−4.6075	0.6997	**0.9978**
Mean	**0.9457**	−4.7569	0.6999	**0.9851**
Standard deviation	0.04	1.08	0.08	0.02

Best results in bold. [a] N.R. not-reportable due to model non-convergence. [b] Mean (SD) power exponent, *n*, of 1.3 ± 0.29.

4. Discussion

While injectable LA PrEP, cabotegravir, helps address limitations to oral PrEP regimens, cost and bi-monthly clinical visits to administer the intramuscular injection by a trained professional may limit its utility among populations of lower socioeconomic status or living far from health infrastructures such as in rural areas [28]. The subtherapeutic PK tail of injectable LA PrEP is another disadvantage that poses a risk of breakthrough infection with ARV resistance in patients who switch therapy or fall out of care [8]. SC implants have been used clinically to deliver therapeutic hormonal contraception for up to 3 years since Nexplanon's FDA approval in 2006 [29]. A similar delivery platform could improve upon injectable LA PrEP by extending the dosing interval, offering retrievability for those who choose to discontinue treatment, and facilitating multiple preventative indications such as contraceptive care. In clinical trials, an ISL-eluting implant was safe and achieved active metabolite (ISL-tp) concentrations in PBMCs above the PrEP efficacy threshold (0.05 pmol/million cells) for 12 weeks with projections that therapeutic exposure could be sustained for up to 16 months [17]. A downstream clinical trial of this implant recapitulated the clinical PK with no dose-dependent adverse events or discontinuations [18]. ISL remains a safe, potent and favorable model drug to develop as an implantable platform for LA HIV PrEP.

First-order release implants such as Implanon® [9] are clinically useful; however, a zero-order implant would improve the predictability of drug release and may offer greater confidence in therapeutic concentrations near the end of the PK interval which may reduce risk of ARV resistance. Li et al. previously published on our reservoir-style, PCL biodegradable implant where zero-order release was directly observed for 13–17 months under in vitro sink conditions with demonstrable batch to batch consistency in raw material, formulation and end-product characteristics [22]. This release profile optimized implant delivered ISL for 6 months after subdermal implantation, was safe and biocompatible in a rat preclinical study [22]. While the original implant study was designed to explore whether dual implants had an influence on plasma PK, herein we sought to utilize a PK modeling approach with an additional independent SC bolus study to fully characterize in vivo absorption kinetics and guide optimization of drug release rates and dose translation efforts.

To estimate PK parameters of ISL in rats, we conducted a dose-ranging, SC bolus PK study of ISL solution that leveraged a sparse design with a population PK modeling approach. Single-dose pharmacokinetics indicated dose proportionality of ISL parent and its metabolite ISL-tp for all doses in rats [12]. ISL and ISL-tp halflife characterized in our study was limited by sample number and short duration of samples in the elimination phase. We observed a significant difference in rat PBMC ISL-tp: plasma ISL ratios compared to that reported in clinical trials of an ISL-eluting implant (~2390) [17]. However, our observation of lower ISL-tp:ISL ratios in rats relative to humans is consistent with the between species differences reported in Sykes et al. [24]. These findings indicate that allometric scaling of certain preclinical animal models will have limited utility in identifying clinical doses to achieve ISL's PBMC efficacy target.

Based on the deconvolution model, only 3 of the 10 animals did not achieve an absorption rate of 50 µg/day for at least 50% of the analyzed profile. Graphical analysis of Figure 7c,d in Barrett et al. demonstrated that a SC ISL-eluting implant releasing approximate 250 µg/day was projected to achieve 2 nM ISL plasma concentrations in humans 60 days post-implantation [30]. During first in human trials of their down-selected implant formulation, steady-state ISL plasma concentrations of 1 nM were associated with ISL-tp PBMC concentrations of 0.272 pmol/million cells, which is 5-fold higher than target exposure [17]. Given our average dose of 50 µg/day, and assuming no between-species differences in drug delivery mechanisms, we predict that our implant will result in human PBMC concentrations at or exceeding the PrEP threshold of 0.05 pmol/million cells. Furthermore, all estimated absorption rates throughout the 154 day observation window were below 0.20 mg/day, with the exception of one timepoint that approached, but was still below the 0.25 mg/day upper dosing limit for oral ISL. In examining the PK curve associated with this observation it is likely that impaired mechanical integrity of the implant contributed to this isolated observation of abnormally high release [22]. Cumulative drug release estimated by deconvolution in our study showed that less than 15% of the mean loaded dose (98.6 mg) was absorbed during the 154 day observation window. Consistent with that estimation, analysis of residual ISL in recovered implants from three animals at a 3-month interim necropsy demonstrated 3–7% (min–max) of ISL had been absorbed. Fine-tuning an ARV-eluting implant to deliver a constant rate of the lowest effective dose within an in vivo system is a worthwhile developmental goal to conserve ARV payload and extend the dosing interval. Herein, we provide in vivo evidence that a reservoir-style implant comprised of PCL can deliver ISL at a constant rate that falls within the safe and potentially effective dosing range for humans for up to 5 months. Furthermore, this implant can be loaded with sufficient ISL to sustain this release profile for well over a year.

ISL's in vivo absorption was consistent with a constant rate drug release as the zero-order equation explained a high proportion of variance in our data with r^2 values ranging from 0.87–0.99. Because biofluids transport into our implant architecture via passive diffusion, swelling is possible. Thus, we also tested a Ritger–Peppas model of mass transport intended to deal with the moving boundaries problem that swelling presents for Fickian diffusion. Importantly, the estimated power exponent term within our Ritger–Peppas model fits approximated 1 with a mean (SD) of 1.3 (0.29). This finding further supports our conclusion of time-independent constant rate transport (i.e., zero-order) [31]. For cylindric drug delivery systems such as ours, Ritger–Peppas exponential terms of 2 are associated with the time-dependent Super Case-II release, terms of 0.45–0.89 with non-Fickian anomalous release and terms of 1 with zero-order release [31]. Furthermore, the Higuchi equation explained a relatively low proportion of variance in our data with r^2 values ranging from 0.59 to 0.83. Finally, within our first-order model fits, a constant function better fit the data than a linear one leading to negative r^2 values. Ultimately, the overall zero-order controlled release mechanism characterized by our PCL, biodegradable, reservoir-style implant may improve the predictability of estimating PK beyond the observation period, assuming polymer and physiologic conditions are optimized. While implant constant rate kinetics may become less predictable when physiochemical conditions change, i.e., the implant polymer degrades, and/or the payload decreases, the current polymer has been adjusted for persistent physiological conditions up to 1 year. Historically, reservoir-style architecture provides the greatest predictability of drug PK for implantable polymeric systems [26]. Our zero-order controlled release implant has leveraged this advantage to improve prediction of in vivo ISL absorption and guide drug development of a LA-implantable PrEP device.

While unquantifiable ISL concentrations at 24 h following the 0.1 mg/kg SC dose increased residual model error, downstream analyses were determined by population estimates that met assumptions of linearity and normality per the diagnostics plots (Figure 3b–d). Differences in T_{max}, 10 min vs. 30 min, are likely attributed to the study size and random variability due to sparse sampling methods. While an individual UIR estimation for each rat used in the implant study would be an ideal study design, blood volume constraints precluded the intensive sampling necessary to accurately estimate ISL's UIR in individual animals. The Wistar rats utilized in the two separate studies were sourced from a reputable vendor and the studies were conducted within the same animal facility using identical sample collection processes. There was <20% difference in the range of animal weights in the subcutaneous bolus and implant studies [22]. Deconvolution profiles (Figure 4a,b) elucidated between group differences for mean absorption rate (p = 0.03) and cumulative mass absorbed at 90 days (p = 0.03), that were not statistically different at 154 days, p = 0.12 and p = 0.89, respectively. Since implant PCL fabrication, ISL formulation and animal cohorts were matched across both animal groups and interrogated across several in vitro studies, no biopharmaceutical reasons have been identified for these differences and are likely attributed to mismatched numbers in groups [22]. The parallel in vitro ISL dissolution study conducted by Li et al. reported a mean in vitro release rate of 0.066 ± 0.008 mg/day, which compares well to our deconvolved mean in vivo absorption rate of 0.072 ± 0.024 μg/day indicating a high correlation of drug release kinetics from in vitro-to-in vivo [22]. This PK modeling approach identified a direct relationship between in vitro and in vivo absorption rates.

The predictive applicability of deconvolution is associated with estimating a suitable UIR. Because our UIR is derived from subcutaneous PK data, our model relies on an assumption of near instantaneous absorption. Our estimated absorption (Ka) value of 12 hr^{-1} provides evidence of instantaneous absorption of ISL into the plasma from the subcutaneous space. While sampling was limited in the SC bolus rat study, the observed data was developed into a fit-for-purpose model describing the mean population parameters necessary for deconvolution. Additionally, dose level half-life estimation was limited by sparse sampling in the elimination phase and was not utilized as a PK parameter for deconvolution. The sparse observed data prior to C_{max} indicated rapid absorption of ISL 10–30 min after SC bolus administration for all doses. Future studies present an opportunity to intensely sample the absorption and terminal phases to improve robustness for applications that require a full PK model. Though limited literature is available on the individual acceptability of UIRs estimated from extravascular routes of administration, various extended release models have been reported that leverage extravascular routes of administration for deconvolution purposes [32,33]. Additionally, European and U.S. agencies accept UIR derived from extravascular routes of administration for regulatory review purposes [34,35]. The applicability of our model to inform dose-translation efforts also relies on an assumption that SC absorption rates do not differ between species. This approach to derive in vivo absorption has been successful in translating dosing from preclinical to clinical phases [17,30]. Altogether, we found the deconvolution methods conducted herein to be highly applicable to this platform and do not expect dose translation to be inhibited by use of a SC dose-ranging design for PK analysis and UIR determination.

Novel LA PrEP platforms with simplified dosing are critical for addressing adherence challenges in HIV prevention. The PK modeling approach applied herein demonstrates that our reservoir-style biodegradable ISL-eluting implant can sustain zero-order release at least 30 μg/day for 5 months in rats. We project that this implant could achieve clinical exposure exceeding ISL's PrEP efficacy threshold. Li et al. previously demonstrated the utility of this implant design for sustained zero-order hormonal contraceptive delivery [22]. This deconvolution approach estimated the daily dose of a prototype implant designed to deliver extended-duration antivirals. Together, these data support the utility of PK modeling as a guide to preclinical development of this implant device as a promising candidate for co-formulation and downstream applications in multipurpose prevention technologies (MPT) including LA HIV PrEP.

Supplementary Materials: The following supporting information can be downloaded at: https://www.mdpi.com/article/10.3390/pharmaceutics16020201/s1, Figure S1: Model code for regression analysis of 4 kinetic models of drug distribution in SAS^{TM} 9.4.; Table S1: Population pharmacokinetic (PopPK) model development and comparison in Phoenix NLME^{TM}.

Author Contributions: Conceptualization, L.M.J. and M.L.C.; methodology, M.L.C., T.S.K. and G.R.; formal analysis, T.S.K., M.L.C., S.Y. and G.R.; resources, L.M.J., G.G., E.L., A.D.M.K. and M.L.C.; data curation, B.V.H. and L.L.; writing—original draft preparation, T.S.K. and M.L.C.; writing—review and editing, L.M.J., A.D.M.K., J.B.D., A.V.D.S., G.G., A.K., A.V.D.S., T.S.K., C.N. and R.B.; supervision, A.D.M.K. and M.L.C.; funding acquisition, L.M.J., A.V.D.S. and M.L.C. All authors have read and agreed to the published version of the manuscript.

Funding: This research was made possible through the support of the U.S. President's Emergency Plan for AIDS Relief (AID-0AA-A-14-00012 and AID-OAA-A-17-00011) in collaboration with the United States Agency for International Development (USAID). The contents are the responsibility of the authors and do not necessarily reflect the views of RTI, USAID, PEPFAR, or the United States Government. This work was also supported by the University of North Carolina at Chapel Hill Center For AIDS Research (CFAR), an NIH funded program (P30AI050410).

Institutional Review Board Statement: The animal study protocol was approved by the Institutional Review Board (or Ethics Committee) of the University of Pittsburgh (protocol code 20056840 on 5 May 2020 for studies involving animals.

Informed Consent Statement: Not applicable.

Data Availability Statement: The data presented in this study are openly available online in the National Library of Medicine, PubMed at [https://doi.org/10.1016/j.jconrel.2021.10.021, accessed on 12 October 2023], reference number [22].

Acknowledgments: We thank RTI and CPAC members for their contributions to coordinating work in this study.

Conflicts of Interest: Authors L.M.J., L.L., A.K. and A.V.D.S. are the inventors of pending patent applications for the implant technology presented in this manuscript. Ariane Van Der Straten is an individual consultant under ASTRA consulting. The authors declare that the research was conducted in the absence of any commercial or financial relationships that could be construed as a potential conflict of interest.

References

1. Grant, R.M.; Lama, J.R.; Anderson, P.L.; McMahan, V.; Liu, A.Y.; Vargas, L.; Goicochea, P.; Casapía, M.; Guanira-Carranza, J.V.; Ramirez-Cardich, M.E.; et al. iPrEx Study Team Preexposure chemoprophylaxis for HIV prevention in men who have sex with men. *N. Engl. J. Med.* **2010**, *363*, 2587–2599. [CrossRef] [PubMed]
2. Mayer, K.H.; Molina, J.-M.; Thompson, M.A.; Anderson, P.L.; Mounzer, K.C.; De Wet, J.J.; DeJesus, E.; Jessen, H.; Grant, R.M.; Ruane, P.J.; et al. Emtricitabine and tenofovir alafenamide vs. emtricitabine and tenofovir disoproxil fumarate for HIV pre-exposure prophylaxis (DISCOVER): Primary results from a randomised, double-blind, multicentre, active-controlled, phase 3, non-inferiority trial. *Lancet* **2020**, *396*, 239–254. [CrossRef] [PubMed]
3. Van Damme, L.; Corneli, A.; Ahmed, K.; Agot, K.; Lombaard, J.; Kapiga, S.; Malahleha, M.; Owino, F.; Manongi, R.; Onyango, J.; et al. FEM-PrEP Study Group Preexposure prophylaxis for HIV infection among African women. *N. Engl. J. Med.* **2012**, *367*, 411–422. [CrossRef] [PubMed]
4. Marrazzo, J.M.; Ramjee, G.; Richardson, B.A.; Gomez, K.; Mgodi, N.; Nair, G.; Palanee, T.; Nakabiito, C.; van der Straten, A.; Noguchi, L.; et al. VOICE Study Team Tenofovir-based preexposure prophylaxis for HIV infection among African women. *N. Engl. J. Med.* **2015**, *372*, 509–518. [CrossRef]
5. Landovitz, R.J.; Donnell, D.; Clement, M.E.; Hanscom, B.; Cottle, L.; Coelho, L.; Cabello, R.; Chariyalertsak, S.; Dunne, E.F.; Frank, I.; et al. HPTN 083 Study Team Cabotegravir for HIV prevention in cisgender men and transgender women. *N. Engl. J. Med.* **2021**, *385*, 595–608. [CrossRef]
6. Delany-Moretlwe, S.; Hughes, J.P.; Bock, P.; Ouma, S.G.; Hunidzarira, P.; Kalonji, D.; Kayange, N.; Makhema, J.; Mandima, P.; Mathew, C.; et al. HPTN 084 study group Cabotegravir for the prevention of HIV-1 in women: Results from HPTN 084, a phase 3, randomised clinical trial. *Lancet* **2022**, *399*, 1779–1789. [CrossRef] [PubMed]
7. APRETUDE (Cabotegravir Extended-Release Injectable Suspension) Prescribing Information, ViiV Healthcare. Available online: https://www.accessdata.fda.gov/drugsatfda_docs/label/2021/215499s000lbl.pdf (accessed on 18 November 2023).

8. Landovitz, R.J.; Li, S.; Eron, J.J.; Grinsztejn, B.; Dawood, H.; Liu, A.Y.; Magnus, M.; Hosseinipour, M.C.; Panchia, R.; Cottle, L.; et al. Tail-phase safety, tolerability, and pharmacokinetics of long-acting injectable cabotegravir in HIV-uninfected adults: A secondary analysis of the HPTN 077 trial. *Lancet HIV* **2020**, *7*, e472–e481. [CrossRef]
9. Adams, K.; Beal, M.W. Implanon: A review of the literature with recommendations for clinical management. *J. Midwifery Women's Health* **2009**, *54*, 142–149. [CrossRef]
10. Brown, M.S.; Hanif, H.; Little, K.M.; Clark, M.R.; Thurman, A.R.; Flomen, L.; Doncel, G.F. End-user research in support of long-acting systemic antiretroviral delivery systems: Insights from qualitative research with providers and target users in South Africa. *BMC Infect. Dis.* **2022**, *22*, 919. [CrossRef]
11. Humphries, H.; Upfold, M.; Mahlase, G.; Mdladla, M.; Gengiah, T.N.; Abdool Karim, Q. Implants for HIV prevention in young women: Provider perceptions and lessons learned from contraceptive implant provision. *PLoS ONE* **2022**, *17*, e0262043. [CrossRef]
12. Markowitz, M.; Sarafianos, S.G. 4′-Ethynyl-2-fluoro-2′-deoxyadenosine, MK-8591: A novel HIV-1 reverse transcriptase translocation inhibitor. *Curr. Opin. HIV AIDS* **2018**, *13*, 294–299. [CrossRef] [PubMed]
13. Markowitz, M.; Gettie, A.; St Bernard, L.; Andrews, C.D.; Mohri, H.; Horowitz, A.; Grasperge, B.F.; Blanchard, J.L.; Niu, T.; Sun, L.; et al. Once-Weekly Oral Dosing of MK-8591 Protects Male Rhesus Macaques from Intrarectal Challenge with SHIV109CP3. *J. Infect. Dis.* **2020**, *221*, 1398–1406. [CrossRef] [PubMed]
14. Schürmann, D.; Rudd, D.J.; Zhang, S.; De Lepeleire, I.; Robberechts, M.; Friedman, E.; Keicher, C.; Hüser, A.; Hofmann, J.; Grobler, J.A.; et al. Safety, pharmacokinetics, and antiretroviral activity of islatravir (ISL, MK-8591), a novel nucleoside reverse transcriptase translocation inhibitor, following single-dose administration to treatment-naive adults infected with HIV-1: An open-label, phase 1b, consecutive-panel trial. *Lancet HIV* **2020**, *7*, e164–e172. [CrossRef] [PubMed]
15. Matthews, R.P.; Jackson Rudd, D.; Zhang, S.; Fillgrove, K.L.; Sterling, L.M.; Grobler, J.A.; Vargo, R.C.; Stoch, S.A.; Iwamoto, M. Safety and Pharmacokinetics of Once-Daily Multiple-Dose Administration of Islatravir in Adults without HIV. *J. Acquir. Immune Defic. Syndr.* **2021**, *88*, 314–321. [CrossRef] [PubMed]
16. Patel, M.; Zang, X.; Matthews, R.; Plank, R.; Sklar, P.; Grobler, J.; Robertson, M.; Vargo, R. Islatravir PK threshold & dose selection for monthly oral HIV-1 PrEP 2021. In Proceedings of the Conference on Retroviruses and Opportunistic Infection, Virtual, 7–10 March 2021.
17. Matthews, R.P.; Patel, M.; Barrett, S.E.; Haspeslagh, L.; Reynders, T.; Zhang, S.; Rottey, S.; Goodey, A.; Vargo, R.C.; Grobler, J.A.; et al. Safety and pharmacokinetics of islatravir subdermal implant for HIV-1 pre-exposure prophylaxis: A randomized, placebo-controlled phase 1 trial. *Nat. Med.* **2021**, *27*, 1712–1717. [CrossRef] [PubMed]
18. Matthews, R.P.; Zang, X.; Barrett, S.E.; Koynov, A.; Goodey, A.; Heimbach, T.; Weissler, V.L.; Leyssens, C.; Reynders, T.; Xu, Z.; et al. A Randomized, Double-Blind, Placebo-Controlled, Phase 1 Trial of Radiopaque Islatravir-Eluting Subdermal Implants for Pre-exposure Prophylaxis Against HIV-1 Infection. *J. Acquir. Immune Defic. Syndr.* **2023**, *92*, 310–316. [CrossRef] [PubMed]
19. Matthews, R.P.; Ankrom, W.; Friedman, E.; Jackson Rudd, D.; Liu, Y.; Mogg, R.; Panebianco, D.; De Lepeleire, I.; Petkova, M.; Grobler, J.A.; et al. Safety, tolerability, and pharmacokinetics of single- and multiple-dose administration of islatravir (MK-8591) in adults without HIV. *Clin. Transl. Sci.* **2021**, *14*, 1935–1944. [CrossRef]
20. Merck Announces Clinical Holds on Studies Evaluating Islatravir for the Treatment and Prevention of HIV-1 Infection—Merck.com. Available online: https://www.merck.com/news/merck-announces-clinical-holds-on-studies-evaluating-islatravir-for-the-treatment-and-prevention-of-hiv-1-infection/ (accessed on 9 September 2023).
21. Squires, K.; Correll, T.; Robertson, M.; Klopfer, S.; May Tan Hwang, P.; Zhou, Y.-P.; Rhee, E. Effect of Islatravir on Total Lymphocyte and Lymhocyte Subset Counts. Available online: https://www.natap.org/2023/CROI/croi_35.htm (accessed on 10 September 2023).
22. Li, L.; Gatto, G.J.; Brand, R.M.; Krovi, S.A.; Cottrell, M.L.; Norton, C.; van der Straten, A.; Johnson, L.M. Long-acting biodegradable implant for sustained delivery of antiretrovirals (ARVs) and hormones. *J. Control. Release* **2021**, *340*, 188–199. [CrossRef]
23. National Research Council. The National Academies Guide for the Care and Use of Laboratory Animals: Eighth Edition, Copyright 2011, NAS | OLAW. Available online: https://olaw.nih.gov/resources/publications/guide-care-2011.htm (accessed on 20 February 2023).
24. Sykes, C.; Van Horne, B.; Jones, J.; Kashuba, A.D.M.; Gatto, G.; Van Der Straten, A.; Johnson, L.; Cottrell, M.L. Intracellular islatravir pharmacology differs between species in an in vitro model: Implications for preclinical study design. *J. Antimicrob. Chemother.* **2022**, *77*, 1000–1004. [CrossRef]
25. Beal, S.L. Ways to fit a PK model with some data below the quantification limit. *J. Pharmacokinet. Pharmacodyn.* **2001**, *28*, 481–504. [CrossRef]
26. Bruschi Mathematical models of drug release. In *Strategies to Modify the Drug Release from Pharmaceutical Systems*; Elsevier: Amsterdam, The Netherlands, 2015; pp. 63–86. ISBN 9780081000922.
27. Costa, P.; Sousa Lobo, J.M. Evaluation of mathematical models describing drug release from estradiol transdermal systems. *Drug Dev. Ind. Pharm.* **2003**, *29*, 89–97. [CrossRef] [PubMed]
28. Spinelli, M.A.; Grinsztejn, B.; Landovitz, R.J. Promises and challenges: Cabotegravir for preexposure prophylaxis. *Curr. Opin. HIV AIDS* **2022**, *17*, 186–191. [CrossRef] [PubMed]
29. Organon & Co.; Jersey City, NJ, U.S.; Food and Drug Administration. *Nexplanon (Etonogestrel Implant) Prescribing Information*. Available online: https://www.organon.com/product/usa/pi_circulars/n/nexplanon/nexplanon_pi.pdf (accessed on 18 February 2023).

30. Barrett, S.E.; Teller, R.S.; Forster, S.P.; Li, L.; Mackey, M.A.; Skomski, D.; Yang, Z.; Fillgrove, K.L.; Doto, G.J.; Wood, S.L.; et al. Extended-Duration MK-8591-Eluting Implant as a Candidate for HIV Treatment and Prevention. *Antimicrob. Agents Chemother.* **2018**, *62*, e01058-18. [CrossRef] [PubMed]
31. Ritger, P.L.; Peppas, N.A. A simple equation for description of solute release II. Fickian and anomalous release from swellable devices. *J. Control. Release* **1987**, *5*, 37–42. [CrossRef]
32. Macha, S.; Yong, C.-L.; Darrington, T.; Davis, M.S.; MacGregor, T.R.; Castles, M.; Krill, S.L. In vitro-in vivo correlation for nevirapine extended release tablets. *Biopharm. Drug Dispos.* **2009**, *30*, 542–550. [CrossRef] [PubMed]
33. Yang, Y.; Manda, P.; Pavurala, N.; Khan, M.A.; Krishnaiah, Y.S.R. Development and validation of in vitro-in vivo correlation (IVIVC) for estradiol transdermal drug delivery systems. *J. Control. Release* **2015**, *210*, 58–66. [CrossRef]
34. FDA Duidance for Industry: Extended Release Oral Dosage Forms: Development, Evaluation, and Application of In Vitro/In Vivo Correlations. Available online: https://www.fda.gov/regulatory-information/search-fda-guidance-documents/extended-release-oral-dosage-forms-development-evaluation-and-application-vitroin-in-vivo-correlations (accessed on 27 September 2023).
35. European Medicines Agency: Guideline on the Pharmacokinetic and Clinical Evaluation of Modified Release Dosage Forms. Available online: https://www.ema.europa.eu/en/documents/scientific-guideline/guideline-pharmacokinetic-clinical-evaluation-modified-release-dosage-forms_en.pdf (accessed on 26 September 2023).

pharmaceutics

MDPI

Article

Impact of LS Mutation on Pharmacokinetics of Preventive HIV Broadly Neutralizing Monoclonal Antibodies: A Cross-Protocol Analysis of 16 Clinical Trials in People without HIV

Bryan T. Mayer [1,*,†], Lily Zhang [1,†], Allan C. deCamp [1], Chenchen Yu [1], Alicia Sato [1], Heather Angier [1,‡], Kelly E. Seaton [2], Nicole Yates [2], Julie E. Ledgerwood [3,‡], Kenneth Mayer [4], Marina Caskey [5], Michel Nussenzweig [5], Kathryn Stephenson [6], Boris Julg [6,7], Dan H. Barouch [7], Magdalena E. Sobieszczyk [8], Srilatha Edupuganti [9], Colleen F. Kelley [9], M. Juliana McElrath [1], Huub C. Gelderblom [1], Michael Pensiero [3], Adrian McDermott [3,‡], Lucio Gama [3], Richard A. Koup [3], Peter B. Gilbert [1,10], Myron S. Cohen [11], Lawrence Corey [1,12], Ollivier Hyrien [1], Georgia D. Tomaras [2] and Yunda Huang [1,13,*]

1 Vaccine and Infectious Disease Division, Fred Hutchinson Cancer Center, Seattle, WA 98109, USA; yzhang2@scharp.org (L.Z.); adecamp@scharp.org (A.C.d.); cyu@scharp.org (C.Y.); asato@scharp.org (A.S.); hangier@fredhutch.org (H.A.); jmcelrat@fredhutch.org (M.J.M.); hgelderb@fredhutch.org (H.C.G.); pgilbert@fredhutch.org (P.B.G.); lcorey@fredhutch.org (L.C.); ohyrien@fredhutch.org (O.H.)

2 Duke University Medical Center, Durham, NC 27705, USA; kelly.seaton@duke.edu (K.E.S.); nicole.yates@duke.edu (N.Y.); georgia.tomaras@duke.edu (G.D.T.)

3 Vaccine Research Center, National Institute of Allergy and Infectious Diseases, Bethesda, MD 20892, USA; mpensiero@niaid.nih.gov (M.P.); adrian.mcdermott@sanofi.com (A.M.); lucio.gama@nih.gov (L.G.); richard.koup@nih.gov (R.A.K.)

4 The Fenway Institute, Boston, MA 02215, USA; kmayer@fenwayhealth.org

5 Laboratory of Molecular Immunology, The Rockefeller University, New York, NY 10065, USA; mcaskey@rockefeller.edu (M.C.); nussen@mail.rockefeller.edu (M.N.)

6 Ragon Institute of Mass General, MIT and Harvard, Cambridge, MA 02139, USA; kstephen@bidmc.harvard.edu (K.S.); bjulg@mgh.harvard.edu (B.J.)

7 Center for Virology and Vaccine Research, Beth Israel Deaconess Medical Center, Boston, MA 02215, USA; dbarouch@bidmc.harvard.edu

8 Columbia University Irving Medical Center, New York, NY 10032, USA; mes52@cumc.columbia.edu

9 Department of Medicine, Division of Infectious Diseases, Emory University School of Medicine, Atlanta, GA 30322, USA; sedupug@emory.edu (S.E.); colleen.kelley@emory.edu (C.F.K.)

10 Department of Biostatistics, University of Washington, Seattle, WA 98195, USA

11 Institute for Global Health and Infectious Diseases, University of North Carolina at Chapel Hill, Chapel Hill, NC 27514, USA; mscohen@med.unc.edu

12 Departments of Medicine and Laboratory Medicine, University of Washington, Seattle, WA 98195, USA

13 Department of Global Health, University of Washington, Seattle, WA 98195, USA

* Correspondence: bmayer@fredhutch.org (B.T.M.); yunda@fredhutch.org (Y.H.)

† These authors contributed equally to this work.

‡ Location at the time of the work.

Abstract: Monoclonal antibodies are commonly engineered with an introduction of Met428Leu and Asn434Ser, known as the LS mutation, in the fragment crystallizable region to improve pharmacokinetic profiles. The LS mutation delays antibody clearance by enhancing binding affinity to the neonatal fragment crystallizable receptor found on endothelial cells. To characterize the LS mutation for monoclonal antibodies targeting HIV, we compared pharmacokinetic parameters between parental versus LS variants for five pairs of anti-HIV immunoglobin G1 monoclonal antibodies (VRC01/LS/VRC07-523LS, 3BNC117/LS, PGDM1400/LS PGT121/LS, 10-1074/LS), analyzing data from 16 clinical trials of 583 participants without HIV. We described serum concentrations of these monoclonal antibodies following intravenous or subcutaneous administration by an open two-compartment disposition, with first-order elimination from the central compartment using non-linear mixed effects pharmacokinetic models. We compared estimated pharmacokinetic parameters using the targeted maximum likelihood estimation method, accounting for participant differences. We observed lower clearance rate, central volume, and peripheral volume of distribution for all LS variants compared to parental monoclonal antibodies. LS monoclonal antibodies showed several improvements in pharmacokinetic parameters, including increases in the elimination half-life by 2.7- to 4.1-fold, the dose-normalized area-under-the-curve by 4.1- to 9.5-fold, and the predicted concentration at 4 weeks post-administration by 3.4- to 7.6-fold. Results suggest a favorable pharmacokinetic profile of LS variants regardless of HIV epitope specificity. Insights support lower dosages and/or less frequent dosing of LS variants to achieve similar levels of antibody exposure in future clinical applications.

Keywords: monoclonal antibodies; LS mutation; two-compartment model; population pharmacokinetics modeling; HIV prevention; targeted maximum likelihood estimation (TMLE)

1. Introduction

Broadly neutralizing monoclonal antibodies (mAbs) are being considered as a potential strategy to prevent and treat human immunodeficiency virus (HIV) disease [1–6]. Particularly, the recent Antibody Mediated Prevention (AMP) studies of VRC01, an HIV broadly neutralizing mAb intravenously administered every 8 weeks, demonstrated no overall protection but a prevention efficacy of 75% against VRC01-neutralization-sensitive HIV viruses [7,8]. The AMP studies also established that prevention efficacy was related to the predicted neutralization ID80 titer (PT80) of VRC01 as an integrated biomarker of VRC01 serum concentration at the time of HIV acquisition, as well as the in vitro VRC01 neutralization sensitivity against the acquired virus [7–9]. These findings suggest that greater mAb-mediated prevention efficacy could be achieved by improving the neutralization activity of a mAb against circulating viruses, and/or by prolonging elevated concentrations of the mAb over time [10]. The next generation of HIV mAbs showed promising improvements compared to VRC01 for prevention, therapy, and potential for long-term virologic control or remission; in vitro analyses of a new series of mAbs highlight more target options for binding sites on the HIV trimer and generally improved neutralization potential [11–13]. To improve the pharmacokinetic (PK) profiles, the most common approach has been the use of the LS mutation (M428L/N434S) in the constant fragment crystallizable (Fc) region of immunoglobin G (IgG)-based mAbs [14–16], which could potentially also improve in vivo activity [17].

The LS mutation enhances the Fc region's binding affinity for the neonatal Fc receptor (FcRn) at a low pH (e.g., lysosomal compartments and vaginal lumen) without affecting its release at pH 7.4 [14–16]. The FcRn is expressed in endothelial cells of the vasculature and some myeloid cells, where it can rescue IgG from lysosomal degradation by binding these molecules within this low pH compartment and then releasing them back into circulation [15,18]. Importantly, the tissue distribution of FcRn enables the delivery of IgG mAbs to relevant mucosal compartments and may thereby contribute to HIV protection at the site of exposure [14]. Increased persistence of anti-HIV neutralizing antibodies in mucosal tissue associated with the LS mutation could have significant advantages in preventing HIV acquisition, as increased maintenance at the portals of entry might allow for consistent protective levels with less frequent administrations and/or a reduction in dosing.

So far, clinical studies of several anti-HIV LS variant mAbs demonstrated extended durability as measured by improved elimination half-life [19,20]. However, less is known about how the LS modification could improve other aspects of PK profile, including clearance rate, distributional volumes, or average concentrations over a dosing interval. In addition, there have not been systematic PK analyses of available HIV mAbs to understand whether the benefits of the LS modification are consistent for mAbs targeting different epitope sites on the HIV trimer.

Here, we hypothesized that across multiple mAbs the LS modification would significantly and predictably improve PK profiles over the parental mAb as determined by measures of extended elimination rates and reduced clearance. To that end, we analyzed data from 16 clinical trials of 11 HIV-1 mAbs with and without the LS mutation. Using these data, we developed population PK models for each of the 11 HIV-1 mAbs and investigated differences in PK parameters between five pairs of parental and LS mutation versions. We used the targeted maximum likelihood estimation (TMLE) [21] method to examine how key PK parameters changed between parental and LS variants. TMLE allows for rigorous adjustment for demographic and clinical factors that affect PK using an ensemble of learners, estimated under the counterfactual scenario that the two versions of mAbs (parental and LS) were assessed in the same study population. In addition, we accounted for variabilities and correlations of PK parameter estimates in assessing the statistical precision of the comparisons. Overall, we found that the LS modification reliably improved the PK profiles for each assessed anti-HIV mAb pair in a largely consistent manner.

2. Materials and Methods

2.1. Data

We analyzed PK data for 5 parental HIV-1 mAbs targeting three different epitopes on the HIV trimer: VRC01 (CD4 binding site), 3BNC117 (CD4 binding site), PGDM1400 (V1V2 glycan), PGT121 (V3 glycan), 10-1074 (V3 glycan), and their LS variants (Supplemental Table S1). We also analyzed data from VRC07-523LS, an engineered LS variant derived from VRC01 [22]. We included individual-level data collected from 16 phase 1 clinical trials of these mAbs [19,20,23–29], conducted in compliance with all relevant ethical requirements, including obtaining participant consent. All trials aimed to evaluate the given study product(s), administered via subcutaneous (SC) or intravenous (IV) route, in healthy adult participants without HIV. Details of the clinical trials and assays used to generate the PK data can be found in Table 1.

Table 1. Description of the 16 phase 1 clinical trials included in the study. Note: only data from intravenous and subcutaneous delivery routes were included in the analyses. mAb, monoclonal antibody; IV, intravenous; SC, subcutaneous; PK, pharmacokinetics.

mAb	HIV-1 Epitope Specificity	Clinical Trial (clinicaltrial.gov Registration #)	Study Country(ies)	Single or Combination mAb Regimens	IV Administration (n)	SC Administration (n)	PK Timepoints after Administration through the First 4 Weeks of the Study **	Assay	References
VRC01	CD4 binding site	VRC602 (#NCT01993706)	USA	Single	18	5	End of infusion, 1, 2, 4, 8, 12, 24, and 72 h, and 2, 7, 14, 21, and 28 days	ELISA (VRC)	[25]
		HVTN 104 (#NCT02165267)	USA	Single	64	20	1 h, 3 days, and 2 and 4 weeks	ELISA (VRC)	[26]
		HVTN 116 (#NCT02797171)	USA/RSA	Single	62	0	Weeks 1-2	Singulex (VRC)	[30]
VRC01LS		VRC606 (#NCT02599896)	USA	Single	21	18	1, 2, 6, 24, 48, and 72 h, and 1, 2, 3, and 4 weeks	ELISA (VRC)	[19]
		HVTN 116 (#NCT02797171)	USA/RSA	Single	17	0	Weeks 1-2	Singulex (VRC)	[30]
VRC07-523LS		VRC605 (#NCT03015181)	USA	Single	17	8	End of infusions, 1, 2, 6, 24, 48, and 72 h, and 1, 2, 3, and 4 weeks	ELISA (VRC)	[20]
		HVTN 127/HPTN 087 (#NCT03387150)	USA/SUI	Single	59	41	3, 6, and 28 days	ELISA (VRC)/ BAMA (Duke)	[31,32]
		HVTN 130/HPTN 089 (#NCT03928821)	USA	Combination	26	0	3, 6, 14, and 28 days	BAMA (Duke)	[28]
		HVTN 136/HPTN 092 (#NCT04212091)	USA	Combination	10	10	1 h, and 1, 2, 3, 6, 14, and 28 days	BAMA (Duke)	[33]
3BNC117		MCA-0835/Nussenzweig 446 * (#NCT02018510)	USA	Single	22	0	End of infusion, 0. 5, 3 6, 9, and 12 h; and 1, 2, 4, 7, 14, 21, and 28 days	ELISA (Celldex)	[34]
		YCO-0899/Nussenzweig 583 (#NCT02824536)	USA	Combination	18	0	End of infusion, 1, 2, 7, 14, and 28 days	ELISA (Duke, transferred from Celldex)	[23]
3BNC117-LS		YCO-0946/Nussenzweig 684 * (#NCT03254277)	USA	Single	3	12	End of infusion/injection, 1 (SC), 3, 7, 14, and 28 days	ELISA (Duke, transferred from Celldex)	[27]
		YCO-0971/Nussenzweig 739 * (#NCT03554408)	USA	Combination	5	30	End of infusion/injection, 1, 3, 7, 14, and 28 days	ELISA (Duke, transferred from Celldex)	[35]

Table 1. *Cont.*

mAb	HIV-1 Epitope Specificity	Clinical Trial (clinicaltrial.gov Registration #)	Study Country(ies)	Single or Combination mAb Regimens	IV Administration (n)	SC Administration (n)	PK Timepoints after Administration through the First 4 Weeks of the Study **	Assay	References
PGDM1400	V1V2 Glycan	IAVI T002/Barouch 693 * (#NCT03205917)	USA	Single	9	0	End of infusion, 3, 6, and 24 h, and 1, 2, 3, 7, and 12 days	BAMA (Duke)	[24]
		IAVI T002/Barouch 693 * (#NCT03205917)	USA	Combination	12	0	End of infusion, 3, 6, and 24 h, and 1, 2, 3, 7, and 12 days	BAMA (Duke)	[24]
		HVTN 130/HPTN 089 (#NCT03928821)	USA	Combination	15	0	3, 6, 14, and 28 days	BAMA (Duke)	[28]
PGDM1400 LS		HVTN 140/HPTN101 (Part A) (#NCT05184452)	USA/KEN/RSA/ZIM	Single	9	6	1 h, and 3, 6, and 28 days	BAMA (Duke)	[36]
PGT121	V3 Glycan	IAVI T001/Barouch 628 * (#NCT02960581)	USA	Single	12	4	End of infusion, 30 min, 3, 6, 9, and 12 h, and 1, 2, 3, 7, 14, 21, and 28 days	BAMA (Duke)	[29]
		IAVI T002/Barouch 693 * (#NCT03205917)	USA	Combination	12	0	End of infusion, 3, 6, and 24 h, and 1, 2, 3, 7, and 12 days	BAMA (Duke)	[24]
		HVTN 130/HPTN 089 (#NCT03928821)	USA	Combination	15	0	3, 6, 14, and 28 days	BAMA (Duke)	[28]
PGT121.414.LS		HVTN 136/HPTN 092 (# NCT04212091)	USA	Single and combination	20	13	0, 1, 2, 3, 6, 14, and 28 days	BAMA (Duke)	[33]
10-1074		MCA-0885/Nussenzweig 514 * (#NCT02511990)	USA	Single	14	0	End of infusion, 30 min, 3, 6, 9, 12 h, and 1, 2, 4, 7, 14, 21, and 28 days	ELISA (Celldex)	[37]
		YCO-0899/Nussenzweig 583 (#NCT02824536)	USA	Combination	18	0	End of infusion, 2 days, 1, 2, and 4 weeks	ELISA (Duke, transferred from Celldex)	[23]
10-1074-LS		YCO-0971/Nussenzweig 739 * (#NCT03554408)	USA	Single	9	12	End of infusion/injection, 1, 3, 7, 14, and 28 days	ELISA (Duke, transferred from Celldex)	[35]
		YCO-0971/Nussenzweig 739 * (#NCT03554408)	USA	Combination	5	30	End of infusion/injection, 1, 3, 7, 14, and 28 days	ELISA (Duke, transferred from Celldex)	[35]

* Only PK data from groups without HIV were included in all presented analyses. ** All available timepoints were used for analysis; for brevity only time points collected in the first 4 weeks were listed here. For more information, please review documentation for each study publicly available, or by request. All studies complied with relevant ethical requirements including obtaining participant consent and were registered with ClinicalTrials.gov.

2.2. PK Assays

2.2.1. Enzyme-Linked Immunosorbent Assay (ELISA)

ELISA was used to quantify the mAb concentrations in human serum for VRC01 [25,26], VRC01LS [19], VRC07-523LS [20,31], 3BNC117 [23,34], 3BNC117LS [27], 10-1074 [37], and 10-1074LS [35] in 10 of the 16 clinical trials. Briefly, for a series of diluted samples, serum antibody levels were measured using optical density readings based on antibody-specific anti-idiotype binding. Serum concentrations were then determined from a standard curve using a 5PL curve fit. For VRC01 [25,26], VRC01LS [19], and VRC07-523LS [20,31,32], the specific anti-idiotype antibodies were developed and purified by the Vaccine Research Center (VRC), National Institute of Allergy and Infectious Diseases (NIAID) at the National Institutes of Health (NIH). For 3BNC117 [23,34], 10-1074 [37], and their LS variants [27,35], the specific anti-idiotype antibodies were developed and purified by the Duke Vaccine Immunogenicity Program [23]. For this assay, the lower limits of quantification (LLoQ) varied over the range of 0.12–1.1 mcg/mL depending on the specific mAb and study (Supplemental Table S2).

2.2.2. Singulex Assay

The Singulex assay was used to quantify concentrations in human serum for VRC01/VRC01LS [30] in the HIV Vaccine Trials Network (HVTN) 116 phase 1 clinical trial. Briefly, concentrations in the (diluted) samples were measured using Singulex paramagnetic microparticle (MP) beads using the Singulex Erenna® Capture Antibody Labeling Kit (Singulex/EMD Millipore) according to the manufacturer's protocol. The Erenna detects single molecules of VRC01/VRC01LS as photons over a read time. Each measurement of photons above a threshold was considered a detected event and the sum of all photons from these events over the read time (i.e., signal intensity) is called the event photon (EP) measurement for the sample or standard. The EP curve of the VRC01 standard was used to calibrate the VRC01/VRC01LS concentrations in serum samples. The median LLoQ across runs was 2.31×10^{-4} mcg/mL.

2.2.3. PK Binding Antibody Multiplex Assay (BAMA)

The BAMA assay derived on a Bio-Plex instrument was used to quantify concentrations in human serum for VRC07 523LS [20,28], PGDM1400 [24,36], and PGT121 [29,33,38] under the oversight of the Quality Assurance for Duke Vaccine Immunogenicity Programs. This anti-idiotype (anti-ID) assay measured concentrations by its ability to bind anti-idiotype antibody captured on fluorescent magnetic beads. The Bio-Plex software (https://www.bio-rad.com/en-us/product/bio-plex-data-pro-software?ID=LSVAC015, accessed on 26 March 2024) provides 2 readouts: a background-subtracted median fluorescent intensity (MFI), where background refers to a plate level control (i.e., a blank well containing antigen-conjugated beads run on each plate), and a concentration based on a standard curve using a 5PL curve fit. For this assay, the LLoQ varied over the range of 0.023–0.50 mcg/mL depending on the specific mAb and study (Supplemental Table S2).

2.3. Population PK Modeling

We modeled serum mAb concentrations over time using non-linear mixed effects modeling with the Monolix software system (Version 2021 R2) (https://lixoft.com/products/monolix/, accessed on 26 March 2024). We estimated population parameters, including the population mean and variance of each PK parameter and parameters of the error term variance by fitting the non-linear mixed effects models using the method of maximum likelihood implemented with the Stochastic Approximation Expectation Maximization (SAEM) algorithm as implemented in Monolix. We described PK following intravenous (IV) or subcutaneous (SC) mAb administration by an open 2-compartment disposition model with first-order elimination from the central compartment. A 2-compartment model assumes that antibody clearance is bi-phasic, which has been routinely used to describe the PK of IgG mAbs [39]. The model was parameterized in terms of the central volume

(Vc, L), clearance from the central compartment (CL, L/day), peripheral volume (Vp, L), and intercompartmental clearance (Q, L/day). For SC administrations, a depot compartment was added to characterize first-order absorption via the SC route with 2 parameters: bioavailability (F, %) and absorption rate (ka, day^{-1}).

2.4. Error Model and Random Effects Specification

For each population PK model, we considered population-level (average) parameters as fixed effects and inter-participant variation of these parameters as random effects with standard deviations (ω). We did not include random effects for absorption rate and bioavailability due to limited data availability to estimate these parameters across all mAbs. We fitted PK models to serum concentrations, assuming normally distributed error terms. Concentrations below the LLoQ were treated as left-censored (Supplemental Table S2). We assessed two structures for the conditional variance of the error term, given the random effects, assuming the following: (1) the error was proportional to the conditional expectation of the concentration; or (2) the residual variance was the sum of 2 terms, one constant term and another term proportional to the conditional expectation of the concentration. For each mAb, we selected the final error model structure based on the model with the lowest corrected Bayesian Information Criteria (BIC) [40]. In addition, we assumed the variance–covariance matrix of the random effects to be diagonal with heterogeneous variances for Vc, CL, Vp, and Q. Individual-level PK parameters were assumed to follow log-normal distributions. For PGDM1400LS, there were limited data to estimate the distribution phase and hence the random effects were not included for Vp and Q.

We predicted participant-specific model parameters using their empirical Bayes estimates (EBEs), defined as the most probable value given the estimated population parameters and the data from the given participant. We predicted these (random) individual-specific parameters by the mode of their conditional posterior distribution. They were subsequently used to predict the most probable trajectory of the serum concentration for the corresponding participant.

In addition to the model parameters (CL, Vc, Vp, and Q), we derived four individual-level PK features from the base population PK model without covariate adjustment: distribution half-life, elimination half-life, dose-normalized area under the curve (AUC), and predicted concentrations at 4 weeks post-infusion following a single IV bolus administration of 1.4 g of given mAb. A fixed dose of 1.4 g was considered to approximate a body-weight-based dose of 20 mg/kg for a 70 kg individual. The dose-normalized AUC values and predicted concentrations were computed for each individual participant using their individual-level parameters (i.e., EBEs) as predicted from the fitted PK model for the given mAb.

We used the Lin's concordance correlation coefficient (CCC) [41] to assess the concordance between individual-level parameter estimates via fitting a model to each trial separately vs. a model fitted to the pooled trials for each mAb. The CCC is a composite measure of precision (Pearson's correlation coefficient) and accuracy (bias). The quality of concordance is interpreted as poor when CCC < 0.9, moderate when $0.9 <$ CCC< 0.95, and substantial when CCC > 0.95 [42].

2.5. TMLE Methods

Although all 16 clinical trials were carried out in similar study populations supporting overlapping distributions of baseline characteristics between the comparison groups, the parental vs. LS variant groups were not randomized. Therefore, to formally compare PK features between the parental and LS variants, we used the TMLE method [21]. TMLE is an alternative to standard linear or non-linear regression that can improve robustness and efficiency and reduce confounding bias, especially for comparisons between non-randomized groups. Using this method, for each mAb pair we estimated the mean of each PK feature for each mAb in the pair, and the average treatment effect (ATE) of LS vs. parental of the pair, adjusted for potential confounders and predictors of PK variability: baseline body weight (kg), creatinine clearance (mL/min), age (years), and sex at birth (female vs. male), as implemented via the *tmle* package (version 1.5.0.2) in R (version 4.2.1) [43]. Notably, because it was challenging to precisely and accurately estimate these two features without richly sampled kinetic data proximal to product administration, the Q parameter and the distribution half-life were not compared between the parental and LS variants. A sensitivity analysis was also performed to account for single vs. co-administration for parental/LS mAb pairs. All TMLE estimation results of means and ATEs were averaged over 20 runs with a fixed random seed on top of the 10-fold cross-validation estimation procedure to ensure stability of the estimates. The set of learning algorithms used by TMLE for estimating the mean PK feature and ATE conditional on baseline covariates included the following: SL.glm, SL.step, SL.ranger, SL.earth, SL.glmnet, and SL.mean [44]. In addition, to account for variability and co-variability of the individual-level estimates for each PK feature, due to the fact that they were derived from a common population PK model, a bootstrap procedure based on 500 sampled-with-replacement datasets was used to estimate the variance of the estimates for mAb (parental or LS), derive the 95% confidence intervals (CIs), and test for a non-zero mean difference between the two mAbs in each parental/LS pair. We used the Holm procedure [45] to adjust the resulting p-values for comparisons of the multiple PK features within a given mAb pair.

3. Results

3.1. Study Population Characteristics

Here we analyzed the PK properties of anti-HIV mAbs targeting three different epitopes on the HIV-1 trimer: CD4 binding site (VRC01, 3BNC117), V3 glycan (PGT121, 10-1074), and V1V2 glycan (PGDM1400) (Table 1 and Supplementary Table S1). For each mAb, we evaluated both the parental formulation and its LS-formulated variant. For VRC01, we also compared the PK features with VRC07-523LS, the LS variant of another CD4-binding-site-targeting mAb that was engineered by the same developer and is a clonal relative of VRC01 from the same donor [22].

We leveraged data from a total of 16 phase 1 clinical trials, with 95% of the 583 participants enrolled from clinical sites in the US (Table 1). In terms of the study populations, the 10-1074 and 3BNC117 studies included participants who were slightly older (medians 41–49 years), weighed more (medians 79–83 kgs), and comprised fewer females (20–46%). For the remaining mAbs, participants' age ranged over medians of 26–31 years, weights ranged over medians of 71–76 kgs, and the proportion of female participants ranged from 48–67% (Supplemental Table S3).

3.2. Population PK Model Fitting

Serum concentrations of all the mAbs exhibited bi-phasic clearance following administration (Supplemental Figure S1). For each mAb variant assessed in multiple clinical trials, we fitted models both separately by trial and then with a single model combining the data across trials to determine the suitability of pooling data across studies for a given mAb. For PK features, except Vc and Vp, there was at least moderate concordance (CCC > 0.9) between the trial-level and combined models. For Vc, the estimates in the SC population tended to be more deviant between the trial-level and combined models (triangles in the Vc row of Supplemental Figure S2). The Vc and bioavailability (F) parameters were mathematically non-identifiable in SC groups without precise estimation of Vc from groups receiving IV administration, and thus pooling the data across studies could improve the accuracy and precision of estimates of Vc in the SC groups by increasing the sample size for the IV data. Subsequently, we used the combined model for the remaining analyses.

3.3. Covariate-Unadjusted Base PK Models for Each mAb

The final, covariate-unadjusted 2-compartment models generally exhibited good fit to all the data. Goodness of fit at the individual and population levels was demonstrated for all mAbs based on analysis of the residuals and model-predicted concentrations (Supplemental Figures S3–S5, and Supplemental File S1). The population PK parameters for all the mAbs are depicted in Supplemental Table S4. Population-level bioavailability after SC administration compared to IV estimates ranged from 39–78%. Figures 1 and 2 show the distributions of the PK parameter estimates obtained from the base population PK model without covariate adjustment, which are the basis for the subsequent comparisons. The unadjusted individual parameters for Vc and CL were consistent across the parental and LS variants across the mAbs. The CL was notably slower for LS variants compared to parental variants, an expected trend for the LS mutation. The Vc tended to be slightly lower for the LS variants as well, but not uniformly, potentially indicating a smaller distribution space (i.e., higher maximum concentration) in the central compartment for LS variants. The Q and Vp parameters showed notable variation across the mAbs (Figure 1). Variation in these parameters may be attributable to varying sampling schemes where first-in-human trials of a mAb would have more frequent early sampling to precisely determine the distribution phase than subsequent trials of a product. For PGDM1400LS, limited data precluded estimation of the inter-individual variation for Vp and Q, and therefore all individual-level predicted parameters were equivalent to the population estimate. Overall, there was a general pattern that the LS variant had both lower Vp and Q (except for VRC01), suggesting a different distribution process to the peripheral compartment with the LS mutation. In addition, the LS mutation extended elimination half-lives in all mAb pairs. Consistent with the improved elimination half-life and differences in central compartment distribution, drug exposure also appeared much higher for the LS variants as determined by dose-adjusted AUC and predicted week 4 concentration with 1.4 g IV dose (Figure 2).

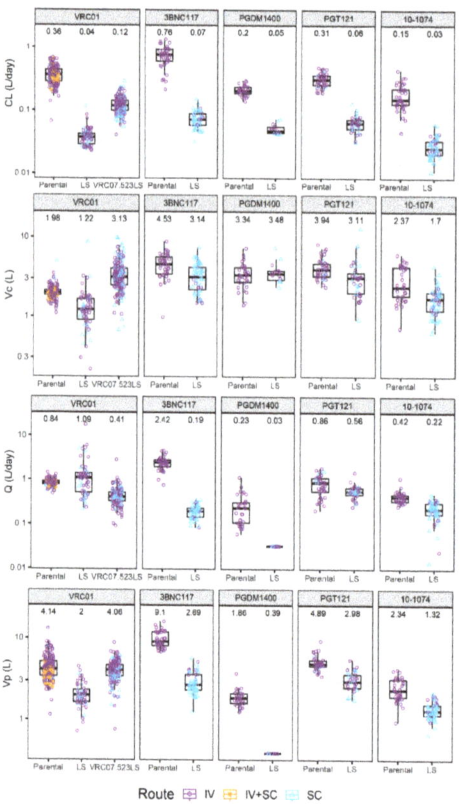

Figure 1. Distributions of population-PK-model-estimated covariate-unadjusted individual-level pharmacokinetic (PK) parameter estimates (points) for each monoclonal antibody pair of parental vs. LS

by PK parameter. The following PK parameters (top to bottom) are displayed: clearance (CL) (L/day), central volume (Vc) (L), inter-compartmental clearance (Q) (L/day), and peripheral volume (Vp) (L). Colors of the points indicate different administration routes. In all plots, box plots indicate median, interquartile range (IQR) (box), and 1.5× IQR (whiskers). The median value is listed above each box plot.

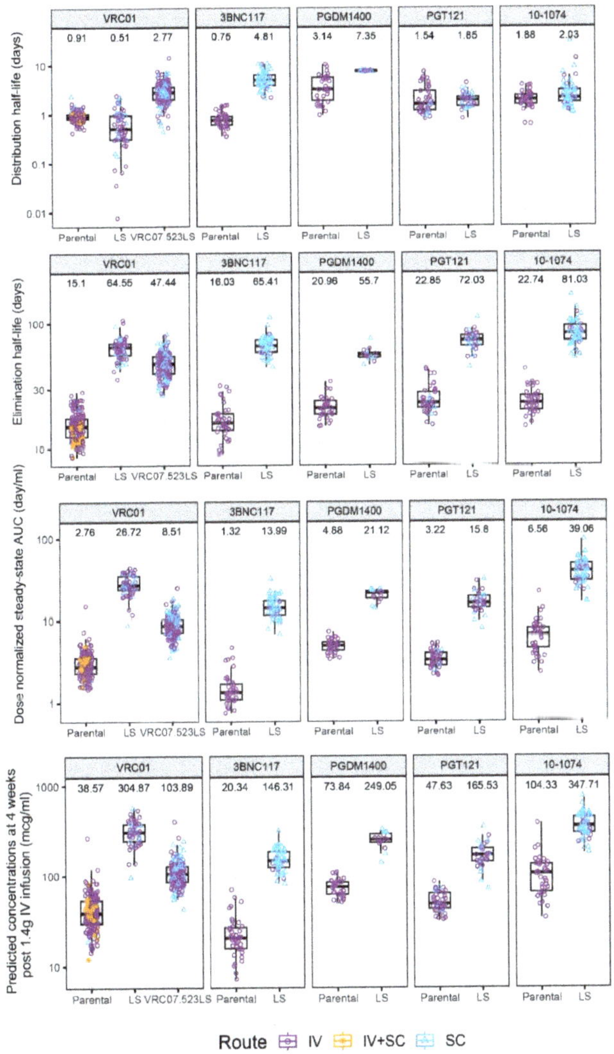

Figure 2. Distributions of population-PK-model-estimated covariate-unadjusted individual-level PK features for each monoclonal antibody pair of parental vs. LS. The following PK features (top to bottom) are displayed: distribution half-life (days), elimination half-life (days), dose-normalized steady-state area under the curve (AUC) (day/mL), and estimated concentrations at 4 weeks post-infusion (mcg/mL). Colors of the points indicate different administration routes. In all plots, box plots indicate median, interquartile range (IQR) (box), and 1.5× IQR (whiskers). The median value is listed above each box plot.

3.4. Covariate-Adjusted Differences in LS and Parental PK

We next used a combination of bootstrapping and TMLE to estimate the PK parameters and features for each mAb pair (parental vs. LS) after adjusting for baseline covariates. The individual-level PK parameter estimates obtained using TMLE depict the counterfactual value of these parameters/features, providing an unbiased comparison of these values as if the same group of individuals received the parental and LS mAbs (Figures 3 and 4). Likely due to the largely comparable characteristics between participants who received the parental vs. LS mAbs (Supplemental Table S3), the direction of the LS benefit in these TMLE-adjusted estimates, as well as the average estimates, was consistent with those observed in the unadjusted estimates as shown in Figure 2.

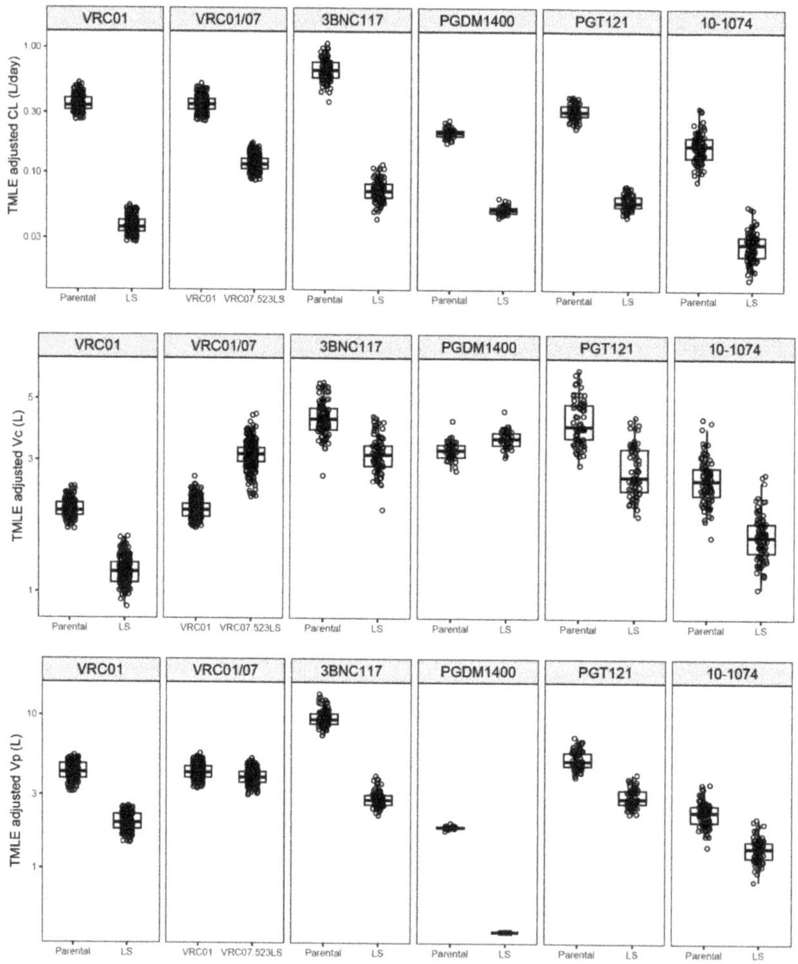

Figure 3. Distributions of targeted maximum likelihood estimation (TMLE)-adjusted individual-level pharmacokinetic (PK) parameter estimates (points) for each monoclonal antibody pair of parental vs. LS by PK parameter. The following PK parameters (**top** to **bottom**) are displayed: clearance (CL) (L/day), central volume (Vc) (L), peripheral volume (Vp) (L). In all plots, box plots indicate median, interquartile range (IQR) (box), and 1.5× IQR (whiskers). Each dot indicates the TMLE-adjusted PK feature estimate for each participant as if all received the parental mAb (left) or all LS mAb (right) under a causal framework.

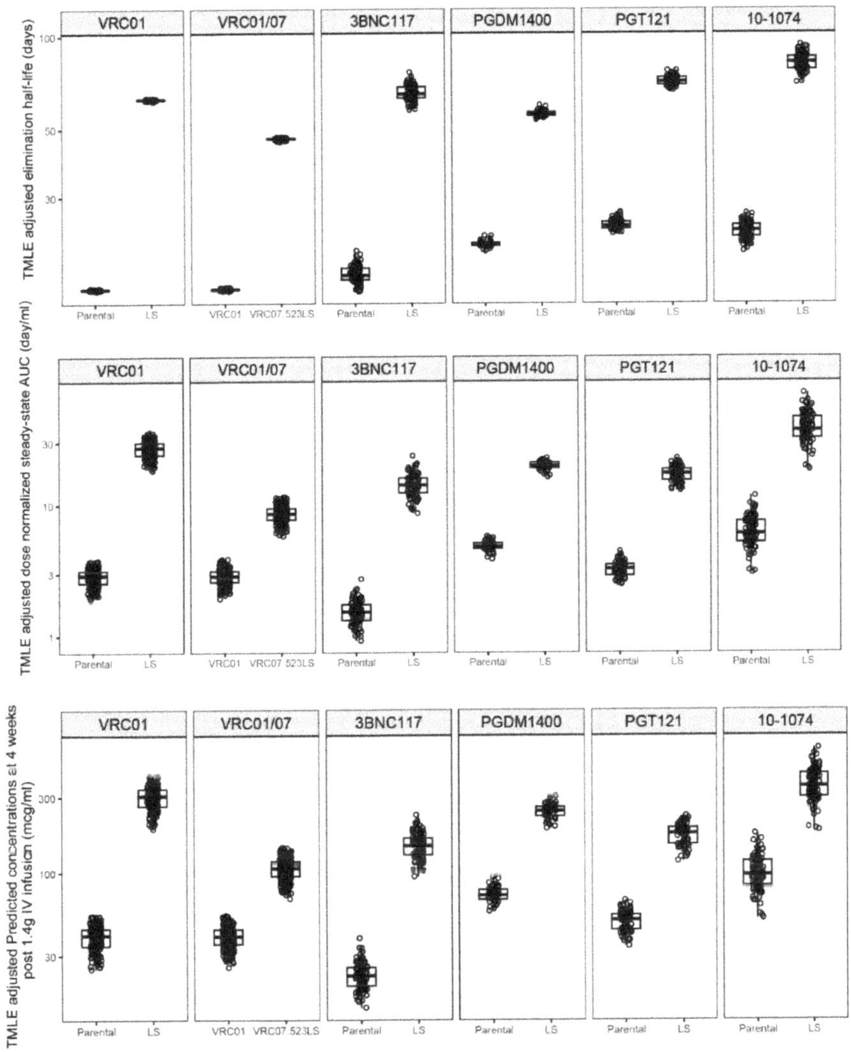

Figure 4. Distributions of targeted maximum likelihood estimation (TMLE)-adjusted individual-level PK features for each mAb pair of parental vs. LS by PK feature. The following PK features (**top** to **bottom**) are displayed: distribution half-life (days), elimination half-life (days), dose-normalized steady-state area under the curve (AUC) (day/mL), and estimated concentrations at 4 weeks post-infusion (mcg/mL). In all plots, box plots indicate median, interquartile range (IQR) (box), and 1.5× IQR (whiskers). Each dot indicates the TMLE-adjusted PK feature estimate for each participant as if all received the parental mAb (left) or all LS mAb (right) under a causal framework.

Specifically, we found significant differences in CL, Vc, and Vp between parental and LS variants for each mAb pair, except for Vc and Vp for PGDM1400/LS (Figure 5, Supplemental Table S5). Among the significant differences, the changes in parameters were all in the same direction, to the left of 1.0 or to the right of 1.0 in terms of the LS/parental ratio, as shown in the forest plot of Figure 5. Specifically, the mean CL decreased in a range of 4.2- to 9.1-fold ($p < 0.001$ for all), the mean Vc decreased in a range of 1.4- to 1.7-fold

($p \leq 0.03$, except PDGM1400, $p = 0.157$), and the mean Vp decreased in a range of 1.7- to 3.3-fold ($p < 0.001$ all except PGDM1400 not tested).

CL (L/day)

mAbs	Parental Mean (95% CI)	LS Mean (95% CI)	LS/Parental (95% CI)
VRC01	0.35 (0.33, 0.37)	0.04 (0.03, 0.04)	0.11 (0.09, 0.12)
3BNC117	0.66 (0.59, 0.75)	0.07 (0.06, 0.08)	0.11 (0.09, 0.13)
PGDM1400	0.21 (0.2, 0.22)	0.05 (0.05, 0.06)	0.24 (0.21, 0.28)
PGT121	0.31 (0.29, 0.34)	0.06 (0.05, 0.07)	0.19 (0.16, 0.22)
10-1074	0.16 (0.15, 0.19)	0.03 (0.02, 0.03)	0.16 (0.14, 0.19)

Vc (L)

mAbs	Parental Mean (95% CI)	LS Mean (95% CI)	LS/Parental (95% CI)
VRC01	1.98 (1.83, 2.14)	1.18 (0.92, 1.51)	0.6 (0.46, 0.77)
3BNC117	4.25 (3.82, 4.73)	3.13 (2.42, 4.06)	0.74 (0.56, 0.98)
PGDM1400	3.27 (2.86, 3.73)	3.62 (3.28, 3.99)	1.11 (0.96, 1.28)
PGT121	4.18 (3.65, 4.77)	2.76 (2.31, 3.3)	0.66 (0.53, 0.82)
10-1074	2.56 (2.27, 2.89)	1.6 (1.39, 1.85)	0.63 (0.52, 0.75)

Vp (L)

mAbs	Parental Mean (95% CI)	LS Mean (95% CI)	LS/Parental (95% CI)
VRC01	4.22 (3.95, 4.51)	1.98 (1.67, 2.36)	0.47 (0.39, 0.56)
3BNC117	9.53 (8.74, 10.41)	2.83 (2.24, 3.59)	0.3 (0.23, 0.38)
PGT121	5.19 (4.73, 5.69)	2.96 (2.63, 3.33)	0.57 (0.5, 0.66)
10-1074	2.32 (2.03, 2.66)	1.37 (1.07, 1.74)	0.59 (0.45, 0.77)

Elimination half-life (day)

mAbs	Parental Mean (95% CI)	LS Mean (95% CI)	LS/Parental (95% CI)
VRC01	15.14 (14.5, 15.82)	62.47 (58.8, 66.37)	4.13 (3.83, 4.45)
3BNC117	16.91 (15.31, 18.67)	65.35 (61.77, 69.14)	3.86 (3.44, 4.34)
PGDM1400	21.08 (19.05, 23.33)	55.81 (31.25, 99.66)	2.65 (1.48, 4.74)
PGT121	24.26 (22.19, 26.52)	70.92 (66.08, 76.11)	2.92 (2.6, 3.29)
10-1074	23.17 (21.53, 24.95)	81.42 (76.02, 87.19)	3.51 (3.18, 3.88)

Dose normalized steady-state AUC (day/L)

mAbs	Parental Mean (95% CI)	LS Mean (95% CI)	LS/Parental (95% CI)
VRC01	2.84 (2.69, 3)	26.91 (24.09, 30.06)	9.46 (8.41, 10.64)
3BNC117	1.51 (1.34, 1.7)	13.91 (12.49, 15.48)	9.23 (7.84, 10.86)
PGDM1400	4.79 (4.5, 5.09)	19.7 (17.6, 22.05)	4.11 (3.62, 4.68)
PGT121	3.19 (2.94, 3.46)	16.82 (15.15, 18.69)	5.27 (4.56, 6.1)
10-1074	6.07 (5.39, 6.84)	37.35 (32.85, 42.47)	6.15 (5.22, 7.24)

Predicted concentrations at 4 weeks post 1.4g IV infusion (mcg/ml)

mAbs	Parental Mean (95% CI)	LS Mean (95% CI)	LS/Parental (95% CI)
VRC01	39.05 (36.59, 41.67)	297.67 (266.39, 332.64)	7.62 (6.76, 8.6)
3BNC117	22.16 (19.39, 25.32)	145.82 (130.22, 163.29)	6.58 (5.55, 7.8)
PGDM1400	71.78 (66.43, 77.56)	240.12 (219.25, 262.97)	3.35 (2.99, 3.74)
PGT121	48.34 (44.31, 52.73)	169.23 (152.67, 187.59)	3.5 (3.02, 4.06)
10-1074	96.53 (85.68, 108.74)	345.37 (304.63, 391.56)	3.58 (3.05, 4.19)

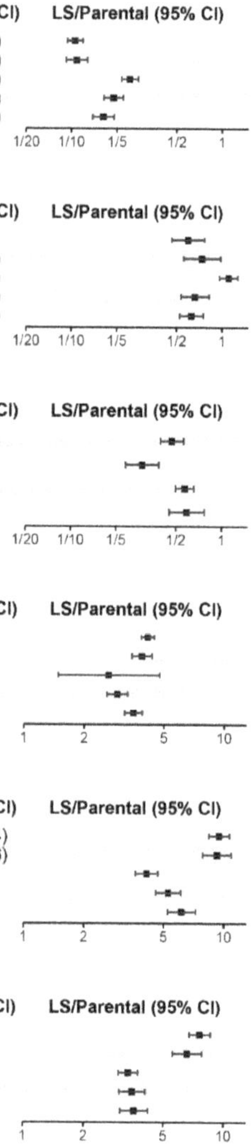

Figure 5. Mean and fold-change estimates between parental and LS variants for each pharmacokinetic feature. For each PK feature, the means for the parental and for the LS variant, LS versus parental (LS/parental) fold change and the associated 95% CIs were estimated using the target maximum likelihood estimation (TMLE) method. The fold-change was computed by exponentiating the average treatment effect estimated on log-normally distributed PK features. In the last column, the points indicate the estimated fold-change of the given parameter comparing the LS to the parental variant for the given monoclonal antibody with the error bars indicating the 95% CI obtained via a bootstrap procedure as described in the Materials and Methods section.

In addition, the direction of the changes for the PK parameters was associated with both extended elimination clearance and higher observed drug exposure for the LS variant compared to the parental for each mAb pair, all statistically significant at $p < 0.001$ (Figure 5 and Supplemental Table S5). The elimination half-life for the LS variant was extended by a mean range of 2.7- to 4.1-fold, the dose-normalized AUC was increased by a mean range of 4.1- to 9.5-fold, and the predicted concentration at 4 weeks post-administration was increased by a mean range of 3.4- to 7.6-fold.

To illustrate the overall impact of the LS mutation on the PK profiles, we simulated PK curves for each mAb administered intravenously at 1.4 g using the mean TMLE-estimated parameters (Figure 6). A general visual pattern emerged comparing the parental to the LS variant: the LS variant concentration was initially higher (i.e., higher peak serum concentration), then all mAbs had a sharp drop in serum concentration during the distribution phase that was less pronounced with the LS mutation. Finally, the LS variant remained at higher concentration with slower decay throughout the elimination phase. A slight exception to this pattern was observed for the PGDM1400 pair, wherein the parental and LS had similar values immediately following administration. In addition, PDGM1400LS exhibited no prominent distribution phase, likely due to the limited available data for this LS variant. Consequently, for PGDM1400LS there appeared to be one single decay rate over time, and it remained at higher serum concentrations than the parental variant throughout.

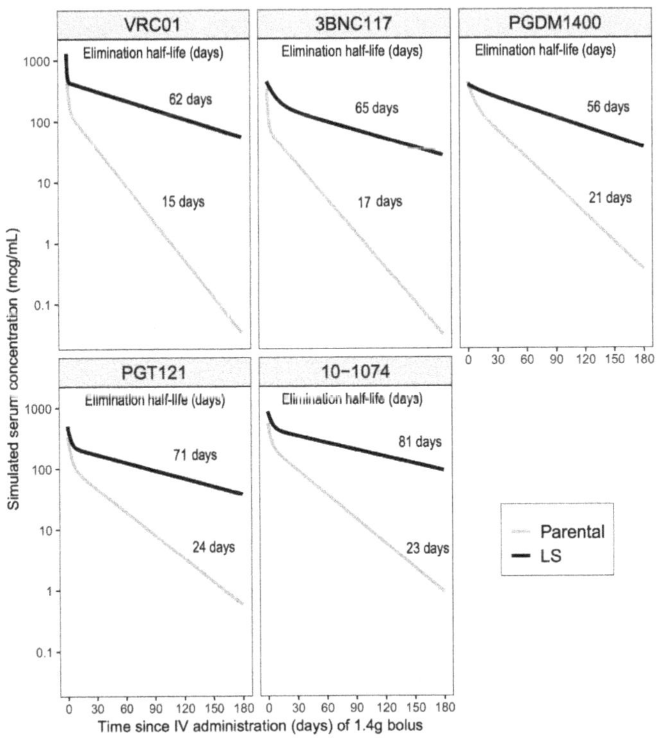

Figure 6. Simulated serum concentrations (mcg/mL) for each monoclonal antibody parental and LS variant pair. Gray curves are for parental mAbs and black curves for LS variants. Values shown on top of each curve are the estimated mean elimination half-life (days) for the corresponding mAb. The mean pharmacokinetic (PK) parameters estimated via targeted maximum likelihood estimation (TMLE) were used in the simulations, assuming a 1.4 g dose administered via bolus intravenous infusion. When a PK parameter was not estimated with TMLE (inter-compartmental clearance (Q) for all antibodies and peripheral volume (Vp) for PGDM1400LS), the mean parameter estimates from the covariate-unadjusted model were used.

3.5. Covariate-Adjusted Differences in VRC01 vs. VRC07-523LS PK

Lastly, we examined the differences between VRC01 and VRC07-523LS to gain more insights into when a mAb has additional modifications other than the LS mutation. While VRC01 is not the parental formulation of VRC07-523LS—the parental VRC07 has never been tested in a clinical setting—VRC07 was engineered from VRC01. Therefore, PK differences between VRC01 and VRC07-523LS cannot be solely attributed to the LS mutation. The covariate-unadjusted individual parameter estimates for VRC07-523LS were generally different from VRC01 (Figure 1). Compared to VRC01, VRC07-523LS also exhibited a longer elimination half-life, a higher does-normalized AUC, and a higher predicted concentration 4 weeks post-administration (Figure 2). Using TMLE, we found significant differences for all parameters except Vp between VRC01 and VRC07-523LS (Figures 3 and 4 and Supplemental Table S5). The mean CL decreased 2.9-fold ($p < 0.001$) between VRC01 and VRC07-523LS, a smaller decrease than that observed among the five pairs of parental/LS mAbs. The mean central volume for VRC07-523LS increased 1.6-fold ($p < 0.001$) compared to VRC01, which is the opposite direction of the trend observed comparing LS to parental variants where the volumes decreased. No significant difference was detected in Vp ($p = 0.247$ d). The elimination half-life for VRC07-523LS was extended over VRC01 by 3.1-fold ($p < 0.001$), similar to the improvement observed with LS mutation for the parental/LS variant mAb pairs. The drug exposure was also higher for VRC07-523LS compared to VRC01, with a 3.0-fold increase ($p < 0.001$) in dose-normalized AUC and a 2.7-fold increase ($p < 0.001$) in serum concentration at 4 weeks post-administration, but these increases were below the ranges observed in the five pairs of parental/LS mAbs. Overall, the PK changes of VRC07-523LS compared to VRC01 appeared somewhat different from what was observed by directly comparing parental to LS among the other mAbs, including VRC01LS vs. VRC01.

4. Discussion

To our knowledge, this is the first unified pharmacokinetics analysis of multiple anti-HIV mAbs from multiple clinical trials to examine the effect of LS mutation on their PK properties upon IV or SC administration in people without HIV. In this cross-protocol analysis of mAb serum concentrations over time, we leveraged data for five pairs of parental vs. LS variant anti-HIV mAbs evaluated in 16 different phase I clinical trials involving a total of 583 participants from the US, Europe, and Africa. Importantly, our comparison analyses adjusted for participants' clinical and demographic characteristics across independent clinical trials. Overall, we found consistent and relatively predictable PK improvement associated with the LS modification compared to the parental mAbs. These findings are important because no other meta-analyses of anti-HIV mAbs have been previously performed to understand systematically the effect of LS mutation; the learned knowledge could be applied to the research and development of LS mAbs, and in the design of clinical trials evaluating LS mAbs, given the knowledge of their parental formulation.

Specifically, we observed bi-phasic clearance of serum concentrations across all pairs of parental/LS mAbs, supporting the use of an open two-compartment disposition model with first-order elimination as the best fitting PK model to describe these data, in agreement with previous analyses for most of these mAbs individually [19,20,23–29]. In addition, we successfully fit population two-compartment models to each mAb using non-linear mixed effects modeling. The parameter estimates of these anti-HIV parental mAbs from our models were consistent with those reported in a previous meta-analysis of non-HIV parental mAbs [39]; to our knowledge, no meta-analysis of anti-HIV or non-HIV LS mAbs has been reported. For each pair of these parental/LS mAbs, we observed that LS mAbs had a reduced CL rate resulting in a 2.7- to 4.1-fold extended elimination half-life and 4.1- to 9.5-fold higher drug exposure in serum. We also observed that the distributional properties were different in the LS mAbs, generally with 1.4- to 3.3-fold lower central and peripheral volumes (Vc and Vp), also contributing to increased dose-normalized concentrations in serum. These results exemplify the complex biological implications of the LS mutation. It is believed that the LS mutation extends mAb persistence in serum, primarily because the FcRn on various cells helps to reduce degradation and clearance and to improve recycling of the mAb back into circulation. Consequently, as hypothesized, all the LS mAbs here have extended elimination half-lives and reduced CL. Coupled with the observed uniform reductions in the volume parameters with the LS mutation, these results suggest that the general distribution of the mAbs in the serum and tissue was likely altered by FcRn binding. Overall, the increased serum drug exposure and overall improvement to the PK profile seem to result from a combination of both the improved half-life and the distributional differences. Critically, as applied to future clinical designs, these improvements suggest reduced dosing frequency needed for LS variants; for example, a 2-month interval for parental mAbs could be reduced to a 6-month interval for their LS variants.

Importantly, we also found that the LS mutation, when compared to the parental formulation, had a similar effect on PK regardless of the specific epitopes on the HIV trimer that the mAbs target. This could potentially be attributed to the fact that the LS modification is a relatively small change of two amino acids (AAs) in the constant Fc region of a relatively large molecule of approximately one thousand five hundred AAs. The quantitative differences in the parameters were not widely different across the mAbs, generally falling within 0.5 $\log_{10}$ changes of each other. It is important to note that these estimated differences between each parental/LS pair adjusted for potential imbalances in participants' characteristics that might influence PK. The relative consistency of these parental/LS differences across mAbs indicates that prediction of PK for a mAb with an untested LS mutation may be performed with limited uncertainty based on the PK of the parental formulation. This is important knowledge for the development and planning of future trials of newly discovered HIV-1 mAbs—and potentially for mAbs targeting other pathogens—and their LS modifications. Phase 1 studies remain critical to assess safety and PKs, specifically as the manufacturing process could introduce unexpected variables even though the LS mutation is a relatively small change in the context of the size of the antibody. This analysis helps to provide a baseline for what should be expected when assessing PKs for an LS-modified anti-HIV IgG mAb.

Lastly, we found that PK features were largely consistent across trials for the same mAb. This highlights the general conservation of the PK biological properties and the reproducibility of these phase 1 studies. Particularly, the estimates of CL, half-lives, and simulated drug exposures were remarkably concordant whether the trials were modeled in combination or separately for a given mAb. We did, however, observe some variation in the volume parameters. The Vc parameter is meant to approximate serum blood volume (~3L) [46] but may exhibit variation, as it also functions as a scaling parameter between the administered dose mass and the observed serum concentration. Thus, the Vc differences observed between the trials could be attributable to varying sampling schemes or differing assays rather than differences in demographic or biological factors.

There were several limitations to this analysis. First, the included trials had different PK sampling time points because some were first-in-human trials of the mAbs, with more detailed and intensive sampling especially after the single administration, and others were follow-on trials of the same mAb, with different dosing levels, administration routes, or co-administration with other mAbs. PK studies may be particularly sensitive to sampling schemes, specifically for estimation of the distributional phase (i.e., early kinetics). For PGDM1400, this challenge was observed by the difficulty in estimating the inter-subject variation for distribution PK parameters (Q and Vp). Despite this limitation, the differences in mean adjusted estimates of Vp between parental and LS mutations were in a tight range (1.7- to 3.3-fold).

Second, because the context of this research is for the prevention of HIV acquisition, as opposed to therapeutic treatment, limited sampling time points immediately following the SC administrations were available from the 16 phase 1 trials. Therefore, our ability to discern the effect of the LS mutation on first-order absorption properties (i.e., the absorption rate from the SC space and bioavailability) of the mAbs was limited. To precisely estimate the absorption rate, sampling schemes with finer granularity are required generally within the first week of product administration. However, since serum concentrations are expected to provide the highest protection during this early interval, PK studies in the context of prevention typically do not focus on these early time points as opposed to later time points. Consequently, there were no pairs of parental and LS variants in PK studies that were run with SC groups under a common design to precisely compare this parameter. The limited SC data also prevented precise estimation of bioavailability, as that requires studies with both SC and IV groups for a given LS/parental mAb pair. When we were able to estimate bioavailability, we observed a large range of estimates across the mAbs (i.e., 38% to 76%) that is consistent with a wide range of bioavailability observed generally with mAbs [47]. However, we have limited available data to conclude with certainty whether the range observed here reflects true biological variability across these mAbs, or whether it is a combined result of limited sample size to precisely estimate Vc and F together with different PK sampling time points across the mAb pairs.

Third, the use of different binding antibody assays may have contributed to PK variability in the analysis, particularly when comparing across studies. Since concentration measurements on the same set of sera were only done by a given assay for each mAb, it is infeasible to estimate an assay effect. Therefore, the generalizability of results may be limited when trying to compare parameters or concentrations across the mAbs when different assays were used.

Fourth, while this analysis illustrated common changes to the PKs of mAbs comparing parental to LS variants, it is limited to conclusions made using serum concentrations. In a recent example using a mAb LS modification to prevent COVID-19 [48] in cynomolgus monkeys, the LS modification increased standardized uptake of sotrovimab but did not change the tissue–blood ratios compared to its parental counterpart. Regarding HIV prevention mAbs, preliminary data showed that VRC01LS and VRC07-523LS have different tissue distribution in SIV/SHIV-free naïve/uninfected macaques, and VRC01 and VRC01LS have different rectal tissue to blood ratios in concentrations [49]. Yet it is unclear how much of this distinction is due to possible non-specific binding (Fab) or unequal interaction with FcRn receptors. Additionally, while the LS mutation predictably increases drug persistence via mutation only to the Fc region and not to the Fab regions, it should always be confirmed that the neutralization potential of the LS-mutated mAb has not been altered compared to the parental version.

Fifth, currently available data were collated from trials generally conducted in the US, with relatively limited data from other regions. Therefore, future updates of this analysis are needed when data of the studied or new anti-HIV mAbs become available from on-going and planned trials conducted in other regions. On the other hand, we do not anticipate a significant regional effect as demonstrated, for example, in the analysis of PK data from the AMP trials that were conducted globally in the Americas, Europe, and South Africa [9,44].

Lastly, our analysis was limited to three classes of anti-HIV mAbs and may not translate to mAbs that target other binding sites on the HIV trimer (e.g., 10E8 targeting the MPER). Translation of the presented findings to non-HIV mAbs would also require additional research.

Credited to favorable tolerability, safety, effectiveness, and durability [50], there has been a rapid increase of mAb use for disease treatment and prevention. As of January 2023, the Food and Drug Administration approved more than 165 mAb products for use across a variety of diseases, including rheumatoid arthritis, cancer, Crohn's disease, psoriasis, and hyperlipidemia [51], and to prevent infections such as respiratory syncytial virus (RSV) [52] and severe acute respiratory syndrome coronavirus 2 (SARS-CoV-2) [53]. As more mAbs are being developed, and increasingly used for prevention and treatment of infections, cancer, and other diseases, our investigation provides important knowledge based on a systematic analysis of five pairs of anti-HIV mAbs for the research and development of LS-formulated mAbs, especially those against HIV. Our findings provide benchmarking for what could be expected from LS mutations on emerging new mAbs regardless of indication. More research is needed to understand whether the impact of the LS modification on PKs holds in mucosal compartments and across mAbs against other non-HIV infections or diseases.

Supplementary Materials: The following supporting information can be downloaded at: https://www.mdpi.com/article/10.3390/pharmaceutics16050594/s1, Supplemental Materials contain additional tables and figures, and an additional file with diagnostic plots for each model for each monoclonal antibody. Table S1: List of mAbs; Table S2: Lower limits of quantification (LLoQ) of each assay by monoclonal antibody and study; Table S3: Demographics of included study participants from each study; Table S4: Population pharmacokinetic parameter estimates; Table S5: LS/Parental mean value and fold change for each pharmacokinetic; Figure S1: Serum concentrations (mcg/mL) for each monoclonal antibody; Figure S2: Individual-level predictions for each pharmacokinetic (PK); Figure S3: Diagnostic plots of all pharmacokinetic models pooled over monoclonal antibodies; Figure S4: Scatter visual predictive check (VPC) plot for each mAb by subcutaneous (SV) and intravenous (IV) administration. Lines depict the median trajectory, shaded region depicts 90% prediction interval, and the points depict individual participant observation (DV). Values are dose normalized for comparison; Figure S5: Scatter visual predictive check (VPC) plot for each monoclonal antibody pooled over administration type; Supplemental File S1: Diagnostic plots for each model for each monoclonal antibody.

Author Contributions: Conceptualization, Y.H. and B.T.M.; methodology, Y.H. and B.T.M.; software, L.Z.; validation, L.Z. and B.T.M.; formal analysis, L.Z. and B.T.M.; investigation, Y.H. and B.T.M.; resources, Y.H. and O.H.; data curation, C.Y., A.S., K.E.S., N.Y., M.J.M., R.A.K. and G.D.T.; writing—original draft preparation, B.T.M., H.A. and Y.H.; writing—review and editing, all authors; visualization, L.Z., H.A. and B.T.M.; supervision, Y.H.; project administration, H.A.; funding acquisition, Y.H. and O.H. All authors have read and agreed to the published version of the manuscript.

Funding: This research was funded by the National Institute of Allergy and Infectious Diseases (NIAID) U.S. Public Health Service Grant UM1 AI068635 [HVTN SDMC] to Peter B. Gilbert and Yunda Huang, and the Bill & Melinda Gates Foundation grant INV-032929 to Ollivier Hyrien. Overall support for the HIV Vaccine Trials Network (HVTN) and the HIV Prevention Trials Network (HPTN) is provided by the National Institute of Allergy and Infectious Diseases (NIAID), Office of the Director (OD), National Institutes of Health (NIH), National Institute on Drug Abuse (NIDA), the National Institute of Mental Health (NIMH), and the Eunice Kennedy Shriver National Institute of Child Health and Human Development (NICHD) under Award Numbers UM1AI068614/UM1AI068619 (HVTN/HPTN Leadership and Operations Center), UM1AI068635/UM1AI068617 (HVTN/HPTN Statistical and Data Management Center), and UM1AI068618/UM1AI068613 (HVTN/HPTN Laboratory Center). The content is solely the responsibility of the authors and does not necessarily represent the official views of the National Institutes of Health.

Institutional Review Board Statement: Ethical review and approval were waived for this study because all studies were reviewed and approved individually, and participants agreed to use of their anonymized samples for future research.

Informed Consent Statement: Informed consent was obtained from all participants involved in each of the studies included.

Data Availability Statement: Please refer to the study-specific publications for information on which data are available for each of the 16 clinical trials. The combined dataset can be made available upon request.

Acknowledgments: We acknowledge the study team members of VRC602, VRC605, VRC606, HVTN104, HVTN116, HVTN127/HPTN087, HVTN130/HPTN089, HVTN136/HPTN092, HVTN140/HPTN101, YCO-0899, YCO-0946, YCO-0971, IAVIT001, IAVIT002, MCA-0835, MCA-0885.

Conflicts of Interest: C.F.K.: research grants to my institution from Moderna, Novavax, Gilead, Viiv. B.J. is a part-time employee of Leyden Labs B.V., the Netherlands. M.C.: Scientific Advisory Board—Gilead Sciences. All other authors have no conflicts of interest to declare.

References

1. Walker, L.M.; Phogat, S.K.; Chan-Hui, P.-Y.; Wagner, D.; Phung, P.; Goss, J.L.; Wrin, T.; Simek, M.D.; Fling, S.; Mitcham, J.L.; et al. Broad and Potent Neutralizing Antibodies from an African donor reveal a new HIV-1 vaccine target. *Science* **2009**, *326*, 285–289. [CrossRef] [PubMed]
2. Wu, X.; Yang, Z.-Y.; Li, Y.; Hogerkorp, C.-M.; Schief, W.R.; Seaman, M.S.; Zhou, T.; Schmidt, S.D.; Wu, L.; Xu, L.; et al. Rational Design of Envelope Identifies Broadly Neutralizing Human Monoclonal Antibodies to HIV-1. *Science* **2010**, *329*, 856–861. [CrossRef] [PubMed]
3. Wu, X.; Wang, C.; O'Dell, S.; Li, Y.; Keele, B.F.; Yang, Z.; Imamichi, H.; Doria-Rose, N.; Hoxie, J.A.; Connors, M.; et al. Selection Pressure on HIV-1 Envelope by Broadly Neutralizing Antibodies to the Conserved CD4-Binding Site. *J. Virol.* **2012**, *86*, 5844–5856. [CrossRef] [PubMed]
4. Caskey, M.; Klein, F.; Lorenzi, J.C.C.; Seaman, M.S.; West, A.P., Jr.; Buckley, N.; Kremer, G.; Nogueira, L.; Braunschweig, M.; Scheid, J.F.; et al. Viraemia suppressed in HIV-1-infected humans by broadly neutralizing antibody 3BNC117. *Nature* **2015**, *522*, 487–491. [CrossRef] [PubMed]
5. Caskey, M.; Schoofs, T.; Gruell, H.; Settler, A.; Karagounis, T.; Kreider, E.F.; Murrell, B.; Pfeifer, N.; Nogueira, L.; Oliveira, T.Y.; et al. Antibody 10-1074 suppresses viremia in HIV-1-infected individuals. *Nat. Med.* **2017**, *23*, 185–191. [CrossRef]
6. Mouquet, H.; Scharf, L.; Euler, Z.; Liu, Y.; Eden, C.; Scheid, J.F.; Halper-Stromberg, A.; Gnanapragasam, P.N.P.; Spencer, D.I.R.; Seaman, M.S.; et al. Complex-type *N* -glycan recognition by potent broadly neutralizing HIV antibodies. *Proc. Natl. Acad. Sci. USA* **2012**, *109*, E3268–E3277. [CrossRef] [PubMed]
7. Corey, L.; Gilbert, P.B.; Juraska, M.; Montefiori, D.C.; Morris, L.; Karuna, S.T.; Edupuganti, S.; Mgodi, N.M.; Decamp, A.C.; Rudnicki, E.; et al. Two Randomized Trials of Neutralizing Antibodies to Prevent HIV-1 Acquisition. *N. Engl. J. Med.* **2021**, *384*, 1003–1014. [CrossRef]
8. Gilbert, P.B.; Huang, Y.; Decamp, A.C.; Karuna, S.; Zhang, Y.; Magaret, C.A.; Giorgi, E.E.; Korber, B.; Edlefsen, P.T.; Rossenkhan, R.; et al. Neutralization titer biomarker for antibody-mediated prevention of HIV-1 acquisition. *Nat. Med.* **2022**, *28*, 1924–1932. [CrossRef]
9. Seaton, K.E.; Huang, Y.; Karuna, S.; Heptinstall, J.R.; Brackett, C.; Chiong, K.; Zhang, L.; Yates, N.L.; Sampson, M.; Rudnicki, E.; et al. Pharmacokinetic serum concentrations of VRC01 correlate with prevention of HIV-1 acquisition. *EBioMedicine* **2023**, *93*, 104590. [CrossRef]
10. Mayer, B.T.; Decamp, A.C.; Huang, Y.; Schiffer, J.T.; Gottardo, R.; Gilbert, P.B.; Reeves, D.B. Optimizing clinical dosing of combination broadly neutralizing antibodies for HIV prevention. *PLoS Comput. Biol.* **2022**, *18*, e1010003. [CrossRef]
11. Wagh, K.; Bhattacharya, T.; Williamson, C.; Robles, A.; Bayne, M.; Garrity, J.; Rist, M.; Rademeyer, C.; Yoon, H.; Lapedes, A.; et al. Optimal Combinations of Broadly Neutralizing Antibodies for Prevention and Treatment of HIV-1 Clade C Infection. *PLoS Pathog.* **2016**, *12*, e1005520. [CrossRef] [PubMed]
12. Wagh, K.; Seaman, M.S.; Zingg, M.; Fitzsimons, T.; Barouch, D.H.; Burton, D.R.; Connors, M.; Ho, D.D.; Mascola, J.R.; Nussenzweig, M.C.; et al. Potential of conventional & bispecific broadly neutralizing antibodies for prevention of HIV-1 subtype A, C & D infections. *PLoS Pathog.* **2018**, *14*, e1006860.
13. Sok, D.; van Gils, M.J.; Pauthner, M.; Julien, J.-P.; Saye-Francisco, K.L.; Hsueh, J.; Briney, B.; Lee, J.H.; Le, K.M.; Lee, P.S.; et al. Recombinant HIV envelope trimer selects for quaternary-dependent antibodies targeting the trimer apex. *Proc. Natl. Acad. Sci. USA* **2014**, *111*, 17624–17629. [CrossRef] [PubMed]
14. Ko, S.-Y.; Pegu, A.; Rudicell, R.S.; Yang, Z.-Y.; Joyce, M.G.; Chen, X.; Wang, K.; Bao, S.; Kraemer, T.D.; Rath, T.; et al. Enhanced neonatal Fc receptor function improves protection against primate SHIV infection. *Nature* **2014**, *514*, 642–645. [CrossRef] [PubMed]
15. Roopenian, D.C.; Akilesh, S. FcRn: The neonatal Fc receptor comes of age. *Nat. Rev. Immunol.* **2007**, *7*, 715–725. [CrossRef] [PubMed]
16. Pyzik, M.; Rath, T.; Lencer, W.I.; Baker, K.; Blumberg, R.S. FcRn: The Architect Behind the Immune and Nonimmune Functions of IgG and Albumin. *J. Immunol.* **2015**, *194*, 4595–4603. [CrossRef] [PubMed]
17. Zalevsky, J.; Chamberlain, A.K.; Horton, H.M.; Karki, S.; Leung, I.W.L.; Sproule, T.J.; A Lazar, G.; Roopenian, D.C.; Desjarlais, J.R. Enhanced antibody half-life improves in vivo activity. *Nat. Biotechnol.* **2010**, *28*, 157–159. [CrossRef] [PubMed]
18. Akilesh, S.; Christianson, G.J.; Roopenian, D.C.; Shaw, A.S. Neonatal FcR Expression in Bone marrow-derived cells functions to protect serum IgG from catabolism. *J. Immunol.* **2007**, *179*, 4580–4588. [CrossRef] [PubMed]
19. Gaudinski, M.R.; Coates, E.E.; Houser, K.V.; Chen, G.L.; Yamshchikov, G.; Saunders, J.G.; Holman, L.A.; Gordon, I.; Plummer, S.; Hendel, C.S.; et al. Safety and Pharmacokinetics of the Fc-Modified HIV-1 human monoclonal antibody VRC01LS: A Phase 1 open-label clinical trial in healthy adults. *PLoS Med.* **2018**, *15*, e1002493. [CrossRef]
20. Gaudinski, M.R.; Houser, K.V.; A Doria-Rose, N.; Chen, G.L.; Rothwell, R.S.S.; Berkowitz, N.; Costner, P.; A Holman, L.; Gordon, I.J.; Hendel, C.S.; et al. Safety and pharmacokinetics of broadly neutralising human monoclonal antibody VRC07-523LS in healthy adults: A phase 1 dose-escalation clinical trial. *Lancet HIV* **2019**, *6*, e667–e679. [CrossRef]
21. van der Laan, M.J. and D. Rubin, Targeted Maximum Likelihood Learning. *Int. J. Biostat.* **2006**, *2*, 11. [CrossRef]
22. Rudicell, R.S.; Kwon, Y.D.; Ko, S.-Y.; Pegu, A.; Louder, M.K.; Georgiev, I.S.; Wu, X.; Zhu, J.; Boyington, J.C.; Chen, X.; et al. Enhanced potency of a broadly neutralizing HIV-1 antibody *In Vitro* improves protection against lentiviral infection *In Vivo*. *J. Virol.* **2014**, *88*, 12669–12682. [CrossRef] [PubMed]

23. Cohen, Y.Z.; Butler, A.L.; Millard, K.; Witmer-Pack, M.; Levin, R.; Unson-O'brien, C.; Patel, R.; Shimeliovich, I.; Lorenzi, J.C.C.; Horowitz, J.; et al. Safety, pharmacokinetics, and immunogenicity of the combination of the broadly neutralizing anti-HIV-1 antibodies 3BNC117 and 10-1074 in healthy adults: A randomized, phase 1 study. *PLoS ONE* **2019**, *14*, e0219142. [CrossRef] [PubMed]
24. Julg, B.; Stephenson, K.E.; Wagh, K.; Tan, S.C.; Zash, R.; Walsh, S.; Ansel, J.; Kanjilal, D.; Nkolola, J.; Walker-Sperling, V.E.K.; et al. Safety and antiviral activity of triple combination broadly neutralizing monoclonal antibody therapy against HIV-1: A phase 1 clinical trial. *Nat. Med.* **2022**, *28*, 1288–1296. [CrossRef] [PubMed]
25. Ledgerwood, J.E.; Coates, E.E.; Yamshchikov, G.; Saunders, J.G.; Holman, L.; Enama, M.E.; DeZure, A.; Lynch, R.M.; Gordon, I.; Plummer, S.; et al. Safety, pharmacokinetics and neutralization of the broadly neutralizing HIV-1 human monoclonal antibody VRC01 in healthy adults. *Clin. Exp. Immunol.* **2015**, *182*, 289–301. [CrossRef] [PubMed]
26. Mayer, K.H.; Seaton, K.E.; Huang, Y.; Grunenberg, N.; Isaacs, A.; Allen, M.; Ledgerwood, J.E.; Frank, I.; Sobieszczyk, M.E.; Baden, L.R.; et al. Safety, pharmacokinetics, and immunological activities of multiple intravenous or subcutaneous doses of an anti-HIV monoclonal antibody, VRC01, administered to HIV-uninfected adults: Results of a phase 1 randomized trial. *PLoS Med.* **2017**, *14*, e1002435. [CrossRef] [PubMed]
27. National Library of Medicine. 3BNC117-LS First-in-Human Phase 1 Study. ClinicalTrials.gov #NCT03254277 2021 Available online: https://www.clinicaltrials.gov/study/NCT03254277?term=NCT03254277&rank=1 (accessed on 20 December 2023).
28. E Sobieszczyk, M.; Mannheimer, S.; A Paez, C.; Yu, C.; Gamble, T.; A Theodore, D.; Chege, W.; Yacovone, M.; Hanscom, B.; Heptinstall, J.; et al. Safety, tolerability, pharmacokinetics, and immunological activity of dual-combinations and triple-combinations of anti-HIV monoclonal antibodies PGT121, PGDM1400, 10-1074, and VRC07-523LS administered intravenously to HIV-uninfected adults: A phase 1 randomised trial. *Lancet HIV* **2023**, *10*, e653–e662. [CrossRef] [PubMed]
29. Stephenson, K.E.; Julg, B.; Tan, C.S.; Zash, R.; Walsh, S.R.; Rolle, C.-P.; Monczor, A.N.; Lupo, S.; Gelderblom, H.C.; Ansel, J.L.; et al. Safety, pharmacokinetics and antiviral activity of PGT121, a broadly neutralizing monoclonal antibody against HIV-1: A randomized, placebo-controlled, phase 1 clinical trial. *Nat. Med.* **2021**, *27*, 1718–1724. [CrossRef]
30. National Library of Medicine. Evaluating the Safety, Pharmacokinetics, and Anti-Viral Activity of VRC01 and VRC01LS in the Serum and Mucosa of Healthy, HIV-Uninfected Adults. ClinicalTrials.gov #NCT02797171 2021. Available online: https://www.clinicaltrials.gov/study/NCT02797171?term=NCT02797171&rank=1 (accessed on 20 December 2023).
31. National Library of Medicine. Evaluating the Safety and Serum Concentrations of a Human Monoclonal Antibody, VRC-HIVMAB075-00-AB (VRC07-523LS), Administered in Multiple Doses and Routes to Healthy, HIV-Uninfected Adults. ClinicalTrials.gov #NCT03387150 2023. Available online: https://www.clinicaltrials.gov/study/NCT03387150?term=NCT03387150&rank=1 (accessed on 20 December 2023).
32. Walsh, S.R.; Gay, C.L.; Karuna, S.T.; Hyrien, O.; Skalland, T.; Mayer, K.H.; Sobieszczyk, M.E.; Baden, L.R.; Goepfert, P.A.; Del Rio, C.; et al. A Randomised Clinical Trial of the Safety and Pharmacokinetics of VRC07-523LS Administered via Different Routes and Doses (HVTN 127/HPTN 087). *medRxiv* **2024**, 2024.01.10.23299799. [CrossRef]
33. National Library of Medicine. Evaluating the Safety, Tolerability, Pharmacokinetics, and Antiviral Activity of the Monoclonal Antibody PGT121.414.LS Administered Alone and in Combination With VRC07-523LS Via Intravenous or Subcutaneous Infusions in Healthy, HIV-uninfected Adult Participants. ClinicalTrials.gov ID#NCT04212091 2023. Available online: https://www.clinicaltrials.gov/study/NCT04212091?term=NCT04212091&rank=1 (accessed on 20 December 2023).
34. Schoofs, T.; Klein, F.; Braunschweig, M.; Kreider, E.F.; Feldmann, A.; Nogueira, L.; Oliveira, T.; Lorenzi, J.C.C.; Parrish, E.H.; Learn, G.H.; et al. HIV-1 therapy with monoclonal antibody 3BNC117 elicits host immune responses against HIV-1. *Science* **2016**, *352*, 997–1001. [CrossRef]
35. National Library of Medicine. First-in-Human Study of 10-1074-LS Alone and in Combination With 3BNC117-LS. ClinicalTrials.gov #NCT03554408 2021. Available online: https://www.clinicaltrials.gov/study/NCT03554408?term=NCT03554408&rank=1 (accessed on 20 December 2023).
36. National Library of Medicine. A Clinical Trial to Evaluate the Safety, Tolerability, and Pharmacokinetics of PGDM1400LS Alone and in Combination With VRC07-523LS and PGT121.414.LS in Healthy, HIV-uninfected Adult Participants. ClinicalTrials.gov #NCT05184452 2023. Available online: https://www.clinicaltrials.gov/study/NCT05184452?term=%20NCT05184452&rank=1 (accessed on 20 December 2023).
37. National Library of Medicine. A Study of the Safety, Pharmacokinetics and Antiretroviral Activity of 10-1074. ClinicalTrials.gov #NCT02511990 2018. Available online: https://www.clinicaltrials.gov/study/NCT02511990?term=NCT02511990&rank=1 (accessed on 20 December 2023).
38. Wesley, M.S.; Chiong, K.T.; Seaton, K.E.; Arocena, C.A.; Sawant, S.; Hare, J.; Hernandez, K.; Rojas, M.; Heptinstall, J.; Beaumont, D.; et al. Validation of a Triplex Pharmacokinetic Assay for Simultaneous Quantitation of HIV-1 Broadly Neutralizing Antibodies PGT121, PGDM1400, and VRC07-523-LS. *Front. Immunol.* **2021**, *12*, 709994. [CrossRef] [PubMed]
39. Davda, J.P.; Dodds, M.G.; A Gibbs, M.; Wisdom, W.; Gibbs, J.P. A model-based meta-analysis of monoclonal antibody pharmacokinetics to guide optimal first-in-human study design. *mAbs* **2014**, *6*, 1094–1102. [CrossRef] [PubMed]
40. Delattre, M.; Lavielle, M.; Poursat, M.-A. A note on BIC in mixed-effects models. *Electron. J. Stat.* **2014**, *8*, 456–475. [CrossRef]
41. Lin, L.I.-K. A Concordance correlation coefficient to evaluate reproducibility. *Biometrics* **1989**, *45*, 255–268. [CrossRef] [PubMed]
42. McBride, G.B. A proposal for strength-of-agreement criteria for Lins Concordance Correlation Coefficient. NIWA Client Rep: HAM2005–062, 2005. Available online: https://www.medcalc.org/download/pdf/McBride2005.pdf (accessed on 19 April 2024).

43. Gruber, S.; Laan, M.v.d. Tmle: An R Package for Targeted Maximum Likelihood Estimation. *J. Stat. Softw.* **2012**, *51*, 1–35. [CrossRef]
44. Huang, Y.; Naidoo, L.; Zhang, L.; Carpp, L.N.; Rudnicki, E.; Randhawa, A.; Gonzales, P.; McDermott, A.; Ledgerwood, J.; Lorenzo, M.M.; et al. Pharmacokinetics and predicted neutralisation coverage of VRC01 in HIV-uninfected participants of the Antibody Mediated Prevention (AMP) trials. *EBioMedicine* **2021**, *64*, 103203. [CrossRef] [PubMed]
45. Holm, S. A Simple Sequentially Rejective Multiple Test Procedure. *Scand. J. Stat.* **1979**, *6*, 65–70.
46. Dong, J.Q.; Salinger, D.H.; Endres, C.J.; Gibbs, J.P.; Hsu, C.P.; Stouch, B.J.; Hurh, E.; Gibbs, M.A. Quantitative prediction of human pharmacokinetics for monoclonal antibodies: Retrospective analysis of monkey as a single species for first-in-human prediction. *Clin. Pharmacokinet.* **2011**, *50*, 131–142. [CrossRef] [PubMed]
47. Sánchez-Félix, M.; Burke, M.; Chen, H.H.; Patterson, C.; Mittal, S. Predicting bioavailability of monoclonal antibodies after subcutaneous administration: Open innovation challenge. *Adv. Drug Deliv. Rev.* **2020**, *167*, 66–77. [CrossRef]
48. Aweda, T.A.; Cheng, S.-H.; Lenhard, S.C.; Sepp, A.; Skedzielewski, T.; Hsu, C.-Y.; Marshall, S.; Haag, H.; Kehler, J.; Jagdale, P.; et al. In vivo biodistribution and pharmacokinetics of sotrovimab, a SARS-CoV-2 monoclonal antibody, in healthy cynomolgus monkeys. *Eur. J. Nucl. Med.* **2022**, *50*, 667–678. [CrossRef]
49. Lemos, M.P.; Astronomo, R.D.; Andersen-Nissen, E.; Huang, Y.; Shao, M.; Mann, P.; Czartoski, J.; Orrell, C.; Naidoo, A.; Mgaga, Z.; et al. "Rectal Biopsies from Men and Women Infused Intravenously with VRC01 mAb Show Partial Protection that Wanes Over Time in Ex Vivo HIV-1 Challenges." (Oral Poster), Abstract of the HIV Research for Prevention Meeting, HIVR4P, 21–25 October, 2018, Madrid. *AIDS Res. Hum. Retroviruses* **2018**, *34*.
50. Rajewsky, K. The advent and rise of monoclonal antibodies. *Nature* **2019**, *575*, 47–49. [CrossRef] [PubMed]
51. Mieczkowski, C.A. The Evolution of Commercial Antibody Formulations. *J. Pharm. Sci.* **2023**, *112*, 1801–1810. [CrossRef]
52. Harris, E. FDA Approves RSV Monoclonal Antibody for Infants and Young Children. *JAMA* **2023**, *330*, 586. [CrossRef] [PubMed]
53. Herman, G.A.; O'Brien, M.P.; Forleo-Neto, E.; Sarkar, N.; Isa, F.; Hou, P.; Chan, K.C.; Bar, K.J.; Barnabas, R.V.; Barouch, D.H.; et al. Efficacy and safety of a single dose of casirivimab and imdevimab for the prevention of COVID-19 over an 8-month period: A randomised, double-blind, placebo-controlled trial. *Lancet Infect. Dis.* **2022**, *22*, 1444–1454. [CrossRef] [PubMed]

pharmaceutics

MDPI

Review

Polymer Delivery Systems for Long-Acting Antiretroviral Drugs

Mohammad Ullah Nayan [1,†], Sudipta Panja [1,†], Ashrafi Sultana [1], Lubaba A. Zaman [1], Lalitkumar K. Vora [2], Brady Sillman [1], Howard E. Gendelman [1,*] and Benson Edagwa [1,*]

1 Department of Pharmacology and Experimental Neuroscience, University of Nebraska Medical Center, Omaha, NE 68198-5880, USA; mohammadullah.nayan@unmc.edu (M.U.N.); sudipta.panja@unmc.edu (S.P.); asultana@unmc.edu (A.S.); lubaba.zaman@unmc.edu (L.A.Z.); brady.sillman@unmc.edu (B.S.)

2 School of Pharmacy, Queen's University Belfast, Medical Biology Centre, Belfast BT9 7BL, UK; l.vora@qub.ac.uk

* Correspondence: hegendel@unmc.edu (H.E.G.); benson.edagwa@unmc.edu (B.E.); Tel.: +1-402-559-4044 (H.E.G. & B.E.); Fax: +1-402-559-3733 (H.E.G. & B.E.)

† These authors contributed equally to this work.

Abstract: The success of long-acting (LA) drug delivery systems (DDSs) is linked to their biocompatible polymers. These are used for extended therapeutic release. For treatment or prevention of human immune deficiency virus type one (HIV-1) infection, LA DDSs hold promise for improved regimen adherence and reduced toxicities. Current examples include Cabenuva, Apretude, and Sunlenca. Each is safe and effective. Alternative promising DDSs include implants, prodrugs, vaginal rings, and microarray patches. Each can further meet patients' needs. We posit that the physicochemical properties of the formulation chemical design can optimize drug release profiles. We posit that the strategic design of LA DDS polymers will further improve controlled drug release to simplify dosing schedules and improve regimen adherence.

Keywords: long-acting formulations; polymer; antiretroviral therapy; HIV; chronic infectious diseases; implants; vaginal rings; prodrug nanoformulations; microarray patches

1. Introduction

Long-acting (LA) drug delivery systems (DDSs) harness polymer properties to achieve spatiotemporal control over release and drug biodistribution. This allows the LA dosage regimens to extend from days to months. Such DDSs significantly reduce the burden of chronic disease. Treatment success parallels medication adherence. The development of LA medicines is traced back to the 1930s with the discovery of hydrophobic drug release from implants [1]. However, the discovery of low-molecular-weight (MW) dye diffusion through silicone tubing laid the groundwork for rate-controlling LA DDS polymers [2]. Biopharmaceutical research activities directed their efforts to convert biocompatible silicone tubing into versatile drug delivery materials. These efforts included improved uses of atropine, histamine, anesthetic, steroid, antimalarial, and antischistosomal agents [2]. Since the 1960s, the work of Drs. Folkman, Langer, Higuchi, Roseman, Peppas, Heller, Ringsdorf, and Speiser have pioneered the LA DDSs advancements seen today. Widespread research resulted in the development of LA DDSs which extended the dosage intervals for a broad range of drug regimens. This resulted in improved treatment outcomes for contraception and for treatments of psychosis, diabetes, osteoporosis, and ocular diseases.

A key example of a treatable chronic disease that would benefit from LA DDSs is human immune deficiency virus type one (HIV-1) infection. Here, prior studies demonstrated that non-adherence to therapeutic regimens was a significant limiting factor in achieving successful treatment outcomes. One notable point is the development of viral resistance. Another is simply remembering to take the medicines according to daily prescribed requirements. Indeed, patients have repeatedly expressed treatment preferences for infrequent dosing regimens as, without a viable cure or preventive vaccine, daily oral antiretroviral therapy (ART) is a principal means for HIV-1 treatment and prevention. Over the past two decades, considerable efforts have been focused on the development of LA ARTs that include a specific focus on controlled-release LA formulations. Such formulations include solid implants, vaginal rings (VRs), and surfactant-stabilized aqueous nanocrystal suspensions. In these depot-forming DDSs, the drug is either encapsulated in biodegradable polymers for implantation in subcutaneous (SC) space or in reproductive organs or formulated as drug nanocrystals for either intramuscular (IM) or SC injections. Extensive tests have culminated in the approval of surfactant-stabilized aqueous nanocrystal LA cabotegravir and rilpivirine (CAB and RPV LA) suspensions by the US Food and Drug Administration (FDA). CAB and RPV LA offer the convenience of monthly or bimonthly dosing for HIV-1 treatment [3–5].

Multiple randomized clinical trials have demonstrated comparable efficacy between CAB and RPV LA and a standard daily oral regimen [4,6,7]. Based on the data that 91% of prior clinical trial participants preferred LA ART over daily oral medicines, CAB LA proved more effective in preventing HIV-1 infection when compared to daily oral Truvada [8]. This propelled its US FDA approval in December 2021 [5]. The remarkable efficacy of CAB LA was linked to improved treatment satisfaction and therapeutic adherence. However, the limitations associated with CAB LA include the need for frequent clinical visits, injection site reactions, variable pharmacokinetics (PK) profiles, prolonged terminal phase tailings, high costs, and comorbid conditions.

Alternative approaches that include in-situ-forming implants (ISFIs), solid implants (SIs), and prodrug nanocrystals are being developed to further improve the PK profiles and other listed limitations associated with CAB LA. Notably, surfactant-coated stearoylated CAB prodrug nanocrystals achieved sustained therapeutic drug levels for up to a year in preclinical models [9]. Other studies demonstrated that a poly(lactic acid-co-glycolic acid) (PLGA)-based ISFI could potentially extend CAB dosing intervals to every 6 months [10]. Other studies have also shown that solid implants containing highly potent ARVs, such as tenofovir alafenamide (TAF) and islatravir (ISL), can simplify ART dosages [11–14]. However, in November 2021, Merck announced that participants who received once-weekly ISL plus MK-8507 experienced a decline in CD4 T-cells. Further review found that CD4 counts also fell in people taking once-daily ISL plus doravirine for HIV-1 treatment, while those taking ISL alone for pre-exposure prophylaxis (PrEP) experienced a decline in total lymphocyte counts. As a result, studies of ISL for PrEP have been discontinued. These findings underscore the importance of establishing safety profiles of new chemical entities prior to combining them with other therapies. Polymeric TAF implant technologies have demonstrated the zero-order release kinetics of the pharmacologically active tenofovir diphosphate (TFV-DP) metabolite in preclinical models of PK studies [15]. These include non-degradable materials that provide sustained drug delivery. The use of self-administered microbicides can significantly reduce the risk of HIV-1 infection. Vaginal rings (VRs) that consist of non-biodegradable elastomeric polymers are being developed as self-administrable PrEP treatment targeting at-risk populations [16]. Several VRs of TAF and dapivirine (DPV), a non-nucleoside reverse transcriptase inhibitor (NNRTI), have demonstrated that such formulations could extend ART dosing intervals [16–18]. In two separate phase III clinical trials, known as the RING and the ASPIRE studies, DPV VRs demonstrated their effectiveness in reducing HIV-1 transmission. Compared to a placebo ring, the DPV ring reduced transmission rates by 31% in the RING study and 27% in the ASPIRE study [19,20]. These results suggest that the DPV ring could supplement existing HIV prevention practices. Microneedles (MNs), also referred to as microarray patches (MAPs), are another class of LA DDS that use drug-diffusible, rate-controlling, non-biodegradable, or slowly biodegradable polymer membranes to extend the apparent half-lives of drugs [21]. Due to the non-invasive and self-administrable nature, the goal of

MNs is to expand LA ART options and to appeal to a wide variety of users that include pediatric populations. Examples include CAB- and RPV-loaded MNs that are currently being explored as alternative treatment options to daily oral therapy as well as injectables [22,23]. However, further work is required to demonstrate improved or comparable PK and efficacy profiles to the existing CAB and RPV LA injectable formulations.

This review summarizes the physiochemical and biological parameters of the commonly used polymers used for LA DDSs in treatment and prevention regimens for HIV-1 infection. LA ART formulations either in preclinical or clinical development are discussed. For each of these delivery systems, polymer compositions and drug release kinetics are discussed. Additionally, an expert opinion section highlights LA ART DDS design considerations which could potentially encourage widespread acceptance and utilization among the key target populations.

2. LA DDSs

The emergence of polymer-based DDSs can be traced back to the 1930s when implanted pellets with hydrophobic substances were identified for their ability to facilitate sustained drug release [24]. Examples include pellets with estradiol for prostate cancer treatment and testosterone pellets for the treatment of hypogonadism [25]. Later, clinical use of depot-forming formulations loaded with hydrophobic drugs in either water-based or oily mediums, such as procaine penicillin G in water and fluphenazine decanoate in sesame oil, became popular [26,27]. Since then, more drug delivery approaches and models have been developed to improve our understanding of the materials and drug release mechanisms. In the 1960s, T. Higuchi presented his renowned 'Higuchi model' to describe the drug release kinetics from various sustained release matrix systems [28]. The model suggested that the release of the solid drugs dispersed in a matrix varies with the square root of time. At the same time, Folkman discovered that silicone rubber could act as a drug reservoir, allowing for constant drug release after implantation. This breakthrough led to the conceptualization of rate-controlling membranes or reservoir-based implants [29]. Notable examples of this concept in action are Ocuserts®, which are designed to deliver ocular drugs at predictable and controlled rates [30].

The challenges associated with non-biodegradable implants, such as the necessity of surgical removal after product life ends and adverse implant site reactions, spurred the development of biodegradable implants. The development and clinical utilization of biodegradable polymers like polylactic acid (PLA), polyglycolide (PGA), and PLGA date back to the 1970s [31]. At the early stage of development, clinical applications of these biodegradable polymers were restricted to surgical sutures [31]. Later on, PLA and PLGA microparticles and pellet depot systems for delivering contraceptive drugs and luteinizing-hormone-releasing hormone (LHRH) analogs were developed [32]. Other notable examples of commercial products where the biodegradable copolymers were utilized include Lupron Depot and Zoladex® implants for treatment of prostate cancer, breast cancer, and endometriosis-like disease [32]. Furthermore, a polymer-based VR was patented by Upjohn Company for sustained drug release [33]. These developments marked significant milestones in the evolution of polymer-based LA DDSs. Additionally, extensive research was undertaken to develop silicone-based VRs for contraception. In-depth research, primarily sponsored by the World Health Organization (WHO), paved the way for the clinical approval of multiple contraceptive implants in the early 1990s.

The development of LA ARTs began in the early 2000s, with an initial focus on VRs and implants. Later, attention shifted towards nanocrystal formulations [34]. Clinical trials for nanocrystal aqueous suspensions started around 2008 (Figure 1), including phase II trials for HIV-1 prevention (PrEP) with LA RPV [35]. However, the development of this formulation as a single agent for PrEP faced challenges linked to the high prevalence of NNRTI mutations [35,36]. After a decade of research and several clinical trials, the first LA complete therapy for HIV-1 treatment was approved in 2020 [37].

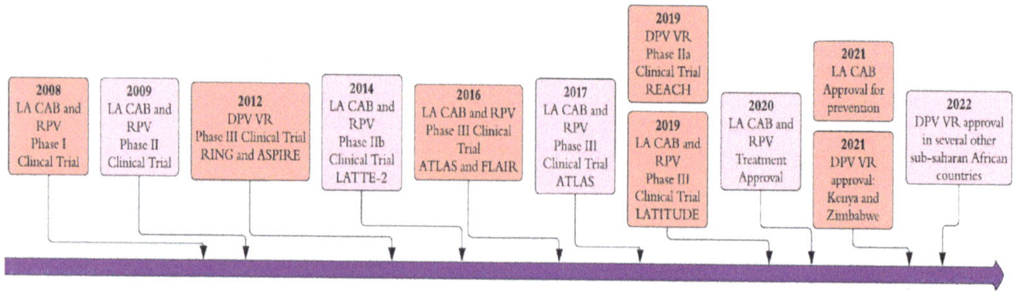

Figure 1. Timeline for LA drug delivery formulations for the treatment and prevention of HIV-1.

3. Drug Release Kinetics from LA DDSs

In polymer-based DDSs, drug release refers to a transfer process in which drug molecules are released from the inner core or matrix to the outer surface of the delivery system and eventually into the surrounding environment or tissue [38]. The rate of drug release from polymer-based DDSs can be modulated by choosing an appropriate polymer with a suitable DDS design. The terminology 'long-acting' refers to the ability to extend the duration of action of a therapeutic agent for a longer period of time. Most often, the terms long-acting, controlled release, sustained release, and extended release are used synonymously [39]. According to the United States Pharmacopeia (USP), the term 'extended release' or 'long-acting' or 'sustained release' is defined as 'a deliberate modification to protract the release rate of an active pharmaceutical ingredient (API) in comparison to an immediate release dosage form' [40]. In this review, the terminology 'long-acting', 'sustained release', or 'controlled release' is used to designate formulations that can extend the apparent half-lives of drugs. The release of drugs from a delivery system can have one or more mechanisms, which can be correlated with a number of existing release kinetics models. The commonly used models are zero-order kinetic, first-order kinetic, Higuchi, Korsmeyer–Peppas, Peppas–Sahlin, and Hixson–Crowell. The zero-order kinetic model considers that there is no drug concentration influence on drug release rates. The zero-order kinetic models are reflected by the mathematical Equation (1), where C_0 and C_t represent the amount of drug at the start and the amount released at time 't'.

$$C_0 - C_t = K_0 t \quad (1)$$

For first-order kinetics, the rate of drug release is proportional to the concentration of the remaining drug in the delivery system and represented by Equation (2), where dc/dt is the rate of drug release, K_1 is the first-order rate constant, and C_t is the concentration of the remaining drug at time t.

$$\frac{dC}{dt} = -K_1 C_t \quad (2)$$

The Korsmeyer–Peppas model (Equation (3)) reflects dissolution-mediated drug release. 'M_t' and 'M_∞' are the amount of eluted drug, 't' is the recorded time '∞'. The ratio of 'M_t/M_∞' denotes the fraction of drug release at the time 't'. K_k is the Korsmeyer rate constant.

$$\frac{M_t}{M_\infty} = Kt^n \quad (3)$$

The Peppas–Sahlin equation (Equation (4)) links to the diffusion- and relaxation-mediated drug release. The diffusion exponent is represented by 'm', and kinetic constants are represented by K_1 and K_2.

$$\frac{M_t}{M_\infty} = K_1 t^m + K_2 t^{2m} \quad (4)$$

The Higuchi model (Equation (5)) is based on diffusion-mediated drug release. Here, 'Q' indicates the amount of drug released per unit area at time t and K_H is the Higuchi constant.

$$Q = K_H\sqrt{t} \tag{5}$$

The Hixson–Crowell model (Equation (6)) applies to uniform-size drug particles, where the rate of drug release is controlled by drug dissolution. This is governed by the surface area or diameter of the drug-encased particles. The rate of drug dissolution is based on the cube root of the drug mass. The M_0 and M_t are the mass of the drug at the initial and at the recorded time 't'. K_{HC} symbolizes the Hixson–Crowell release constant.

$$M_0^{1/3} - M_t^{1/3} = K_{HC}t \tag{6}$$

An ideal LA DDS should exhibit zero-order drug release kinetics where a constant amount of drug is released per unit time. However, maintaining zero-order kinetics is very challenging, and it majorly depends on the physicochemical properties of the drug and excipients [41,42]. Diffusion, osmotic pumping, swelling, degradation, or erosion-induced release are the commonly reported mechanisms for polymer-based DDSs.

Non-degradable polymer-based DDSs can be either reservoir or matrix form. In the reservoir form of DDSs, the release rate is governed by the thickness of the polymer membrane and permeability of the drug through the polymer membrane. Whereas, in the matrix form of DDSs, Fickian diffusion remains the underlying release mechanism [43]. Diffusion refers to the random movement of the drug molecules from the higher- to the lower-concentration region in the matrix. The rate of diffusion in LA DDSs can be described by Fick's law [43]. According to this law, the drug release rate depends on the concentration gradient (ΔC) and diffusibility (D) of the drug through the polymer matrix (Equation (7)). Particularly for slab-like matrix, Equation (7) can be simply transformed to Equation (8).

$$\frac{\partial C}{\partial t} = D\Delta C \tag{7}$$

D is the diffusion coefficient or diffusivity, C is the concentration.

$$\frac{M_t}{M_0} = 4\left(\frac{Dt}{\pi h^2}\right)^2 \tag{8}$$

M_t is the sum of drug released at time t, M_0 is the total of the drug-loaded mass, D is the diffusion coefficient, and h is the thickness of the slab-like matrix.

According to Equation (8), the release rate is directly proportional to the drug's diffusibility (D) through the slab and inversely proportional to the thickness (h) of the polymer [43]. Although Equation (8) can be transformed to other various forms based on the geometry of the DDSs, the parameters, such as the drug's diffusibility and slab thickness, play their role in a similar way. For a particular DDS, the release kinetics can be zero order when 'h' and 'D' remain constant over time (Equation (8)) [43]. The diffusibility is influenced by the size of the drug molecule relative to the pore size of the matrix. Furthermore, matrix pore size and density have been governed by the properties of the polymer used to fabricate the matrix, such as nature of the monomers and the molar composition used to synthesize the polymer [44]. Matrix systems lack a rate-controlling membrane, so the diffusion rate is affected by non-constant drug concentration gradient and diffusion distance. Additionally, the diffusion distance is reliant on the polymer's swelling [44].

In degradable polymer-based DDSs, drug release is majorly controlled by the rate of polymer degradation or erosion and osmotic pumping of the drugs. The chemical degradation of the polymer is influenced by its hydrophilicity. Hydrophilic polymers can absorb water, resulting in an increase in their pore size. This allows them to initiate drug release [45]. Over time, the polymer undergoes degradation, which increases the number and size of pores, eventually leading to continuous release of drug. Unlike degradation, erosion is an alternative method of the drug release process where polymeric chain segments are dissolved by keeping their chemical structure intact [44]. Erosion processes can happen either on the surface or bulk or a combination of both places on the DDS. Surface erosion gradually reduces the size of the DDS from the outside in [46,47]. In bulk erosion, water permeates the entire bulk of the polymer matrix, leading to a uniform degradation with no significant change in their initial size [47]. In addition, bulk erosion may produce faster and unpredictable drug release kinetics, making it less favorable for LA DDSs [48]. Osmotic pumping is another method of drug release, where osmotic pressure drives the influx of water into the non-swelling segment of the DDS, resulting in drug release [49].

Interestingly, the prevalent drug release pattern from LA DDSs is the triphasic rather than monophasic drug release pattern [41]. The first phase of drug release from an LA DDS demonstrates an initial burst release due to the rapid release of drug molecules located near the surface of the DDS or the surrounding tissue [50]. The extent of this burst release is influenced by factors such as the design and morphology of the DDS, polymer properties, fabrication process, storage conditions, and homogeneity of the drug–matrix concentration [50]. While an initial burst release may be desired for rapid drug action, it can significantly reduce the drug depot concentration, consequently impacting the DDS's longevity. Strategies such as hydrophobic polymer coating on the outer surface of the DDS can curtail this challenge [50]. The second phase of drug release involves a slow-release period, where the drug diffuses through the polymer matrix. This phase concurrently occurs with polymer degradation in biodegradable systems [41]. The third phase may exhibit faster release as bulk erosion of the matrix starts. Drug release can also follow a biphasic trend, transitioning from the initial burst release to the zero-order kinetics [51]. Overall, the kinetics of drug release are complex and influenced by multiple factors. Understanding and optimizing the release profile of DDSs are crucial for achieving desired therapeutic outcomes and ensuring the efficacy of DDSs.

Molecular weight (MW) and the molar ratio of the monomeric units of a polymer play a crucial role in determining its physicochemical properties, including solubility, crystallinity, glass transition temperature (T_g), and mechanical strength. Polymers with less elasticity result in non-deformable matrices and smaller pore formation, leading to a slower drug release [52]. Moreover, the degradation of the copolymer is influenced by the nature and molar ratio of monomeric units [53]. For example, a higher molar ratio of hydrophilic monomer, glycolic acid (GA), in PLGA composition leads to its faster biodegradation and subsequent drug release [54].

Polymer crystallinity refers to the proportion of crystalline and amorphous regions within the polymer structure [55]. The ratio of different monomeric units in a copolymer affects the polymer's crystallinity and T_g. Most polymers are semicrystalline in nature, in which the amorphous domains separate the crystalline domains. The water permeability of the polymer is primarily controlled by the percentage of crystallinity. An increased level of polymer crystallinity decreases the overall water permeability, resulting in a slower rate of polymer biodegradation and a decrease in the drug release rate [56]. In high-MW polymers, the influence of crystallinity on drug release depends on the presence of monomer crystallinity [57]. A notable example is the biodegradability of PLGA, where the ratio of its monomers, lactic acid and GA, determines the degree of crystallinity, subsequent biodegradation, and drug release profile. PLLA, composed of L-lactic acid, is highly crystalline, while PDLA, composed of D-lactic acid, is completely amorphous. Like PLLA, PGA is also a highly crystalline polymer. The degree of crystallinity and biodegradability of PLGA copolymer depends on the molar composition of lactic acid and GA comonomers in the copolymer [58]. Furthermore, the ratio of the amorphous and crystalline regions also influences the polymer's T_g, which in turn affects its mechanical strength and water permeability. For example, a polymer with T_g equal or close to physiological temperature can transform into a rubbery state, facilitating water diffusion and promoting drug release [44].

4. Common Polymers in LA DDSs

Advancements in synthetic polymer chemistry have led to the development of numerous biocompatible and biodegradable polymers. Synthetic polymers are derivatized from chemically synthesized monomers and are typically biologically inert, causing minimal adverse immune reactions. They also exhibit more predictable physical and chemical properties compared to natural polymers [59]. The ability to adjust the physicochemical properties of synthetic polymers allows for tailored drug release kinetics, making them more favorable than natural polymers [60]. Polymers used in the fabrication of LA DDSs can be categorized as either biodegradable or non-biodegradable [61]. Injection site reactions and requirement for removal of DDSs after the lifespan spurred the development of biodegradable polymers. The use of biocompatible and biodegradable synthetic polymers in LA DDSs eliminates the need for removal after their lifespan. In the following discussion, we will explore the polymers commonly employed in LA DDSs, along with their physiochemical characteristics and mechanisms of drug release.

4.1. Polyethylene Glycol (PEG)

PEG, also known as macrogol, is synthesized by ring-opening polymerization of ethylene oxide. Its outstanding biocompatibility, hydrophilicity, non-immunogenicity, and antifouling properties towards serum protein contribute to its extensive use in DDSs [62]. Chemical modification with PEG, commonly referred to as PEGylation, enables the conjugation of biomolecules to the surface of micro- or nanoparticles. Surface PEGylation of DDSs substantially reduced its opsonization by decreasing the chances of protein corona formation, which eventually increases the plasma apparent half-life and reduces the immunogenicity in vivo [63,64]. Based on structure, PEG can be classified as linear, branched, and star-shaped [65]. The end hydroxyl group of the PEG can be transformed into various active functional groups, such as amino, maleic amide, and carboxyl moieties, suitable for functionalization with various drugs and other pharmacologically active biomolecules [66,67]. However, the minimal availability of active functional groups in linear PEG offers limited PEG–drug covalent conjugates, resulting in low drug-loading content. Therefore, employing a four- or eight-arm PEG structure can increase the number of active end functional groups, resulting in improved drug loading in DDSs [68,69]. PEG cannot undergo biodegradation; hence, a low-MW PEG has been recommended for use in DDSs so that it can easily be eliminated from the biological system. In addition, PEG can be copolymerized with biodegradable polyesters, such as PLA or PLGA, to improve the overall biodegradability of the copolymer [70].

4.2. Poloxamers

Poloxamers are proprietary polymers of BASF, also known as Pluronics®. These are triblock copolymers and have been synthesized by sequential polymerization of ethylene oxide (EO) and propylene oxide (PO) (Figure 2). The arrangement of blocks in the copolymer can either be PEO-b-PPO-b-PEO or PPO-b-PEO-b-PPO [71]. Poloxamers are amphiphilic in nature because of the hydrophilic and hydrophobic properties of PEO and PPO, respectively. Amphiphilic poloxamers self-assemble into a spherical micelle at a concentration above their critical micelle concentration (CMC) at a specified temperature. These micelles adopt a core–shell morphology with a PPO block in the core and PEO block on the shell. Some poloxamers can transform their morphology from micellar structures to solid micellar cubes, producing thermo-responsive clear and rigid hydrogel at elevated temperatures and concentrations [72]. For example, Poloxamer 407 can form hydrogels in 20% w/w solution [73]. The rate of drug release from the poloxamer-based DDSs primarily depends on the drug's hydrophilicity and relative block lengths of PPO and PEO [74,75]. It has been seen that hydrophilic drugs typically followed diffusion-induced release, whereas hydrophobic drugs followed erosion-induced release [76]. In addition, factors such as poloxamer concentration, lipophilicity, physicochemical properties of the used solute, and the amount of the aqueous channel network in poloxamer gel significantly influence drug release. In an abundant aqueous environment, the structure of poloxamer micelles breaks down quickly, leading to gel matrix degradation [77].

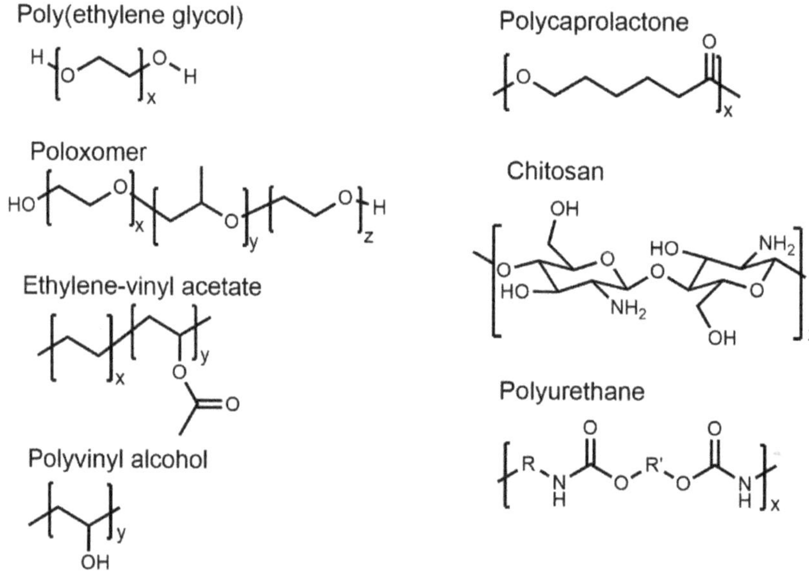

Figure 2. Chemical structures of the polymers used in LA delivery systems.

4.3. *Ethylene-Vinyl Acetate (EVA)*

The random copolymer of ethylene (E) and vinyl acetate (VA) is often referred to as ethylene-vinyl acetate (EVA) (Figure 2). The copolymer has been synthesized by radical copolymerization of ethylene and vinyl acetate with various feed molar ratios of monomers depending on its applicability [78]. EVA has been extensively used in clinical settings, especially for ocular implants, contraceptives, and hormone replacement therapy. NuvaRing®, Progestasert®, and Implanon® are the three most widely used EVA-based contraceptive products, where EVA acts as a release-rate-controlling membrane [79]. EVA has also been used in the delivery of buprenorphine/hydromorphone for treatment of opioid withdrawal.

Like other polymers, EVA's release kinetics is influenced by its physicochemical properties, such as hydrophobicity, crystallinity, etc. Furthermore, EVA's physicochemical properties have been governed by wt% of VA. The commercially available EVA contains 1 to 40 wt% VA, randomly distributed throughout the EVA backbone [80]. The acetate side chain of VA creates steric hindrance that restricts the alignment of the polymeric chain. As a result, an increasing percentage of VA decreases the crystallinity [81]. Eventually, EVA becomes nearly amorphous when the VA content increases to 40 wt% [82]. EVA exhibits a complex combination of amorphous and crystalline phase regions, characterized by at least two T_g (between −35 to −25 °C), which is barely influenced by the overall VA content [83,84]. In addition, the MW distribution, branching, and morphology also have a prominent impact on the physicochemical properties of EVA. In hydrogel reservoir-based delivery systems, EVA is commonly employed as a membrane for controlling the release rate. Additionally, EVA finds extensive application as an excipient in the development of nano- and microparticles as it is compatible with a wide range of excipients that can be used in complex multilayer polymer blends to further tune the release rate [79,85].

4.4. Polyvinyl Alcohol (PVOH)

Polyvinyl alcohol (PVOH) is a highly biocompatible and biodegradable semicrystalline polymer used in LA DDSs [86]. PVOH is synthesized by sodium-hydroxide-mediated hydrolysis of polyvinyl acetate and can be classified as fully or partially hydrolyzed PVOH [87,88]. PVOH is highly water soluble because of its pendant hydroxyl groups (Figure 2). The solubility, mechanical, and adhesion properties of PVOH depend on its MW and percentage of hydrolysis. Partially hydrolyzed PVOH (87–89%) shows improved solubility, flexibility, and adhesion to hydrophobic surfaces, while highly hydrolyzed PVOH (91–99%) exhibits enhanced water stability and increased tensile strength and adhesion to hydrophilic surfaces [89]. The hydroxyl groups in PVOH can form inter- and intramolecular hydrogen bonding, which influences its rheological, viscoelastic, and solution properties. PVOH's characteristics, including high surface stabilization and chelation, make it an excellent candidate for drug delivery [90,91]. PVOH has been used in many clinically approved products over the last few decades. For example, the Iluvien® implant has been fabricated by using PVOH and silicone and used for macular edema [92]. In addition, PVOH has been found to be extensively applied in preclinical research for developing hydrogels, microspheres, and nanoparticles for various therapeutic purposes [86]. PVOH hydrogels can be physically or chemically cross-linked, holding drug molecules in their 3D cross-linked structure [93,94]. These hydrogels have gained popularity as the drug release kinetics from these hydrogels can be predicted by observing water uptake and swelling index [93,95].

PVOH has demonstrated remarkable biocompatibility in various studies conducted since the late 1980s, as evidenced by its successful use in biomedical applications [96,97]. Examples include its inclusion in ophthalmic solutions and tear replacement solutions without causing any discomfort to the eyes [98]. In recent years, PVOH hydrogels have found extensive use in cartilage replacement, tissue sealants for emphysema treatment, wound dressings with healing agents, and as a barrier against secondary infections [99–101]. The toxicity profile of PVOH was extensively evaluated after oral and intravenous injections of PVOH in animal models [102,103]. Oral administration of PVOH to rats demonstrated no reproductive, neurological, or systemic toxicity [104]. Acute oral toxicity of PVOH was found to be very low with poor gastrointestinal absorption and no mutagenic or clastogenic effects [102]. The highest oral dose with no observed adverse effect level (NOAEL) was determined to be 5000 mg/kg/day [102]. After oral administration, PVOH was primarily excreted in feces and minimally in urine [102]. The liver was identified as a principal organ for PVOH deposition, resulting in its slow elimination from the body [105]. Furthermore, genotoxicity tests indicated no evidence of mutagenicity or carcinogenic activity [106].

4.5. Polyurethanes (PUs)

PUs are condensation polymers, synthesized by step-growth polymerization of isocyanates and polyols. Chain extenders are often added during the synthesis process of PUs to enhance their mechanical strength. The stiffer hard segments in PUs are made up of chain extenders and isocyanates, and flexible soft segments of PUs are made up of polyols. The interplay between these segments endows polyurethanes with a balance between rigidity and elasticity [107]. Both aliphatic and aromatic diisocyanates are employed for the synthesis of PU. PUs derived from aliphatic diisocyanates tend to be more biocompatible and show enhanced resistance to biodegradation than those produced from aromatic diisocyanates [108,109]. Polyether and polycarbonate are commonly employed as polyols, with MW typically ranging from 1 to 5 kDa. Notable advancements were made in developing biodegradable polyester-based mechanically strong PUs, widely used in manufacturing VR-like implants [110]. For instance, a biodegradable intravaginal ring was developed using a poly(ester urethane) that released antiretroviral DPV for more than a month [111]. PUs derived from polyethers display enhanced hydrolytic stability and remain elastic at low temperatures. However, they are susceptible to both oxidation and thermal degradation. The addition of antioxidants has further increased their stability. For example, PEG, a widely used hydrophilic polyol, significantly influences the drug release rate from PU-based DDSs [112,113].

The drug release from the PU-based DDSs has three primary mechanisms. These include solute diffusion, polymer swelling, and polymer degradation [114]. PU-based DDSs show an initial burst release which is mostly because of solute diffusion through PUs. In non-degradable PU-based DDSs, the drug release rate is determined by the thickness, swelling capacity, and permeability as well as the drug concentration gradient and polymer–drug compatibility [43]. In biodegradable PU-based DDSs, the release rate is determined by its own swelling capacity and rate of degradation or erosion. The PU can be hydrophilic or hydrophobic depending on the nature of the diol/polyol and chain extender employed during synthesis. Unlike hydrophobic PUs, hydrophilic PUs can easily swell and demonstrate a faster rate of drug release. For example, the rate of DPV release from swellable PU-based DDSs is significantly higher than its non-swellable counterpart [115]. Hydrophobic PUs tend to experience surface erosion, whereas hydrophilic PUs with fewer reactive ester linkages typically undergo bulk erosion [116,117]. The type of diol/polyol used in PU-based DDSs determines its swelling capacity which in turn controls its rate of degradation/erosion. The degradation of PUs entirely depends on the hydrophilicity and biodegradability of diol/polyol together with the chain extender [118,119]. PUs with PEG as diol can easily be hydrated and the rate of degradation is accelerated. Moreover, PUs can also undergo chemical degradation or enzymatic degradation [120].

4.6. Polyesters

Biodegradability of polyesters, such as PGA, PLA, and PLGA, makes them the most extensively used polymers in LA DDSs. These polymers serve as vehicles for delivering various pharmaceutically active biomolecules like proteins, nucleic acid, small drug molecules, etc. PLA is synthesized through lactide ring-opening polymerization. The physiochemical properties and release profile of PLA can be tuned by tailoring the degree of racemization of lactic acid. PLLA, derived from L-lactide, exhibits a semicrystalline structure characterized by its hardness and transparency, with a T_g of 53 °C. In contrast, PDLA, derived from D-lactide, is amorphous in nature, displaying lower mechanical strength and a T_g of 55 °C [121]. Due to the high thermal stability of PLA, it can be fabricated using various thermal processing methods, such as injection molding, extrusion, etc. PLA degrades through the hydrolysis of its ester bond and forms lactic acid as an intermediate degradation product which is further converted to carbon dioxide and water, a biocompatible and non-toxic end product [122]. Moreover, the intermediate lactic acid can accelerate the degradation by catalyzing the degradation process. Biodegradation of PLA is very slow due to its semicrystalline nature. Furthermore, side chain methyl group cause hydrophobicity in PLA which further lowers its rate of degradation. Therefore, depending on the implant size and MW used, PLA can take as long as 2 years to degrade inside the body [121]. The slow biodegradation rate of PLA has been utilized in the development of Brimo DDS®, an intravitreal implant that incorporates PLA for the sustained release of brimonidine to treat geographic atrophy [123].

PGA is another polyester. PGA is synthesized by the ring-opening polymerization of glycolide [46]. In comparison with PLA, PGA is highly crystalline, providing it with notable mechanical properties that include enhanced strength, rigidity, and thermal resilience [124]. However, PGA has limited potential as a standalone polymer for several reasons. First, the hydrolysis of the ester in PGA is faster than PLA due to its high hydrophilicity. Additionally, the degraded byproduct, glycolic acid, can potentially cause inflammation [125]. Second, PGA possesses inherent brittleness and limited solubility in commonly used organic solvents which further restricts its processibility to formulate a desired DDS. Therefore, PGA is frequently employed in combination with other polymers to overcome any usage limitations.

PLGA, one of the extensively utilized polymers in LA formulations, comprises a substantial 46% of the approved LA injectable medications made from polymers [126].

PLGA is synthesized by a combination of ring-opening and condensation polymerization reactions. PLGA can be synthesized in various MWs and lactide-to-glycolide (L/G) ratios, where an increase in the L/G ratio corresponds to a slower rate of polymer degradation. For instance, in aqueous conditions, PLGA polymers with L/G ratios of 50:50, 75:25, and 85:15 have degradation times of approximately 1–2, 4–5, and 5–6 months, respectively [121].

Drug release from PLGA or PLA-like polyester mainly has two mechanisms: diffusion and bulk erosion or degradation [81]. At the early stage of release, diffusion is the main release mechanism which depends on the concentration gradient and the shape of the DDS [41]. Over the time, PLGA erosion becomes the dominating mechanism. The drug release process from PLGA-based DDSs starts with slow water infiltration followed by swelling of polymer. The swelling of the polymer core increases the pore's size and makes way for the drug molecule to diffuse through the polymer matrix or membrane [127]. Soon after, PLGA undergoes degradation, eventually forming lactic acid and GA as byproducts. These acids act as auto-catalysts and further speed up the degradation process [41]. Collectively, the MW, PLA:PGA ratio, and geometry of the PLGA-based DDS influence the water absorption and, eventually, its degradation. Moreover, PLGA–drug interaction, drug solubility, pH of the surrounding environment, osmolarity, and porosity of the DDS have a considerable impact on the release rate. PLGA-based DDSs mostly have triphasic release kinetics [41].

Polycaprolactone (PCL) is a semicrystalline and hydrophobic polyester which is typically synthesized by the ring-opening polymerization of ε-caprolactone. Compared to other hydrophilic polyesters, PCL shows a very slow rate of biodegradation [128]. Microorganisms can slowly degrade PCL by using lipase, an ester hydrolysis enzyme. However, the hydrophobicity limits the interaction between PLA and lipase, therefore chemical degradation remains the primary degradation mechanism [129,130]. PCL-based DDSs can maintain a sustained drug release over several months. PCL is also known for its mechanical strength, which has made it a popular choice in applications like surgical dressing, wound healing, and contraceptive implants [131]. By taking advantage of the compatibility between PCL and various hydrophobic drugs, the drugs can be uniformly dispersed in the PCL matrix [132]. Additionally, it can be blended with an array of both natural and synthetic polymers, including starch, hydroxyapatite, chitosan, PEG, PU, oxazolines, PEO, and PVA. The blending allows fine-tuning of PCL's mechanical strength, crystallinity, and degradation behavior and accomplishes an ideal DDS [133].

For PCL-based DDSs, hydrophobicity and crystallinity are two important factors that dictate the degradation and drug release kinetics. The chemical degradation of PCL-based DDSs happens in two steps. The first step is the rate-limiting step, where water infiltrates and hydrolyzes the amorphous region, which causes mass loss and is followed by an increase in porosity in the DDS. In the second step, a PCL degradation product, 6-hydroxyhexanoic acid, further catalyzes the hydrolysis process which eventually propagates to the crystalline region of the PCL [132]. Interestingly, the degraded form of PCL can be phagocytosed by cells [131]. PCL-based DDSs usually have biphasic release kinetics, where there is an initial burst release of the drugs followed by a sustained release phase. It has been evidently seen that the initial burst release is much higher for hydrophilic drugs than for hydrophobic drugs [109,134,135].

4.7. Chitosan and Hyaluronic Acid (HA)

Chitosan, a natural biopolymer, has widely been used in tissue engineering, bioscaffold creation, and drug delivery application. Chitosan has been derived from chitin (a long-chain natural polysaccharide) by adopting the alkaline or enzymatic deacetylation process [136,137]. Chitosan is a copolymer of D-glucosamine and N-acetyl-D-glucosamine linked through β-(1→4)-glycosidic bonds, giving it a rigid crystalline form. The chitosan backbone contains '–NH_2'- and '–OH'-like active functional groups that can be easily functionalized with active biomolecules, making it popular for nano- or microparticle-based DDSs. Chitosan, a biodegradable polysaccharide, is particularly notable for its versatility in sustained drug delivery [138]. The physicochemical characteristics of chitosan depend on its MW and degree of acetylation. Chitosan exhibits low water solubility, enabling it to remain in the body for extended periods. This property has made it particularly useful for sustained release in oral drug delivery applications [139]. Moreover, inspired by anti-inflammatory, antibacterial, and mucoadhesive properties, chitosan has been widely used in wound dressing, healing, and pulmonary DDSs.

HA is a naturally occurring non-immunogenic and hydrophilic polysaccharide. It is commonly used in the development of DDSs. The HA contains two basic sugar units, namely, D-glucuronic acid and N-acetyl-D-glucosamine, conjugated through β(1–4) and β(1–3) glycoside bonds [140]. HA is known to interact with CD44 cell surface receptors, and exogenous HA can reduce HIV-1 infection of CD4+ T-cells in a CD44-dependent manner [140]. However, endogenous HA failed to do so as it undergoes a rapid degradation at the cell surface by the enzyme called hyaluronidase. Moreover, HA is one of the major components of mucosa, which results in low mucoadhesive properties of exogenous HA. The thiol group (SH) modification on the backbone of HA can further increase its mucoadhesive property. In this context, Agrahari et al. formulated TFV-encapsulated nanofibers (NFs) by using thiolate hyaluronic acid to stop vaginal transmission of HIV [141]. The NFs were demonstrated to be safe toward the genital tract and other major organs of 57BL/6 mice. The NFs show hyaluronidase-responsive TFV release, resulting in the prevention of vaginal HIV transmission.

4.8. Polymers and Polymer Blends

Over the past few decades, research into LA DDSs has shown significant strides forward. The introduction of novel polymers and polymer blends has improved the efficiency of the existing drug delivery systems. This includes better biocompatibility, reductions in local and systemic toxicities, and optimal release profiles. The FDA-approved ISFI is one of them, which employed dimethyl sulfoxide (DMSO) and N-methyl-2-pyrrolidone (NMP)-like organic solvents to lower the solution viscosity of PLGA and induce phase conversion. Compared with DMSO, NMP has more carcinogenic, teratogenic, and mutagenic toxicities [89,142,143]. Therefore, to avoid NMP and minimize the use of PLGA, poly(3-hydroxybutyrate-co-3-hydroxyvalerate (PHBV) and poly (ethylene carbonate) (PEC) were developed.

PHBV is a hydrophobic, thermoplastic, semicrystalline, biocompatible, and biodegradable linear aliphatic polyester that has been synthesized by incorporating 3-hydroxy valerate units into the poly(3-hydroxybutyrate) (PHB) [144]. PHBV is very brittle and has a melting temperature lower than PHB [145]. The thermal properties of PHBV can be improved through blending with PCL or TiO_2 nanoparticles [146]. PHBV was used to formulate an aripiprazole (an antipsychotic drug)-encapsulating nanoporous ISFI [147]. Rheological analyses demonstrated the existence of a highly cross-linked network structure in the PHBV ISFIs. Spectroscopic analysis revealed the intactness of aripiprazole in the network structure. The ISFI demonstrated a burst release followed by a sustained release of aripiprazole over 18 days.

PEC is a biodegradable polymer synthesized by copolymerization of CO_2 with ethylene oxide [148]. Generally, PEC is characterized by high ethylene carbonate content with low polydispersity and high glass transition temperature. PEC undergoes biodegradation through surface erosion, followed by the formation of ethylene glycol as a degradation product [149]. The in vivo biodegradability of PECs is influenced by the presence of reactive oxygen species, which are associated with inflammatory cells. Moreover, the molecular weight and percent of ether contained in PEC also have an immense effect on its degradation rate [150]. An LA ISFI was successfully demonstrated using PEC as a rate-controlling polymer and bovine serum albumin as a model drug. To compare the effects of DMSO and NMP on the microstructure and drug release of PEC-containing ISFIs, a bovine serum-albumin-encasing ISFI containing either PEC/DMSO or PEC/NMP was developed [151]. In comparison with the PEC/NMP ISFI, the PEC/DMSO ISFI demonstrated a quick release of DMSO followed by the formation of a depot with low porosity, which minimizes burst drug release.

The emergence of stimuli-responsive polymeric biomaterials demonstrates improved spatiotemporal drug release. These smart biopolymers are responsive to various internal and external stimuli, including pH, redox potential, temperature, and enzyme activity [152–154]. Phenylboronic acid (PBA)-containing polymers are an example of such smart biomaterials which are sensitive to pH, sugars, and reactive oxygen species [155]. PBA-conjugated polymers have proven to be effective in pH-responsive delivery of anticancer and antiretroviral drugs. For instance, doxorubicin-encapsulating, PBA-modified poly(maleic anhydride)-F127 micelles show a pH-responsive release [156]. The use of PBA-conjugated polymers in LA ARV formulations could offer enhanced pharmacologic benefits. In this circumstance, a PBA-based mucin-like polymer system that blocks HIV-1 migration was developed within the genitourinary system [157,158].

5. LA ARV Delivery

LA DDSs, such as implants, VRs, MNs, and depot-forming nanoformulations, offer a promising avenue for extended release of ARVs beyond two months (Figure 3). In HIV-1-positive patients, the viral load is always persistent in tissue along with systemic circulation. This necessitates achieving continuous effective drug concentration both in the target tissue and plasma without reducing the longevity of the drug depot. Notably, for sexually transmitted diseases like HIV-1, precise drug concentrations within reproductive organs are pivotal for effective disease prevention. While local administration methods, such as VRs, offer a potential solution, they also face the challenges of maintaining the local therapeutic drug concentration without causing any toxicity to the surrounding tissues. Furthermore, the drug can be systemically absorbed, potentially diminishing local drug concentrations. In this context, implants or depot-forming injectables emerge as an alternative strategy as they can deliver the drug to reproductive organs through systemic drug exposure. Yet, achieving optimal drug concentrations without inducing local adverse effects in the implanted or injected site presents a multifaceted obstacle. The intricacies of polymer–drug interactions and device design play a pivotal role in dictating drug exposure and its release kinetics. Here, we will discuss recent advancements in LA ARV formulations, specifically focusing on the interplay between drug and polymer interaction and factors that influence the drug release kinetics. By deciphering the complex dynamics underlying polymer-mediated drug delivery, this exploration aims to pave the way to achieve an efficacious long-acting ARV formulation by fine-tuning their controllable parameters.

5.1. Implants

In 1938, Deansby and Parkes pioneered implantable drug systems by observing the effects of estrone pellet implants on male chickens [159]. Folkman and Long later refined this delivery system with a silicone rubber capsule for continuous drug release. This led to major developments in implantable drug delivery, especially for contraception and ocular treatments [159]. The approval of various contraceptive and ocular implants boosted their acceptance and uptake by patients. This progress influenced implants' development for HIV-1 treatment and PrEP. In the early 2000s, research focused on implants for the sustained release of ARVs. TAF's potency and suitable physiochemical properties, such as hydrophobicity, MW, and solubility in polymer matrices, make it an ideal candidate for delivery through implants. Other drugs like ISL, CAB, DTG, and entecavir are also being explored for implant-based delivery [15,160,161].

The formulations and drug release mechanisms of implantable devices are multifaceted and varied, making their categorization challenging. Nevertheless, those designed for HIV treatment and prevention can be broadly divided into two types. The first includes preformed solid implants, which are made of either biodegradable or non-biodegradable polymers and require surgical or invasive procedures for subdermal placement [15,162]. The second type involves injectable in-situ-forming implants (ISFIs). ISFIs can be injected in the form of suspension or solution in subdermal spaces where solvent-induced phase inversion produces a drug-loaded implant [10,163]. Even though most ARVs and the polymers used in the fabrication of implants are FDA approved with established systemic safety profiles, local toxicity concerns that are either drug specific or associated with the nature of implant materials remain. Improved understanding of how ARVs and their metabolites interact with and are eliminated from the surrounding tissues and extracellular environments is of significance. For instance, high concentrations of TFV and TAF have shown significant implant site toxicities in some preclinical studies and significantly affected rapid development and translation of these devices [164]. Nonetheless, advancements are being made in implant design to regulate TAF release and stability within the device.

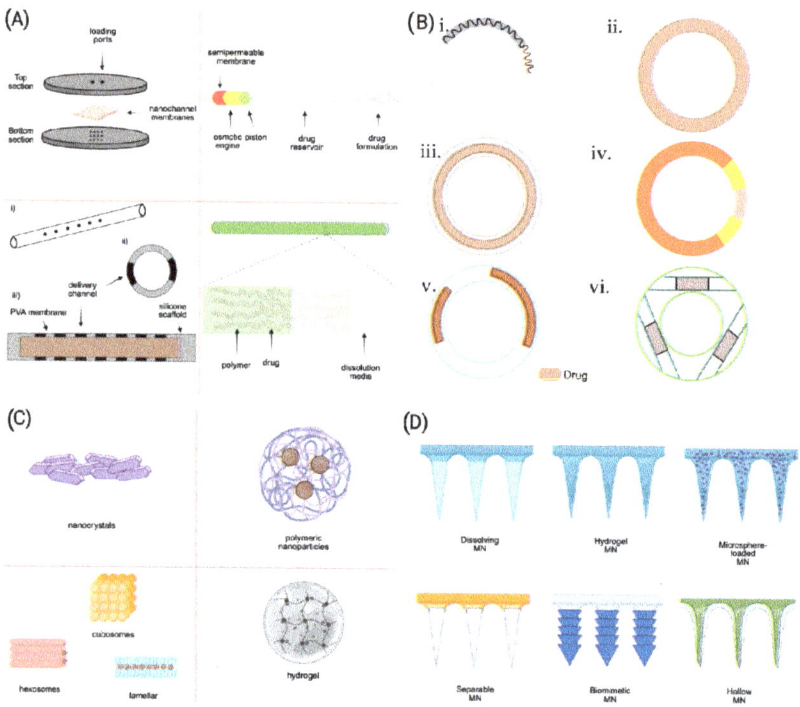

Figure 3. Different types of LA DDSs that are in either clinical or preclinical development. (**A**) Different types of solid implants. Reproduced with permission from Ref [165]. Copyright 2018 Antimicrobial Agents and Chemotherapy. (**B**) Different types of Vaginal rings—(**i**) over-molded metal spring design, (**ii**) matrix-type VR, (**iii**) Full-length reservoir-type VR, (**iv**) sandwich-shaped VR, (**v**) Partial reservoir-type VR, (**vi**) Insertable reservoir-type VR. Images were adapted under the terms of Creative Commons Attribution—Non-commercial from ref. [166]. (**C**) Nanocrystal formulations, polymeric nanoformulations, liposomes, and hydrogels. (**D**) Different types of microarray patches. Reproduced with permission [21].

A wide range of non-biodegradable (such as silicone and PU) and biodegradable polymers (PCL or PLA) have been used for ARV implant generation (Figure 4). Biomedical-grade silicones have been used for implant development for various diseases over six decades and are also the preferred materials for the development of ARVs, such as TAF implants. Irrespective of the extensive use of silicone-based TAF implants, TAF permeability through the silicone implant is low. To improve TAF permeability, implants have been fabricated with nano- or microchannels that aid in improved TAF release kinetics. To further control drug release rates, the implants were either coated with a rate-controlling membrane or the diameter and number of drug release channels were adjusted [15]. Interestingly, TAF release rate from this implant was independent of initial TAF loading. PK evaluation of this device in beagle dogs demonstrated a zero-order release kinetics of TAF (1.07 ± 0.02 mg/day) that sustained therapeutic TFV-DP levels in PBMCs for 40 days. The implant is currently in an early phase I clinical trial to assess its safety and tolerability [167]. The ability to remove the implant can be advantageous in the case of adverse reactions. However, removal of this device requires invasive surgical procedures that may limit broader use. In such cases, transcutaneously refillable silicone-coated nanochannels may offers enhanced pateint convinence and acceptability over conventional solid implants by offering refillability with explantation [11]. This design primarily comprises a transcutaneously refillable medical-grade titanium drug reservoir and a uniquely crafted silicone membrane housing TaN-coated slit nanochannels [11]. These nanochannels are arranged in square arrays and connected to the inlet and outlet of the membrane through a network of microchannels. Drug release rate from the implant is dependent on interactions with the interior surface of the membrane and can be further controlled by adjusting the number of nanochannels.

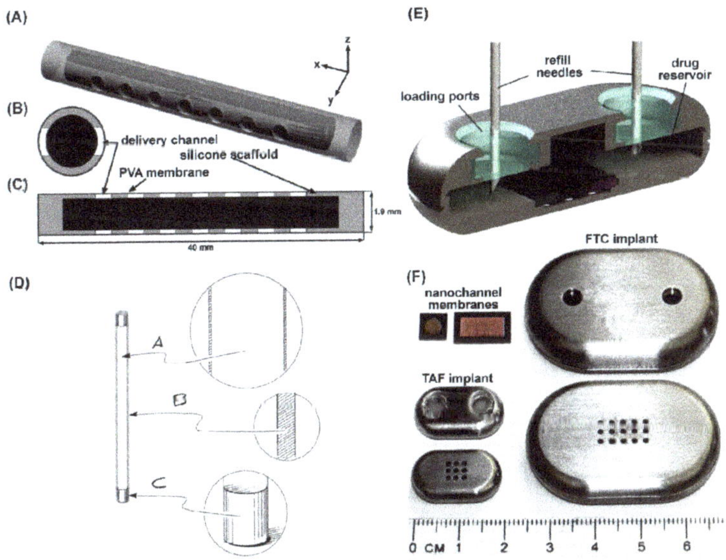

Figure 4. Different types of non-erodible implants are illustrated. (**A–C**) TAF implants contain silicone and PVA. Reproduced with permission from [15]. (**D**) PCL reservoir-style implant for TAF delivery, which comprises a formulated drug core (Reproduced under the Creative Commons CC-BY license from [12]). (**E,F**) Cross-sectional depiction of nanochannel-containing implant showing drug refill needles through the loading ports with resealable silicon plugs. TAF and FTC implants of two different sizes and shapes. Reprinted with permission [15].

Hydrophobic poly(tetramethylene oxide) subunit-containing PUs have also been used to create TAF implants [162]. As hydrophobic PU demonstrates minimal swelling, the drug inside the hydrophobic PU tube forms a suspension encircled by an outer layer of saturated drug. The difference in concentration of these two layers drives the drug release rate, which follows zero-order kinetics. This release rate can be fine-tuned by modifying the membrane wall thickness or composition of PU. PU-based pellets-in-a-tube-type TAF implants where a mixture of TAF, NaCl, and magnesium stearate were converted into a pellet and impulse sealed into PU tubes were created [162]. In this report, the release kinetics of TAF from the implant were not zero order. Interestingly, incorporating a hydrophilic excipient like PEG into the TAF pellet or increasing the crystallinity of the PU membrane had been found to enhance the drug's release from the implant [13]. However, preclinical in vivo, safety, and tolerability studies in different animal models demonstrated that the PU-based TAF implant results in local inflammation and toxicities. The local toxicities observed in those implants may be the result of high TFV and TAF concentrations in the implant site, as observed in those studies. A similar implant design has also been employed to develop CAB implants [168]. In a comparative study, the release rate of CAB from various elastomeric PU membranes, including poly(dimethylsiloxane), poly(ethylene-co-vinyl acetate), and hydrophilic poly(ether) urethanes, was tested. Interestingly, only the segmented hydrophilic poly(ether) urethanes demonstrated a sufficient CAB release rate to reach a pharmacologically relevant concentration. Furthermore, using polyvinylpyrrolidone (PVP) instead of PEG as an excipient caused less mass swelling and higher implant drug release [168].

Biodegradable polymers are widely used for the development of reservoir-type TAF implants. PCL are TAF compatible. When a TAF-containing PCL implant is inserted subcutaneously, the biological fluid from the surroundings slowly penetrates the PCL implant, creating water-filled channels and dissolving TAF to facilitate its release. As more water enters the implant, it induces the bulk erosion of PCL. However, PCL's bulk erosion rate is so slow that it can persist for up to a year [169]. A biodegradable PCL-containing solid TAF implant was recently developed [12]. This implant was 2.5 mm in diameter but of varying thicknesses and produced by a hot-melt extrusion process. The PCL tube was filled with a formulation containing TAF and castor oil and sealed at both ends through heat or injection sealing. The TAF release rate from the PCL implant depended on factors like the implant's surface area, PCL's MW, membrane thickness, and TAF solubility. TAF release was found to correlate with the implant's surface area and inversely with PCL wall thickness [12]. Interestingly, thinner-PCL-wall implants demonstrated a rapid initial TAF release, while high-MW PCL implants released TAF more quickly. The effect of PCL's MW on release rates was common in thinner walls rather than thicker ones. This is possibly due to differences in crystallinity between PCL types [12]. Incorporation of hydrophilic excipients in a PCL implant can increase the TAF release rate. For example, a PCL implant, where TAF was formulated with PEG_{3350} at various weight ratios, increased TAF dissolution from the implant. This leadsto rates increases of TAF [160]. The PEG_{3350}. This eventually increases the drug release rate. It has been reported that the rate of TAF release is proportional to the mass of the PEG_{3350} used and the surface area of the implant.

ISL, a novel nucleoside reverse-transcriptase translocation inhibitor (NRTTI) has gained attention in implant generation due to the potency and extended intracellular apparent half-life of its triphosphate metabolite [14,170]. ISL was incorporated in both biodegradable and non-biodegradable polymer-containing implants for extended release [165]. Two types monolithic matrix-type implant were developed by using either biodegradable PLA and PCL or non-degradable EVA polymer for SC delivery of ISL [165]. The uniform dispersion of crystalline ISL in the polymer matrix resulted in a minimal implant surface porosity, which limited the water infiltration to the implant. Therefore, the implant showed a small transient initial burst release followed by a slow first-order release of ISL. The initial small transient burst release was presumably because of the rapid dissolution of ISL persisting on the implant surface. Furthermore, the rate of ISL release showed an increasing trend with its loading content. Several clinical trials involving ISL implants and oral regimens have been halted as reductions of the participants' CD4+ T-cell and lymphocyte counts have been reported [167].

Given the instability and safety concerns associated with TAF and ISF-based implants, alternative approaches such as the development of stable TFV prodrugs and salts, new ISF analogs, ISFIs, and other potent drugs from different classes such as CAB and DTG (INSTIs) have gained considerable attention. In-situ-forming implants (ISFIs) are injectable drug formulations that turn into solid/semisolid depots after administration [171]. A combination of the drug, excipient or release-rate-limiting agents, and water-miscible organic solvents have been used in the formation of ISFIs. N-methylpyrrolidone (NMP) and dimethyl sulfoxide (DMSO) are the commonly used water-miscible organic solvents used in ISFIs. After SC administration of ISFIs, the water-miscible solvent starts diffusing out from the implant, and surrounding water starts to infiltrate. Subsequently, there is an immediate phase separation of solvents, solidification of formulations, generation of microchannels for water influx, and eventually formation of the ISFI [172]. The drug release rate largely depends on the phase-inversion kinetics of the implant, MW and hydrophobicity of the used drug molecules, and biodegradability of the excipient [173–175].

Biopolymers such as hydroxypropyl methylcellulose, PLGA, PEG, poloxamers, PCL, and poly(N-isopropyl acrylamide) have been employed to develop ISFIs for various indications [176]. The selection of polymer depends significantly on the drug's physicochemical properties and the desired release rate. Generally, a polymer's solubility should be compatible with the drug's solubility characteristics for optimal drug release kinetics [177–179]. PLGA was used as an excipient and NMP/DMSO as water-miscible solvent to formulate DTG and CAB ISFIs [10,180]. The authors investigated the impact of physiochemical properties of the polymer, used solvents, drug–polymer miscibility into the solvent, and polymer-to-solvent ratio on the drug release kinetics.

The rate of drug release from the ISFIs is linked to the microstructure of ISFIs at the administration site. For example, scanning electron microscopy (SEM) imaging of DTG ISFIs demonstrated a central pore surrounded by a highly interconnected porous outer PLGA shell (Figure 5). Over time, the thickness of the outer PLGA shell decreased due to PLGA erotion, leading to an increase in the central pore size. The implant with a thinner shell and larger central pore size facilitated a high rate of drug release from ISFIs. The multispecies tolerability studies of ISFIs support their safety. The acceptability of ISFIs stems from their easy manufacturing and patient-comfortable administration process. Moreover, the clinical availability of leuprolide acetate ISFI (Atrix Laboratories Inc.) and that of Sanofi Aventis (Eligard®) supports the safety of this delivery system [181]. The overall limitations of ISFIs are injection site reaction, injection intervention, minimal control of the size and shape of the implant, and heterogeneous drug release [182,183].

5.2. Vaginal Rings (VRs)

Building on the VR's clinical progress for contraception and hormone therapy, rings containing ARVs have been developed to prevent HIV-1 infection. In addition, VRs are gaining preference over other vaginal products. These include gels or creams based on their efficacy in time-dependent ARV delivery to the vaginal epithelium [33,184,185]. VRs are also widely accepted among women compared to other LA DDSs due to their user-friendly and discreet method of administration and easy reversibility of treatment when needed. The traditional approach to VR design involves uniform dispersion of drugs within polymer matrices. This is accomplished by molding a drug–polymer matrix into a ring-shaped device. Conventional VRs are classified into sandwich (or matrix)-type VRs and core (or reservoir)-type VRs [111]. In the sandwich design, a drug-dispersed polymer matrix is placed between two non-medicated rate-controlling polymer layers [166]. The core-type VR involves creating a small pellet containing a mixture of drugs and polymers. Several cores are combined into a single ring. The ring is covered by a drug-free secondary polymer membrane [166].

The drug release rate in VRs is typically governed by its drug diffusion through the polymer membrane [166]. When the VR is placed vaginally, water enters its wall, leading to polymer swelling and the dissolution of drugs [166]. The dissolved drugs near the matrix's surface are diffused first. This action also forms a drug-depleted layer near the matrix surface. Subsequently, the remaining undissolved drugs undergo dissolution and diffuse through the polymer matrix. Factors such as the drug's solubility in the polymer, the amount of drug loaded into the VR, and the surface area and thickness of the VR membrane influence this diffusion process. DPV, maraviroc, TAF, TDF, UC781, and MC1220 are being utilized in VR fabrication. Among these, DPV, TDF, and TAF, due to their ideal properties, have been predominantly studied for VR development. However, only the DPV VR has progressed to phase III clinical trials and received approval in several African nations for the prevention of HIV-1 [186].

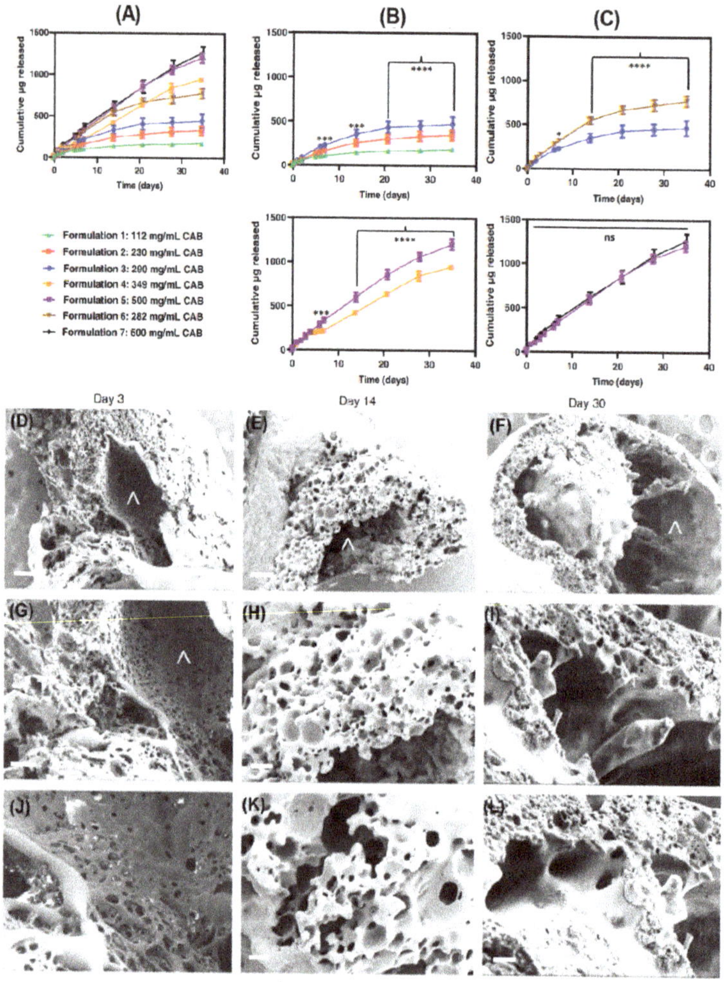

Figure 5. Microstructure in PLGA-based ISFIs: (**A**) Cumulative release of CAB ISFI PLGA formulations. (**B**) Effect of drug loading on CAB release. (**C**) PLGA molecular weight affects CAB release. To compare the drug release from different formulation at different timepoints in (**B**,**C**), two-way ANOVA with tukey's multiple comparison test were performed. * $p < 0.05$, *** $p < 0.001$, **** $p < 0.0001$, and ns (not significant) when $p > 0.05$. (**D–L**) Effect of PLGA degradation on implant microstructure. SEM cross-section images of placebo ISFIs (1:2 w/w PLGA/NMP, PLGA MW 27 kDa) over a 30-day period. (**D–F**) Low-magnification image (100×) of the entire implant (scale bar = 100 µm). (**G–I**) Higher magnification (200×) of the implant shell (shell thickness was measured using SEM scale; scale bar = 50 µm). (**J–L**) Higher (500×) magnification of the center of the implant (scale bar = 20 µm). (**D**,**G**,**J**) Implants imaged at day 3 postincubation in 0.01 M PBS pH 7.4 at 37 °C. (**E**,**H**,**K**) Implants imaged at day 14 postincubation. (**F**,**I**,**L**) Implants imaged at day 30 postincubation. Symbol (*) represents implant shell. Symbols (^) represents the central pore of the implant. Images were reproduced under the terms of the Creative Commons CC-BY license [12,180].

Silicone, EVA, and PUs have been extensively used to develop DPV and TDF VRs [186,187]. They are clinically proven safe polymers for VR development. For instance, five of seven clinically approved VRs are made of silicone [111,188,189]. NuvaRing is the only VR that solely uses EVA, whereas another VR, Ornibel, uses PU as either a drug reservoir or a coating agent to control the drug release [186]. However, these polymers come with drawbacks. For instance, condensation-type silicone elastomers can produce a propanol byproduct. This can increase the drug solubility, prompting more solubilized drugs to move to the polymer surface. This then would lead to an increased initial drug burst release [190]. Furthermore, reports have emerged of biofilm formation due to enhanced bacterial adherence on EVA-based VRs [191–193]. However, biofilm generation can be significantly reduced by surface modifications such as in biocompatible surfactant, argon plasma treatment, or perfluoro-alkyl siloxane grafting on the elastomer [194–197].

The exploration of DPV VR development began nearly 25 years ago. The VRs were then developed using a condensation-cured silicone elastomer system [198,199]. To produce the core- or reservoir-type DPV VRs, DPV was blended with silicone elastomer and reaction injection molded at 50 °C to form an active core. This core was then segmented and enclosed within a non-medicated outer silicone membrane through another reaction-injection-molding process. The resultant VRs had a 5.50 mm and 9.00 mm outer cross-sectional diameter, while the total diameter of the VR was 55.00 mm. DPV release rate from the reservoir VRs depends on the solubility of DPV in the silicone matrix, and the diffusivity of the silicone solubilizes DPV through the non-medicated silicone sheath layer. In contrast, the matrix-type DPV VR was developed through a curing process involving DPV and silicone elastomer with a normal propyl orthosilicate cross-linker mixture. The implants did not show any significant local site reactions and proved to be well tolerated in the phase I clinical trial [198,199]. The PK and release kinetics of these two VR types showed notable differences. Matrix-type VRs exhibited a substantial initial burst release of DPV, reaching peak plasma concentration within 8 h. On the other hand, reservoir-type VRs took 10 days to reach the peak plasma concentration. Moreover, the average plasma concentration in matrix type VRs was almost 50 times higher than in reservoir-type VRs. Interestingly, the matrix-type VRs demonstrated a rapid decline in plasma concentration over time, while the drop in plasma concentration from reservoir-type VRs was relatively slow [199,200]. From the clinical trial findings to date, the matrix-type silicone DPV VR stands out as the most effective one. There are several phase III clinical trials conducted with DPV matrix-type silicone-based VRs, and the DPV VR has been reported to reduce the risk of HIV-1 acquisition by 27% to 37% in African women [201,202].

Apart from DPV, TFV was the only other vaginal microbicide that demonstrated potential as a vaginal PrEP agent during clinical trials. As per the trial result, gel containing 1% TFV exhibited a 39% protection rate against HIV-1 infections in women [203]. Inspired by these findings, considerable research efforts were undertaken towards developing TFV or TFV prodrug-based VRs. While various polymers, including silicone, EVA, and PU, were investigated for creating TFV VRs, it was only the PU-based TFV or TFV prodrug VRs that demonstrated promising results in preclinical studies. For instance, Mesquita et al. developed silicone, EVA, and PU-based TDF VRs and compared the TDF release rate from these VRs [204]. The in vitro release study revealed that the release rate of TDF from a PU-based VR was 5 to 15 times higher in comparison to those created from EVA and silicone [204]. The differences in TDF release rate were attributed to the varying solubility of TDF in employed polymers. Specifically, TDF showed lower solubility in silicone and EVA as opposed to PU, resulting in a decreased diffusion of TDF through the polymer membrane [204].

The higher release rate of TDF from hydrophilic PU-based TDF VR has also been investigated [204]. Additionally, the authors compared both matrix and reservoir types of TDF VRs, made with hydrophilic polyether urethane (PEU). To prepare the matrix-type TDF VR, TDF-loaded hydrophilic PEU was extruded first using a twin-screw extruder and then cut into small pellets. Then, the pelletized segments were re-extruded, cut into 15.5 cm segments, and end-joined using induction welding [204]. For preparing a reservoir-type TDF loaded VR, the hydrophilic PEU was extruded into tubes, and the empty tubes were filled with either TDF only or a formulation of TDF and NaCl (86:14 wt% ratio) [205]. The matrix facilitated an immediate drug release, which then decreased by a factor of at least 20-fold from day 1 to day 28. In contrast, reservoir-type VRs showed a prolonged lag time to start the drug release. Interestingly, when an osmotic agent like NaCl was added to the reservoir-type VR, there was an increased influx of vaginal fluid into the VR, prompting a quicker TDF release. To establish human safety and tolerability, a phase I clinical trial was conducted using a reservoir-type HPEU TDF VR. However, the trial was discontinued due to the observed genital ulcers among the participants. However, no ulceration was observed among the participants who received placebo-containing VRs manufactured using the same technique, indicating the observed local adverse effect is not due to the polymer [17].

Hydrophilic PU (HPU) membrane-based reservoir TFV VRs were also reported [203]. The VR was produced by making an HPU tube by hot-melt extrusion and then filling the tube with either TFV or TFV/glycerol/water-containing semisolid formulation. The end of the tube was sealed with an induction welder. Only the TFV-containing VR showed a slow but steadily increasing TFV release rate in the in vitro release study. However, adding an HPU-permeable substance, such as glycerol with TFV in the polymer matrix, increased its solubility in HPU and release rate [203].

PU-based VRs were also used for the vaginal delivery of IQP-0528 and MIV-150. IQO-0528 is an NNRTI that has physiochemical properties similar to DPV. A matrix-type PU-based VR was developed to deliver IQP-0528 [206]. The VR was prepared by an injection-molding technique after mixing the drug and Tecoflex EG85A PU. In vitro studies with this VR indicated a linear relationship among the drug release rate, VR surface area, and drug loading to the VR. However, in vivo findings indicated that drug release was not solely determined by drug loading or VR surface area. Instead, it was primarily driven by diffusion, contingent on both the VR drug concentration and drug's presence in the surrounding media [206].

5.3. Microneedles (MNs)

Microneedles (MNs), also known as microneedle arrays or patches, are transdermal drug delivery systems featuring tiny projections around 200 μm tall. While initially developed for immediate-release transdermal drug delivery, MNs have now evolved as innovative LA DDSs due to the advancement in polymer chemistry. MNs penetrate through the skin to the dermis, an area rich in blood vessels, ensuring efficient drug absorption while reducing systemic exposure and bypassing first-pass metabolism [207–209]. Their small size and length cause less pain, as pain receptors reside deeper in the skin. The design and composition of MNs, including tip length, spacing, diameter, and overall geometry, can be tuned to optimize drug delivery. Further, MNs can be combined with other drug delivery technologies like nano- or microparticles or hydrogels for extended drug delivery [210,211]. Their benefits have spurred a significant increase in research in recent decades. Particularly in ARV delivery, their discreet, easy, and minimally invasive nature is recognized and widely accepted [212]. For instance, a survey study was conducted to gather opinions from patients and healthcare professionals regarding the use of MNs for ARV delivery. The results showed a positive attitude towards MN technology, with both groups acknowledging the benefits of discreet and self-administrable MNs for administering ARV drugs [213]. Given the constraints related to permeability through the skin's stratum corneum, MN technology proves particularly effective for potent, lipophilic, and low-MW drugs. ARVs such as RPV, CAB, and BIC have demonstrated significant potential when employed with MN technology.

There are several types of MNs based on their design and drug-releasing mechanism. Each MN's drug release kinetics varies based on the MN's geometry, surface area, number of MNs per patch, polymer used, nature of the drugs, and its loading content [214]. In general, MNs are classified into solid MNs, drug-coated MNs, dissolving MNs, and hollow MNs. Solid MNs are usually used to pretreat the skin to create microchannels to facilitate drug absorption before applying any transdermal topical gel or formulation [214]. Polymers with good mechanical strength, such as silicone, PMMA, PCL, PLGA, PLA, and PGA, and metals such as stainless steel, titanium, and ceramics have been used to prepare solid MNs [215]. In drug-coated MNs, the outside of solid inert MN tips is coated with drug formulation. Upon insertion of the drug-coated MNs into the skin, the drug formulation is dissolved in the surrounding medium to release the drugs from MNs. These MNs are less utilized for LA DDSs because of their inability to form a drug depot. However, for LA drug delivery, dissolving MNs containing a nanocrystal formulation of ARVs have utility.

Dissolving MNs are made of biocompatible polymers that can encapsulate drug-containing matrices and, upon administration, dissolve in the biological fluid and release the drug content [216]. Nanocrystal suspension of ARVs, such as CAB, RPV, and bictegravir (BIC), when delivered through dissolving MNs, can easily form dissolution-controlled drug depots (Figure 6). This strategy can allow a less painful way of self-administration of ARVs. The tips of dissolving MNs can be made of biodegradable or non-biodegradable polymer or sugar, and the rate of polymer or sugar degradation determines the MNs' release duration [216–221]. The fast-dissolving MNs use hydrophilic polymers like dextrose, maltose, and chitosan, while slow-dissolving ones employ hydrophobic polymers like PLGA, PVA, and PLA, due to their biocompatibility, mechanical strength, and slow erosion [222–224]. PLGA MNs are mainly loaded with hydrophobic drugs and can be designed to separate from their baseplates upon insertion [223,225,226]. PCL is also used in MN production which allows controlled drug release using near-infrared light as an external trigger [227].

To create a bilayer dissolving MNs for transdermal delivery of RPV, a first layer or MNs tips were developed using varying compositions of PVP and RPV nanocrystal suspension [23]. The drug-free baseplate of the MNs was made using an aqueous solution of PVA and glycerol. The fast-dissolving MNs dissolved within a few minutes after the insertion into the skin and formed an RPV depot in the application site. However, the dissolution testing of MNs in excised skin demonstrated a slower dissolution profile of RPV suspension, which may be due to the hydrophobicity of RPV in the MN tips. Also, the concentration of RPV in the skin tissue was proportional to the MN application time. Although varying compositions of RPV nanocrystal suspension and PVP were used to manufacture MNs, the MNs with 70% RPV, 15% PVP, and 15% water demonstrated better plasma RPV levels compared to other MNs. Due to the discreet nature of the MNs and their ability for localized delivery of drugs, the same type of RPV MN was also generated by the same group for vaginal and CNS delivery of RPV [228,229].

A similar type of bilayer MN has also been developed by the same group for the transdermal delivery of BIC and CAB nanocrystal formulation (Figure 6). For CAB MN generation, bilayer MNs were developed, where the MN tips were generated by using a hydrogel composed of a 20:20:60 PVA:PVP:CAB LA nanoformulation [22]. The baseplate was made using PVA and PVP. The two polymer blends formed a sufficiently strong MAP tip, which could be attributed to the hydrogen bond formed between the carbonyl group of PVP and the hydroxyl group of PVA. This hydrogen bonding also improved the crystallinity and cohesiveness of the polymer. An in vitro MN tip dissolution study showed that CAB MN tips dissolved almost 100% within the first 60 min after administration. Importantly, after a single MN administration in rats, these MNs maintained CAB therapeutic levels for up to a month [22]. For developing BIC-containing MNs, first, a PVP- and PVOH-containing BIC nanosuspension was prepared and lyophilized. Then, the lyophilized powder was reconstituted in a very small amount of deionized water, and MN tips were generated from the resultant suspension. The baseplate of the MNs was made using PVP and glycerol. Similar to CAB MNs, BIC MNs also maintained plasma BIC levels equal to the human therapeutic level for up to a month [230].

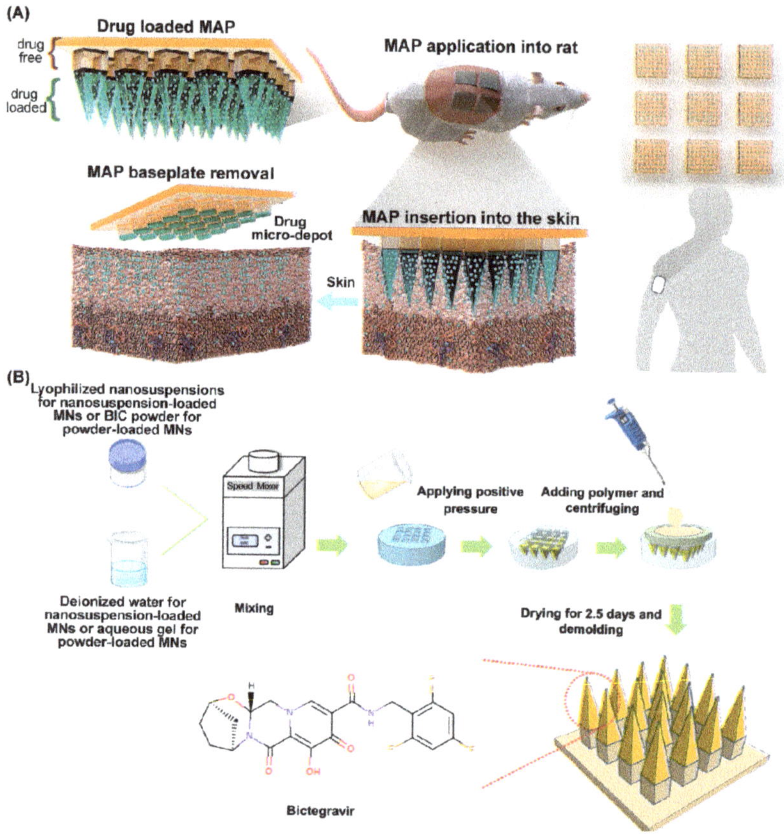

Figure 6. Fast-dissolving MNs have been made for ARV delivery. (**A**) Schematic depiction of programmed dissolving bilayer MAP with a high drug loading of LA microdepots for potential monthly human application. (**B**) Schematic presentation of the MN preparation process. The figure was reproduced under the terms of the Creative Commons CC-BY license [22].

Newer types of MNs, such nano- or microparticle-loaded MNs, hydrogel-forming MNs (HFMNs), and separable MNs, are also becoming very popular and useful in drug delivery [223,231]. HFMNs are particularly important for LA drug delivery and are made from swellable cross-linking polymers. In a dry state, HFMNs are sturdy enough to penetrate the skin. Upon insertion, HFMNs form hydrogels by soaking up water from the interstitial fluid, causing the formation of water-filled channels in the MN tips [232,233]. The drug's release rate from HFMNs is influenced by the dimensions and number of channels, as well as the size of the drug particles [234]. Unlike dissolving MNs, HFMNs have a higher drug-loading capacity [235]. However, this capacity varies based on the polymer type and cross-linker used. Common polymers for HFMNs include PMVE/MA, PVA, EVA, chitosan, and silk fibroin, often cross-linked with hydrophilic polymers or compounds like PEG, dextran, and CMC [231,235]. The drug release rate can be tailored by adjusting the percent of cross-linker and swelling ratio.

5.4. Polymeric Micro- and Nanoparticles

Polymeric microparticles are microscopic particles with sizes of 1 to 1000 μm, where a core substance is encapsulated by one or more layers of the polymer membrane. Microparticles can further be divided into microspheres and microcapsules. While microspheres consist of a uniform matrix with the API dispersed throughout, microcapsules feature a distinct core, which can be liquid, solid, or semisolid, enveloped by a continuous polymer coating [236]. Nanoparticles have a similar feature but a size of 1 to 1000 nm. Micro- and nanoparticles can be prepared through a process called micro- or nanoencapsulation, respectively. This process entails encapsulating solid or liquid drug particles either dispersed or dissolved within a polymeric matrix or as a core surrounded by a polymeric shell. The polymer used in these particulate systems is alginate, chitosan, PLA, or PLGA. However, PLGA is the most widely used polymer for micro- or nanoencapsulation, which is used in around 50% of clinically approved LA injectables [237]. The proven clinical safety and tunable drug release kinetics make PLGA a widely used polymer in particulate drug delivery systems. The encapsulation of drugs in micro- or nanoparticulate systems can be achieved through either physicochemical or mechanical processes.

The physiochemical process of drug entrapment involves coacervation or phase separation and inotropic gelation techniques [238,239]. There are three steps of microencapsulation by the coacervation-phase separation technique, which is carried out under continuous agitation. The first step involves the dispersion of the drugs in the polymer solution or suspension, then the coacervation process starts with a stimulus such as an addition of acid or salt (pH change), an input of non-solvent into the dispersion, or a temperature change in the emulsion which causes the precipitation of the polymer and continuous coating of the polymer on the core drug surface. The last step is the hardening or solidifying of the coated polymer materials on the core drug surface [240–242]. The mechanical process of drug entrapment in micro- or nanoparticulate systems includes fluid bed coating and the solvent evaporation technique [238]. In fluid bed coating, the drug particles are initially suspended in the airflow and then coated with a polymer solution through spraying. Subsequently, the solvent is evaporated to solidify the polymer coating around the drug particles [241,243]. However, the most widely used method for polymeric drug entrapment is the solvent evaporation or extraction method [244,245]. This method involves the emulsification of the polymer and drug solution, followed by the evaporation of the solvent. During the evaporation process, the polymer materials shrink around the core drug particles and produce micro- or nanosized particles [240,241]. The emulsification process of the polymer and drug solution depends on the physiochemical properties of the drug, which can be either a single emulsification process (o/w) for hydrophobic drugs or a double emulsification process (w/o/w) for hydrophilic drugs [241,246,247]. However, challenges still exist in using the above-mentioned processes for nano- or microparticle formulation as controlling size, size distribution, shape, and batch-to-batch variation remains unmanageable. Moreover, encapsulating biologics using these techniques is considerably more challenging due to the inherent nature of proteins and biologics to interact with the polymer surface, as well as concerns about the stability of the biologics [248,249].

5.4.1. PLGA-Based Micro- and Nanoparticles

PLGA has ideal physiochemical properties as a drug carrier, such as biocompatibility and both gel- and capsule-forming ability [51,250]. The superior suitability of PLGA has been acknowledged in clinical settings, and the introduction of PLGA-based depot-forming injectables began with the US FDA's approval of PLGA-based Lupron Depot in 1989. Since then, PLGA microparticles have been used for delivering both small molecules and protein, such as Nutropin Depot.

A lot of preclinical efforts were also employed to develop ARV-loaded PLGA micro- and nanoparticles, as the PLGA particle can encapsulate multiple ARVs in the same formulation. Those particles have shown promise to prolong dosage regimens for both treatment and pre-exposure prophylaxis (PrEP) as they improved the PK and drug release properties of the ARV. Tahir Khuroo et al. developed a PLGA microparticle encapsulating a prodrug of DTG by the emulsification–evaporation method. This PLGA microparticle increased the plasma apparent half-life of the DTG by almost 8.6-fold. However, the PLGA microparticle showed an initial burst release of DTG on day 1. However, the plasma DTG level gradually decreased by 4-fold over a 60-day period [251]. Initial burst release is a common phenomenon for PLGA-based formulations. Several studies on the morphology and drug distribution of PLGA microparticles indicated that PLGA microparticles are very porous and the drugs in the PLGA microparticles are distributed on the surface or the orifice of the large, interconnected pores [252,253]. Initially, water is absorbed through the pores, and then fast dissolution of the neighboring drugs may cause the initial burst release of the drugs. In an alternative scenario, water uptake into PLGA microspheres with a seemingly smooth, thin surface layer can directly initiate drug release through various processes [253]. These processes may involve polymer swelling, leading to the formation of surface pores or the creation of new pore networks. However, after the initial burst release, the PLGA microparticle often undergoes pore rearrangement or a process called pore healing, causing new pore formation and pore closing [253,254]. This may also explain the steady drug release kinetics after the initial burst release from PLGA microparticles. The drug release kinetics from PLGA-based micro- and nanoparticles can be controlled by controlling the MW of PLGA and the glycolic and lactic acid ratio of the PLGA and size and surface porosity of the particle [255–257].

PLGA nanoparticles have been widely employed by Mandal et al. and colleagues for delivering combination ARV (cARV), such as (EVG + TAF + FTC) or (TFV + EVG) or (TAF + FTC) or (BIC + TAF) [258–261]. These nanoparticles were developed by a w/o/w emulsion–solvent evaporation method, where the aqueous solution of the drugs was first emulsified with organic solution of PLGA (75:25 lactide:glycolide ratio) (MW 4000–15,000 Da) and poloxamer, then the resultant emulsion was again emulsified in 1% PVA solution to generate a nanoparticle with a size of 250 nm. The nanoparticle improved the biodistribution of all three encapsulated drugs. Four subsequent injections of these nanoparticles over a 6-week period protected humanized mice from HIV-1 infection for 19 weeks [261]. A similar nanoparticle encapsulating either TAF + FTC or TAF + BIC or TAF + EVG also improved the drug PK and biodistribution of those drugs [258,260,262].

Despite numerous clinically approved depot-forming PLGA-based implants and a significant increase in PLGA-based drug delivery development efforts over the last few decades, there are still notable limitations in PLGA-based drug delivery systems. Such limitations include the degradation of PLGA suspension or solution over the storage period due to its hydrolytic properties. To improve the efficiency and effectiveness of PLGA-based drug delivery, it is crucial to thoroughly understand the factors that influence the degradation rate and drug release kinetics of PLGA-based polymers [263]. In addition to that, injection site reaction, accumulation of immune cells, and deposition of micro- and nanoparticles in the spleen and lung can pose significant challenges for developing polymeric nano- and microparticles [264].

5.4.2. Nanocrystal Formulations

Nanocrystal technology has revolutionized HIV-1 treatment and prevention. This technology facilitates formulation of water-insoluble molecules into LA surfactant-stabilized drug nanosuspensions in aqueous buffers. Using either top-down or bottom-up technology, such as high-pressure homogenization, wet bead milling, microfluidization, and the emulsion template freeze-drying technique, solid drug microparticles are suspended in an aqueous buffer with stabilizing surfactants or water-soluble polymers and nanoformulated into stable solid drug nanoparticles. Compared to other formulation approaches, nanocrystals exhibit superior properties that include high drug loading to affect administration volumes, improve drug dissolution through reduced particle size, protect drug molecules from rapid metabolism to extend their apparent half-lives, and enhance drug stability by keeping it in solid form within the formulation. The use of aqueous buffers to produce drug nanocrystals also minimizes solvent-related toxicities in the final product. The success of this technology in HIV therapy is highlighted by the LA injectable Cabenuva (CAB and RPV LA) and Apretude (CAB LA) which are currently approved for treatment and prevention, respectively (Table 1). LA CAB consists of pure CAB nanocrystals with an average particle size of 200 nm dispersed in a stabilizing aqueous buffer solution of Tween

20, PEG_{3350}, and mannitol [265]. For LA RPV nanocrystals, Poloxamer 338 (P338) is used as the stabilizing surfactant [35].

Table 1. A comprehensive summary of CAB- and RPV-based LA delivery system for the treatment and prevention of HIV-1 infection.

Drugs	(CAB + RPV LA)	CAB LA
Brand name	Cabenuva.	Apretude or Vocabria.
Formulation characteristics	CAB nanocrystals are produced in an aqueous solution of polysorbate 20 (Tween 20), PEG_{3350}, and mannitol with particle size of 200 nm. RPV nanocrystals are produced in an aqueous solution of Poloxamer 338 with an average particle size of 200 nm.	Same as Cabenuva.
Indication	Treatment of HIV-1 infection.	Prevention of HIV-1 infection
Population	Virologically suppressed PLWH.	Adults and adolescents weighing at least 35 kg (77 lbs) who are at risk of sexually acquiring HIV.
Dosage regimen	Two initial injections of CAB 600 mg/3 mL and RPV 900 mg/3 mL given 1 month apart for two consecutive months and then given every two months (CAB 600 mg/3 mL and 900 mg/3 mL) thereafter. Injection is given in separate gluteal muscles.	Two initial injections (600 mg; 3 mL) given one month apart for two consecutive months, followed by maintenance doses (600 mg; 3 mL) given every 2 months thereafter. An OLI of CAB tablets may, or may not, be given for one month prior to starting CAB to assess tolerability.
Approval	First approved by Health Canada in March 2020, followed by the EMA in October 2020 and the US FDA in January 2021.	First approved by the US FDA in December 2021 and the EMA in October 2022.
Efficacy in the clinical trials	Several phase IIb and phase III clinical trials (LATTE-2, FLAIR, ATLAS, ATLAS-2M) proved non-inferiority for maintaining viral suppression compared to standard daily oral therap [3,6,8,266].	Proved superior to approved oral PrEP agent TDF/FTC in 2 phase III clinical trials (HPTN-083 and 084) [267,268].
Side effects	Mild ISRs were the most reported AEs during clinical trials, but none of them caused treatment withdrawal.	Same as Cabenuva.

PLWH—people living with HIV. EMA—European Medicines Agency. US FDA—US Food and Drug Administration. OLI—oral lead in. AEs—adverse events. ISRs—injection site reactions.

Upon IM injection, these nanocrystals form a drug depot at the injection site that slowly dissociates to release the drug for months. Drug release and absorption in biological matrices are dependent on several factors such as nanoparticle agglomeration or dispersion at the injection site, particle size, and surface properties that affect host protein adsorption to the particle surface to facilitate absorption [265]. The longevity of the depot is heavily influenced by the hydrophobicity of the formulated compounds. However, to ensure sufficient dissolution of the nanocrystals into the surrounding medium and eventually demonstrate effective plasma and tissue drug levels, the hydrophobicity of the formulated compound needs to be optimal. The particle size, in addition to the hydrophobicity, also influences the dissolution properties. With a larger surface area, the nano-sized crystal improves the dissolution properties of the drugs, helping maintain a sufficient concentration for efficacy. The benefit of smaller particle size is also evident in the improved syringeability of the formulation [167,265].

To further extend ARV dosage regimens, a prodrug strategy that is coined as 'long-acting slow effective antiretroviral therapy' (LASER ART) was developed (Table 2). The strategy relies on controlled drug release from a tissue macrophage depot. LASER ART was applied to all classes of ARVs with the goal of improving their physicochemical properties to facilitate their formulation into scalable LA agents with improved apparent half-lives. Some are currently in advanced preclinical development with the potential to improve the HIV treatment dosage regimen by up to 6 months or longer [269].

Prodrug nanocrystals of a stearoylated CAB were shown to sustain CAB concentrations for up to one year in rodent and non-human primate (NHP) models [9]. A similar modification of DTG significantly improved the PK profile of the parent drug [269]. However, esterification of CAB and DTG with either shorter or longer than 18-carbon fatty acid failed to maintain plasma CAB and DTG levels beyond two months [269–271]. The prodrug nanocrystal strategy was extended to NRTIs where FTC, 3TC, and TFV were converted into hydrophobic ProTide prodrugs with improved PK profiles [272–274]. Notably, hydrophobic stable TFV ProTide nanocrystals demonstrated effective TFV-diphosphate levels in a two-month PK study in rats after a single IM injection [272]. This formulation demonstrated anti-HBV activity in both HBV-infected transgenic and human hepatocyte-transplanted TK-NOG mice, for at least three months [275,276].

Research across a broad spectrum of hydrophobic ARV prodrugs sheds light on the correlation between prodrug hydrophobicity and drug release. The prolonged PK profile observed in INSTI-esterified prodrugs is influenced by multiple factors that contribute to the observed extended apparent half-lives. These influencing factors encompass (i) the physiochemical attributes of the prodrug, (ii) the chemical and enzymatic conversion of the prodrug to the parent drug, and iii) the absorption of the prodrug and parent drug by the lymphatic systems and tissues (Figure 7). Yet, there is a noticeable gap in research dedicated to understanding the role of polymer/surfactants in drug release and how they interact with cells that undergo mononuclear phagocytosis. Many of these formulations require hydrophilic low-MW polymers like poloxamer or a combination of PEG and polysorbate. This ensures that the particles remain stable in aqueous suspensions. One crucial facet of nanocrystal formulations is their uptake by phagocytic cells, like macrophages. The nature of the polymer in the formulation affects how macrophages interact with the nanoparticles and how stable the nanoparticles remain within phagocytosed cells. Numerous studies highlight the stability of these nanoparticles within macrophages, pointing to the creation of a depot within these cells. Comprehensive research focusing on prodrug nanoparticle size, charge, polymer/surfactant choices, and their effects on macrophage interactions is crucial. Deodhar et al. demonstrate improved stability of surfactant-coated nanoformulated DTG prodrugs in various environments like acidic pH, macrophages, plasma, and various tissues, compared to their counterparts in a free prodrug solution [269]. Nevertheless, the foundational mechanism promoting the heightened stability of the nanoparticle calls for further exploration. Furthermore, various challenges are associated with this technology when the native parent is used in the formulation, including an initial burst release, a prolonged PK tail, and population-level PK variability [265,277]. This variability in CAB LA's PK is also evident between genders. Some females exhibit a smoothed concentration–time curve, presenting a lower peak drug concentration (C_{max}) and a prolonged period to reach this C_{max} compared to males. However, both genders show comparable overall drug exposure ($AUC_{[0-\infty]}$) following a single LA dose of CAB. This suggests that the absorption rate might differ between genders, rather than the total absorption [278]. Additionally, the terminal phase PK tail for CAB LA shows variations between male and female patients.

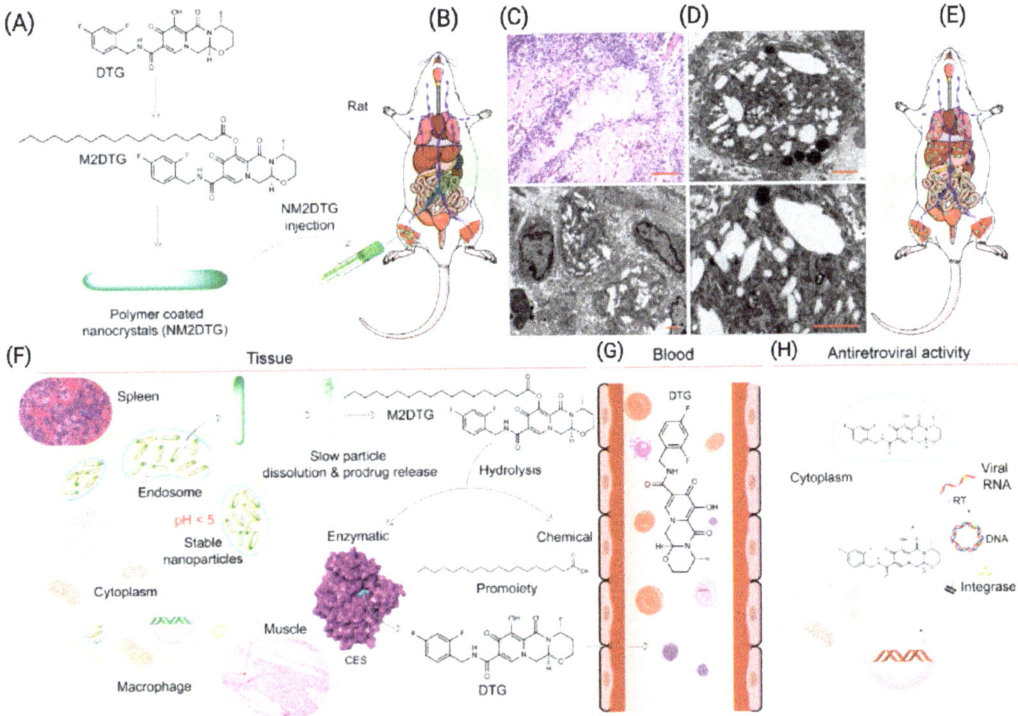

Figure 7. The illustration shows the in vivo fate of LASER ART prodrug nanoformulation. (**A**) Chemical synthesis of prodrug from parent ARV. (**B**) IM injection of LASER ART nanoformulation illustrates the muscle as the primary drug depot. (**C**,**D**) Macrophages phagocytose and store the nanocrystal from the site of injection. (**E**) Biodistribution of prodrug and active drug in tissues including HIV-1 reservoir site. (**F**) Slow prodrug release in the low-pH microenvironment in macrophages and then hydrolysis to release active drug. (**G**,**H**) Slow dissolution and hydrolysis of prodrug in the blood and tissues. The figure was reproduced under the terms of the Creative Commons CC-BY license [269].

Table 2. LA ARV formulations that are either in clinical or preclinical development.

Delivery System	ARV Drug	Polymer	Formulation Details	Predicted Dosage Regimen (Month)	Clinical Trial	Ref.
Implant	TAF	Silicone and PVA	TAF powder is encapsulated in a silicone tube, which has several drug delivery channels coated with PVA. Drug release rate can be controlled by controlling the number of channels or thickness of PVA membrane.	6	Phase I	[15]
	TAF	Silicone	Transcutaneous refillable nanochannel delivery implant, where drug reservoir is made of titanium and drug is released through TaN-coated slit nanochannel.	6	Preclinical	[11]
	TAF	PUs	TAF formulation pellet is encased in PU-based drug reservoir and the drug is released through diffusion through the PU membrane.	3		[162]
	CAB		CAB formulation is converted into pellets and then encased in PU membrane containing drug reservoir. PU membrane acts as RCM for controlling drug release through diffusion.	3		[168]

Table 2. *Cont.*

Delivery System	ARV Drug	Polymer	Formulation Details	Predicted Dosage Regimen (Month)	Clinical Trial	Ref.
Implant	TAF	PCL	TAF formulation containing TAF and castor oil was loaded into PCL tubes and the drug was released through the diffusion and bulk erosion of PCL. Drug release rate can be controlled through controlling MW of PCL.	3	Preclinical	[160]
	ISL	EVA	Crystalline ISL is uniformly dispersed in EVA and the implant is prepared through hot melt extrusion.	12	On hold	[165]
ISFIs	CAB	PLGA	CAB and PLGA were solubilized in NMP:DMSO and injected subcutaneously. Upon injection, the solution went through phase inversion and formed solid implant.	6	Preclinical	[51]
	DTG		Same as CAB PLGA ISFI	6		[180]
VRs	TDF	PUs	TDF formulation was loaded in hydrophilic PU tubes and the PU tubes were end-sealed using induction welding.	1		[203]
	TDF	PUs	TDF and NaCl formulation was loaded onto PU extruded tubes and end-sealed to prepare the VRs.	1		[17]
MNs	CAB	PVP and PVA	MN tips were generated by using a hydrogel composed of 20:20:60 PVA:PVP:CAB and the baseplate of MNs was made using PVP and PVA. The MN tips dissolved quickly to release the CAB and the CAB formed drug depot at the administration site.	1	Preclinical	[22]
	BIC		MNs tips were made of PVP, PVA, and BIC. The baseplate was made of PVP and glycerol.	1		[230]
	RPV		MN tips were made of PVP and RPV. The baseplate was made of PVA and glycerol.	1		[23]
Microparticle	DTG	PLGA	DTG was transformed into a hydrophobic prodrug and encapsulated into PLGA-based microparticles by organic and aqueous solvent emulsification–evaporation.	3		[251]
Prodrug nanocrystal	TFV	Poloxamer, Polysorbate and PEG	TFV was converted into lipophilic ProTide and then formulated as aqueous nanocrystals.	≥3		[272]
	CAB		Lipohilic ester prodrug cabotegravir was synthesized and formulated as aqueous nanocrystals.	12		[9]
	DTG		Fatty acid ester prodrug DTG was synthesized and nanoformulated as aqueous nanocrystals.	≥6		[269]

ISL—islatravir, TFV—tenofovir, TDF—tenofovir disoproxil fumarate, TAF—tenofovir alafenamide fumarate, CAB—cabotegravir, BIC—bictegravir, DTG—dolutegravir, RPV—rilpivirine.

6. Conclusions and Future Perspective of LA Drug Delivery Systems

The prospects for polymer-based LA formulations are bright, given our rapidly expanding knowledge of polymer properties and drug release mechanisms. While current LA drugs for HIV offer improvement over traditional oral HIV medications, there are user- and treatment-related gaps that could be addressed by the emerging LA drug delivery technologies such as prodrug approaches and implants. Ideally, future LA ART medicines should possess three main characteristics.

First, they should be easy to manufacture, scalable, and user-friendly, with a dosing schedule extended at least beyond two months to synchronize dosing with the existing routine patient laboratory test clinic visits. A desirable target product profile is therefore a six-month ART dosage regimen. However, achieving a six-month dosage regimen with the exixting parent drugs is challenging due to their inherent features that include rapid drug metabolism upon absorption from the implant/injection site depot. The constraints on the allowable volume for injections or the size of implantable products, especially for SC or IM delivery, further intensify this challenge. Sometimes, the required drug amount surpasses these permissible limits. A potential solution lies in choosing drugs that are highly potent and have a high resistance to mutations. Prodrug approaches could also be used to achieve this goal by improving intracellular drug delivery and absorption, distribution, metabolism, and excretion (ADME) properties of parent drugs to extend their half-lives. When identifying the optimal drug for LA formulation development, coexisting medical conditions in patients should also be considered. Given that a significant number of HIV patients (~7.5%) are coinfected with hepatitis B or tuberculosis, LA drugs that can concurrently target multiple viruses are especially valuable [279].

The second desired feature of LA formulations is to minimize postmarket complications. This includes curtailing the prolonged PK tail in the terminal phase of drug release, inhibiting the emergence of drug-resistant mutations, and ensuring that the treatment can be reversed, especially for newer drugs with no established safety profiles, if adverse reactions occur. For example, the FDA-approved CAB LA exhibits a long PK tail after the last injection [280]. The CDC recommends that those at risk of HIV continue with oral CAB or PrEP for at least a year after discontinuing CAB LA. Yet, making the shift from LA drugs to oral PrEP is not always straightforward, as many prefer LA treatments over daily oral ART. There are also concerns about how easily LA treatments can be reversed in the event of complications. These have been circumvented through oral lead-ins prior to administering LA injectable agents. Solid LA implantable devices might offer an alterantive reversal solution, but the need for surgical removal postlifespan dilutes this advantage.

The third desired feature of LA formulations is their widespread adoption and utilization by the end users. Factors like cost, storage stability, transportation requirements, and specific storage conditions can significantly influence the acceptance and use of LA medicines. Given that HIV is most common in resource-limited settings with limited financial capacity for medication, inadequate cold storage infrastructure, and a scarcity of medical professionals, an LA medicine that is affordable, stable at room temperature, and self-administrable would greatly boost its adoption and application.

Currently, no LA technology, whether in clinical or preclinical development, embodies all three of these ideal features, suggesting that a 'one size fits all' approach to LA therapy is inappropriate. Given this, there is a push to tailor LA medicines to the unique needs of specific subgroups. Understanding the interactions between the polymers, drugs, and their release mechanisms will enable more precise tuning of drug release rates from these formulations. Additionally, to ensure the clinical success of LA ARVs in such resource-limited settings, it is imperative to involve all stakeholders in the early phase of formulation development. These stakeholders include end users, healthcare professionals, business developers, and policymakers. Collaboration, careful planning, and implementation consideration are essential to prioritize and design the most promising ARVs and LA DDSs that facilitate effective implementation in the most HIV-burdened settings. Moreover, the development of LA ARVs may present regulatory challenges, such as establishing optimal dosing regimens to address the need for combination antiretroviral therapy and managing missed doses and potential drug–drug interactions in patients with comorbidities. Some of these challenges are particularly important for PrEP dosing schedules, as there is no precise biomarker to gauge outcomes. In such situations, conducting thorough in vivo PK studies and evaluating PrEP effectiveness in preclinical animal models becomes imperative. Nonetheless, selecting an appropriate animal model to determine PrEP efficacy remains a significant challenge in preclinical evaluation, given that none of the existing models can fully replicate the human transmission and immunogenicity of HIV infection. Additionally, since LA ARVs can persist in the body for extended periods, ensuring the long-term safety of both the active ingredients and excipients used in LA ARVs in preclinical in vivo models is of significance. While the excipients used in LA ARVs are generally recognized as safe (GRAS), extensive GLP toxicology studies in relevant models are needed to broaden the usage in special populations such as pediatric patients. The requirement for extensive, extended safety study timelines introduces further challenges, including increased overall product development costs and time. In this context, the use of physiologically based pharmacokinetic (PBPK) modeling can be a valuable tool for predicting drug PK. This approach can also streamline formulation screening and expedite the development of LA products.

Overall, advancements in polymer chemistry and insights into their drug release mechanisms have paved the way for promising strategies to extend HIV-1 treatment dosage regimens beyond two months. Nevertheless, there are developmental and translational hurdles that must be overcome for these products to successfully transition to clinical application.

Author Contributions: Conceptualization, B.E. and H.E.G.; literature collection and analysis, M.U.N., S.P., A.S., L.A.Z., L.K.V. and B.S.; figure preparation, M.U.N. and S.P.; writing—original draft preparation, M.U.N., S.P., A.S., L.A.Z., L.K.V. and B.S.; writing—review and editing, B.E. and H.E.G.; funding acquisition, B.E. and H.E.G. All authors have read and agreed to the published version of the manuscript.

Funding: The work is supported by National Institutes of Health grants R01 NS034239; R01 NS036126; R01 MH115860; R01 NS126089; T32NS105594 (HEG); R01 AI145542 (BE and HEG); R01 AI158160 (HEG and BE).

Institutional Review Board Statement: Not applicable.

Informed Consent Statement: Not applicable.

Data Availability Statement: Data sharing is not applicable. No new data were created in this study.

Conflicts of Interest: The authors declare no conflict of interest.

References

1. Deanesly, R.; Parkes, A.S. Testosterone. *Br. Med. J.* **1936**, *1*, 527–528. [CrossRef] [PubMed]
2. Paolini, M.S.; Fenton, O.S.; Bhattacharya, C.; Andresen, J.L.; Langer, R. Polymers for extended-release administration. *Biomed. Microdevices* **2019**, *21*, 45. [CrossRef] [PubMed]
3. Swindells, S.; Andrade-Villanueva, J.-F.; Richmond, G.J.; Rizzardini, G.; Baumgarten, A.; Masiá, M.; Latiff, G.; Pokrovsky, V.; Bredeek, F.; Smith, G. Long-acting cabotegravir and rilpivirine for maintenance of HIV-1 suppression. *N. Engl. J. Med.* **2020**, *382*, 1112–1123. [CrossRef] [PubMed]
4. Smith, G.H.R.; Henry, W.K.; Podzamczer, D.; Masiá, M.D.M.; Bettacchi, C.J.; Arasteh, K.; Jaeger, H.; Khuong-Josses, M.A.; Montes-Ramírez, M.L.; Stellbrink, H.J.; et al. Efficacy, Safety, and Durability of Long-Acting Cabotegravir and Rilpivirine in Adults with Human Immunodeficiency Virus Type 1 Infection: 5-Year Results from the LATTE-2 Study. *Open Forum Infect. Dis.* **2021**, *8*, ofab439. [CrossRef]
5. Parikh, U.M.; Koss, C.A.; Mellors, J.W. Long-Acting Injectable Cabotegravir for HIV Prevention: What Do We Know and Need to Know about the Risks and Consequences of Cabotegravir Resistance? *Curr. HIV/AIDS Rep.* **2022**, *19*, 384–393. [CrossRef]
6. Margolis, D.A.; Gonzalez-Garcia, J.; Stellbrink, H.-J.; Eron, J.J.; Yazdanpanah, Y.; Podzamczer, D.; Lutz, T.; Angel, J.B.; Richmond, G.J.; Clotet, B. Long-acting intramuscular cabotegravir and rilpivirine in adults with HIV-1 infection (LATTE-2): 96-week results of a randomised, open-label, phase 2b, non-inferiority trial. *Lancet* **2017**, *390*, 1499–1510. [CrossRef]
7. Orkin, C.; Oka, S.; Philibert, P.; Brinson, C.; Bassa, A.; Gusev, D.; Degen, O.; García, J.G.; Morell, E.B.; Tan, D.H.S.; et al. Long-acting cabotegravir plus rilpivirine for treatment in adults with HIV-1 infection: 96-week results of the randomised, open-label, phase 3 FLAIR study. *Lancet HIV* **2021**, *8*, e185–e196. [CrossRef]
8. Orkin, C.; Arasteh, K.; Górgolas Hernández-Mora, M.; Pokrovsky, V.; Overton, E.T.; Girard, P.-M.; Oka, S.; Walmsley, S.; Bettacchi, C.; Brinson, C. Long-acting cabotegravir and rilpivirine after oral induction for HIV-1 infection. *N. Engl. J. Med.* **2020**, *382*, 1124–1135. [CrossRef] [PubMed]
9. Kulkarni, T.A.; Bade, A.N.; Sillman, B.; Shetty, B.L.D.; Wojtkiewicz, M.S.; Gautam, N.; Hilaire, J.R.; Sravanam, S.; Szlachetka, A.; Lamberty, B.G. A year-long extended release nanoformulated cabotegravir prodrug. *Nat. Mater.* **2020**, *19*, 910–920. [CrossRef] [PubMed]
10. Young, I.C.; Massud, I.; Cottrell, M.L.; Shrivastava, R.; Maturavongsadit, P.; Prasher, A.; Wong-Sam, A.; Dinh, C.; Edwards, T.; Mrotz, V.; et al. Ultra-long-acting in-situ forming implants with cabotegravir protect female macaques against rectal SHIV infection. *Nat. Commun.* **2023**, *14*, 708. [CrossRef]
11. Chua, C.Y.X.; Jain, P.; Ballerini, A.; Bruno, G.; Hood, R.L.; Gupte, M.; Gao, S.; Di Trani, N.; Susnjar, A.; Shelton, K.; et al. Transcutaneously refillable nanofluidic implant achieves sustained level of tenofovir diphosphate for HIV pre-exposure prophylaxis. *J. Control. Release* **2018**, *286*, 315–325. [CrossRef]

12. Johnson, L.M.; Krovi, S.A.; Li, L.; Girouard, N.; Demkovich, Z.R.; Myers, D.; Creelman, B.; van der Straten, A. Characterization of a Reservoir-Style Implant for Sustained Release of Tenofovir Alafenamide (TAF) for HIV Pre-Exposure Prophylaxis (PrEP). *Pharmaceutics* **2019**, *11*, 315. [CrossRef]
13. Simpson, S.M.; Widanapathirana, L.; Su, J.T.; Sung, S.; Watrous, D.; Qiu, J.; Pearson, E.; Evanoff, A.; Karunakaran, D.; Chacon, J.E.; et al. Design of a Drug-Eluting Subcutaneous Implant of the Antiretroviral Tenofovir Alafenamide Fumarate. *Pharm. Res.* **2020**, *37*, 83. [CrossRef] [PubMed]
14. Markowitz, M.; Grobler, J.A. Islatravir for the treatment and prevention of infection with the human immunodeficiency virus type 1. *Curr. Opin. HIV AIDS* **2020**, *15*, 27–32. [CrossRef]
15. Gunawardana, M.; Remedios-Chan, M.; Miller, C.S.; Fanter, R.; Yang, F.; Marzinke, M.A.; Hendrix, C.W.; Beliveau, M.; Moss, J.A.; Smith, T.J. Pharmacokinetics of long-acting tenofovir alafenamide (GS-7340) subdermal implant for HIV prophylaxis. *Antimicrob. Agents Chemother.* **2015**, *59*, 3913–3919. [CrossRef]
16. Baeten, J.M.; Palanee-Phillips, T.; Mgodi, N.M.; Mayo, A.J.; Szydlo, D.W.; Ramjee, G.; Gati Mirembe, B.; Mhlanga, F.; Hunidzarira, P.; Mansoor, L.E.; et al. Safety, uptake, and use of a dapivirine vaginal ring for HIV-1 prevention in African women (HOPE): An open-label, extension study. *Lancet HIV* **2021**, *8*, e87–e95. [CrossRef]
17. Keller, M.J.; Wood, L.; Billingsley, J.M.; Ray, L.L.; Goymer, J.; Sinclair, S.; McGinn, A.P.; Marzinke, M.A.; Frank, B.; Srinivasan, S.; et al. Tenofovir disoproxil fumarate intravaginal ring for HIV pre-exposure prophylaxis in sexually active women: A phase 1, single-blind, randomised, controlled trial. *Lancet HIV* **2019**, *6*, e498–e508. [CrossRef]
18. Boyd, P.; Variano, B.; Spence, P.; McCoy, C.F.; Murphy, D.J.; Dallal Bashi, Y.H.; Malcolm, R.K. In vitro release testing methods for drug-releasing vaginal rings. *J. Control. Release* **2019**, *313*, 54–69. [CrossRef] [PubMed]
19. Montgomery, E.T.; van der Straten, A.; Chitukuta, M.; Reddy, K.; Woeber, K.; Atujuna, M.; Bekker, L.G.; Etima, J.; Nakyanzi, T.; Mayo, A.J.; et al. Acceptability and use of a dapivirine vaginal ring in a phase III trial. *Aids* **2017**, *31*, 1159–1167. [CrossRef]
20. Nel, A.; van Niekerk, N.; Kapiga, S.; Bekker, L.-G.; Gama, C.; Gill, K.; Kamali, A.; Kotze, P.; Louw, C.; Mabude, Z.; et al. Safety and Efficacy of a Dapivirine Vaginal Ring for HIV Prevention in Women. *N. Engl. J. Med.* **2016**, *375*, 2133–2143. [CrossRef]
21. Vora, L.K.; Moffatt, K.; Tekko, I.A.; Paredes, A.J.; Volpe-Zanutto, F.; Mishra, D.; Peng, K.; Raj Singh Thakur, R.; Donnelly, R.F. Microneedle array systems for long-acting drug delivery. *Eur. J. Pharm. Biopharm.* **2021**, *159*, 44–76. [CrossRef]
22. Tekko, I.A.; Vora, L.K.; Volpe-Zanutto, F.; Moffatt, K.; Jarrahian, C.; McCarthy, H.O.; Donnelly, R.F. Novel Bilayer Microarray Patch-Assisted Long-Acting Micro-Depot Cabotegravir Intradermal Delivery for HIV Pre-Exposure Prophylaxis. *Adv. Funct. Mater.* **2022**, *32*, 2106999. [CrossRef]
23. Mc Crudden, M.T.C.; Larrañeta, E.; Clark, A.; Jarrahian, C.; Rein-Weston, A.; Lachau-Durand, S.; Niemeijer, N.; Williams, P.; Haeck, C.; McCarthy, H.O.; et al. Design, formulation and evaluation of novel dissolving microarray patches containing a long-acting rilpivirine nanosuspension. *J. Control. Release* **2018**, *292*, 119–129. [CrossRef]
24. Deanesly, R.; Parkes, A.S. Biological properties of some new derivatives of testosterone. *Biochem. J.* **1937**, *31*, 1161. [CrossRef]
25. Chien, Y.W. Novel drug delivery systems. *Drugs Pharm. Sci.* **1992**, *50*, 797.
26. Dreyfuss, J.; Ross, J.J., Jr.; Shaw, J.M.; Miller, I.; Schreiber, E.C. Release and elimination of 14C-fluphenazine enanthate and decanoate esters administered in sesame oil to dogs. *J. Pharm. Sci.* **1976**, *65*, 502–507. [CrossRef]
27. Wright, J.C.; Hoffman, A.S. Historical overview of long acting injections and implants. In *Long Acting Injections and Implants*; Springer: Berlin/Heidelberg, Germany, 2011; pp. 11–24.
28. Higuchi, T. Mechanism of sustained-action medication. Theoretical analysis of rate of release of solid drugs dispersed in solid matrices. *J. Pharm. Sci.* **1963**, *52*, 1145–1149. [CrossRef]
29. Folkman, J.; Long, D.M., Jr.; Rosenbaum, R. Silicone rubber: A new diffusion property useful for general anesthesia. *Science* **1966**, *154*, 148–149. [CrossRef]
30. Heilmann, K. *Therapeutic Systems: Rate-Controlled Drug Delivery: Concept and Development*; George Thieme Verlag: Erlangen, Germany, 1984.
31. Frazza, E.; Schmitt, E. A new absorbable suture. *J. Biomed. Mater. Res.* **1971**, *5*, 43–58. [CrossRef]
32. Kent, J.S.; Lewis, D.H.; Sanders, L.M.; Tice, T.R. Microencapsulation of Water Soluble Active Polypeptides. U.S. Patent 4,675,189A, 23 June 1987.
33. Hoffman, A.S. The origins and evolution of "controlled" drug delivery systems. *J. Control. Release* **2008**, *132*, 153–163. [CrossRef]
34. Nowacek, A.S.; Miller, R.L.; McMillan, J.; Kanmogne, G.; Kanmogne, M.; Mosley, R.L.; Ma, Z.; Graham, S.; Chaubal, M.; Werling, J.; et al. NanoART synthesis, characterization, uptake, release and toxicology for human monocyte-macrophage drug delivery. *Nanomedicine* **2009**, *4*, 903–917. [CrossRef]
35. Baert, L.; van't Klooster, G.; Dries, W.; François, M.; Wouters, A.; Basstanie, E.; Iterbeke, K.; Stappers, F.; Stevens, P.; Schueller, L. Development of a long-acting injectable formulation with nanoparticles of rilpivirine (TMC278) for HIV treatment. *Eur. J. Pharm. Biopharm.* **2009**, *72*, 502–508. [CrossRef]
36. Penrose, K.J.; Parikh, U.M.; Hamanishi, K.A.; Else, L.; Back, D.; Boffito, M.; Jackson, A.; Mellors, J.W. Selection of rilpivirine-resistant HIV-1 in a seroconverter from the SSAT 040 trial who received the 300-mg dose of long-acting rilpivirine (TMC278LA). *J. Infect. Dis.* **2016**, *213*, 1013–1017. [CrossRef]
37. Venkatesan, P. Long-acting injectable ART for HIV: A (cautious) step forward. *Lancet Microbe* **2022**, *3*, e94. [CrossRef]
38. Langer, R. New methods of drug delivery. *Science* **1990**, *249*, 1527–1533. [CrossRef]
39. Rahnfeld, L.; Luciani, P. Injectable lipid-based depot formulations: Where do we stand? *Pharmaceutics* **2020**, *12*, 567. [CrossRef]

40. United States Pharmacopeia. <1151> Pharmaceutical dosage forms. In *United States Pharmacopeia*; USP-NF, Ed.; The United States Pharmacopeial Convention: Rockville, MD, USA, 2021.
41. Fredenberg, S.; Wahlgren, M.; Reslow, M.; Axelsson, A. The mechanisms of drug release in poly (lactic-co-glycolic acid)-based drug delivery systems—A review. *Int. J. Pharm.* **2011**, *415*, 34–52. [CrossRef]
42. Berchane, N.; Carson, K.; Rice-Ficht, A.; Andrews, M. Effect of mean diameter and polydispersity of PLG microspheres on drug release: Experiment and theory. *Int. J. Pharm.* **2007**, *337*, 118–126. [CrossRef]
43. Fu, Y.; Kao, W.J. Drug release kinetics and transport mechanisms of non-degradable and degradable polymeric delivery systems. *Expert Opin. Drug Deliv.* **2010**, *7*, 429–444. [CrossRef]
44. Liechty, W.B.; Kryscio, D.R.; Slaughter, B.V.; Peppas, N.A. Polymers for drug delivery systems. *Annu. Rev. Chem. Biomol. Eng.* **2010**, *1*, 149–173. [CrossRef]
45. Webber, W.L.; Lago, F.; Thanos, C.; Mathiowitz, E. Characterization of soluble, salt-loaded, degradable PLGA films and their release of tetracycline. *J. Biomed. Mater. Res. Off. J. Soc. Biomater. Jpn. Soc. Biomater. Aust. Soc. Biomater.* **1998**, *41*, 18–29. [CrossRef]
46. Marin, E.; Briceño, M.I.; Caballero-George, C. Critical evaluation of biodegradable polymers used in nanodrugs. *Int. J. Nanomed.* **2013**, *8*, 3071–3091.
47. Uhrich, K.E.; Cannizzaro, S.M.; Langer, R.S.; Shakesheff, K.M. Polymeric systems for controlled drug release. *Chem. Rev.* **1999**, *99*, 3181–3198. [CrossRef]
48. Maroni, A.; Zema, L.; Cerea, M.; Foppoli, A.; Palugan, L.; Gazzaniga, A. Erodible drug delivery systems for time-controlled release into the gastrointestinal tract. *J. Drug Deliv. Sci. Technol.* **2016**, *32*, 229–235. [CrossRef]
49. Keraliya, R.A.; Patel, C.; Patel, P.; Keraliya, V.; Soni, T.G.; Patel, R.C.; Patel, M. Osmotic drug delivery system as a part of modified release dosage form. *Int. Sch. Res. Not.* **2012**, *2012*, 528079. [CrossRef]
50. Huang, X.; Brazel, C.S. On the importance and mechanisms of burst release in matrix-controlled drug delivery systems. *J. Control. Release* **2001**, *73*, 121–136. [CrossRef]
51. Makadia, H.K.; Siegel, S.J. Poly lactic-co-glycolic acid (PLGA) as biodegradable controlled drug delivery carrier. *Polymers* **2011**, *3*, 1377–1397. [CrossRef]
52. Saltzman, W.M. *Drug Delivery: Engineering Principles for Drug Therapy*; Oxford University Press: Oxford, UK, 2001.
53. Kim, J.M.; Seo, K.S.; Jeong, Y.K.; Lee, H.B.; Kim, Y.S.; Khang, G. Co-effect of aqueous solubility of drugs and glycolide monomer on in vitro release rates from poly (D, L-lactide-co-glycolide) discs and polymer degradation. *J. Biomater. Sci. Polym. Ed.* **2005**, *16*, 991–1007. [CrossRef]
54. Lu, L.; Peter, S.J.; Lyman, M.D.; Lai, H.-L.; Leite, S.M.; Tamada, J.A.; Uyama, S.; Vacanti, J.P.; Langer, R.; Mikos, A.G. In vitro and in vivo degradation of porous poly (DL-lactic-co-glycolic acid) foams. *Biomaterials* **2000**, *21*, 1837–1845. [CrossRef]
55. Ehrenstein, G.W. *Polymeric Materials: Structure, Properties, Applications*; Carl Hanser Verlag GmbH Co KG: Munich, Germany, 2012.
56. Odian, G. *Principles of Polymerization*; John Wiley & Sons: Hoboken, NJ, USA, 2004.
57. Allcock, H.R.; Lampe, F.W.; Mark, J.E.; Allcock, H. *Contemporary Polymer Chemistry*; Prentice-Hall Englewood Cliffs: Upper Saddle River, NJ, USA, 1981.
58. Muller, W.E.; Wang, X.; Schroder, H. *Biomedical Inorganic Polymers*; Springer: Berlin/Heidelberg, Germany, 2016.
59. Ginjupalli, K.; Shavi, G.V.; Averineni, R.K.; Bhat, M.; Udupa, N.; Upadhya, P.N. Poly (α-hydroxy acid) based polymers: A review on material and degradation aspects. *Polym. Degrad. Stab.* **2017**, *144*, 520–535. [CrossRef]
60. Nair, L.S.; Laurencin, C.T. Biodegradable polymers as biomaterials. *Prog. Polym. Sci.* **2007**, *32*, 762–798. [CrossRef]
61. Dash, A.; Cudworth II, G. Therapeutic applications of implantable drug delivery systems. *J. Pharmacol. Toxicol. Methods* **1998**, *40*, 1–12. [CrossRef]
62. Gan, M.; Zhou, Q.; Ge, J.; Zhao, J.; Wang, Y.; Yan, Q.; Wu, C.; Yu, H.; Xiao, Q.; Wang, W. Precise in-situ release of microRNA from an injectable hydrogel induces bone regeneration. *Acta Biomater.* **2021**, *135*, 289–303. [CrossRef]
63. Rahme, K.; Dagher, N. Chemistry Routes for Copolymer Synthesis Containing PEG for Targeting, Imaging, and Drug Delivery Purposes. *Pharmaceutics* **2019**, *11*, 327. [CrossRef]
64. Sanchez-Cano, C.; Carril, M. Recent Developments in the Design of Non-Biofouling Coatings for Nanoparticles and Surfaces. *Int. J. Mol. Sci.* **2020**, *21*, 1007. [CrossRef]
65. Dirisala, A.; Uchida, S.; Toh, K.; Li, J.; Osawa, S.; Tockary, T.A.; Liu, X.; Abbasi, S.; Hayashi, K.; Mochida, Y. Transient stealth coating of liver sinusoidal wall by anchoring two-armed PEG for retargeting nanomedicines. *Sci. Adv.* **2020**, *6*, eabb8133. [CrossRef]
66. Zhang, D.; Zheng, H.; Geng, K.; Shen, J.; Feng, X.; Xu, P.; Duan, Y.; Li, Y.; Wu, R.; Gou, Z. Large fuzzy biodegradable polyester microspheres with dopamine deposition enhance cell adhesion and bone regeneration in vivo. *Biomaterials* **2021**, *272*, 120783. [CrossRef]
67. Liu, S.; Li, K.; Hussain, I.; Oderinde, O.; Yao, F.; Zhang, J.; Fu, G. A Conductive Self-Healing Double Network Hydrogel with Toughness and Force Sensitivity. *Chem. Eur. J.* **2018**, *24*, 6632–6638. [CrossRef]
68. Apostolides, D.E.; Patrickios, C.S.; Sakai, T.; Guerre, M.; Lopez, G.R.; Améduri, B.; Ladmiral, V.; Simon, M.; Gradzielski, M.; Clemens, D. Near-model amphiphilic polymer conetworks based on four-arm stars of poly (vinylidene fluoride) and poly (ethylene glycol): Synthesis and characterization. *Macromolecules* **2018**, *51*, 2476–2488. [CrossRef]
69. Xue, S.; Li, X.; Li, S.; Chen, N.; Zhan, Q.; Long, L.; Zhao, J.; Hou, X.; Yuan, X. Bone fracture microenvironment responsive hydrogel for timing sequential release of cargoes. *Colloids Surf. A Physicochem. Eng. Asp.* **2021**, *629*, 127413. [CrossRef]

70. Le Fer, G.; Dilla, R.A.; Wang, Z.; King, J.; Chuang, S.S.; Becker, M.L. Clustering and Hierarchical Organization of 3D Printed Poly (propylene fumarate)-block-PEG-block-poly (propylene fumarate) ABA Triblock Copolymer Hydrogels. *Macromolecules* **2021**, *54*, 3458–3468. [CrossRef]
71. Roy, D.; Cambre, J.N.; Sumerlin, B.S. Future perspectives and recent advances in stimuli-responsive materials. *Prog. Polym. Sci.* **2010**, *35*, 278–301. [CrossRef]
72. Alexandridis, P.; Zhou, D.; Khan, A. Lyotropic liquid crystallinity in amphiphilic block copolymers: Temperature effects on phase behavior and structure for poly (ethylene oxide)-b-poly (propylene oxide)-b-poly (ethylene oxide) copolymers of different composition. *Langmuir* **1996**, *12*, 2690–2700. [CrossRef]
73. Pozzo, D.C.; Hollabaugh, K.R.; Walker, L.M. Rheology and phase behavior of copolymer-templated nanocomposite materials. *J. Rheol.* **2005**, *49*, 759–782. [CrossRef]
74. Bodratti, A.M.; Alexandridis, P. Formulation of poloxamers for drug delivery. *J. Funct. Biomater.* **2018**, *9*, 11. [CrossRef]
75. Zhang, K.; Shi, X.; Lin, X.; Yao, C.; Shen, L.; Feng, Y. Poloxamer-based in situ hydrogels for controlled delivery of hydrophilic macromolecules after intramuscular injection in rats. *Drug Deliv.* **2015**, *22*, 375–382. [CrossRef]
76. Kamaly, N.; Yameen, B.; Wu, J.; Farokhzad, O.C. Degradable Controlled-Release Polymers and Polymeric Nanoparticles: Mechanisms of Controlling Drug Release. *Chem. Rev.* **2016**, *116*, 2602–2663. [CrossRef]
77. Van Hemelryck, S.; Dewulf, J.; Niekus, H.; van Heerden, M.; Ingelse, B.; Holm, R.; Mannaert, E.; Langguth, P. In vitro evaluation of poloxamer in situ forming gels for bedaquiline fumarate salt and pharmacokinetics following intramuscular injection in rats. *Int. J. Pharm. X* **2019**, *1*, 100016. [CrossRef]
78. Demarteau, J.; Kermagoret, A.; Jérôme, C.; Detrembleur, C.; Debuigne, A. Controlled Synthesis of Ethylene-Vinyl Acetate Based Copolymers by Organometallic Mediated Radical Polymerization. In *Controlled Radical Polymerization: Materials*; ACS Symposium Series; American Chemical Society: Washington, DC, USA, 2015; Volume 1188, pp. 47–61.
79. Schneider, C.; Langer, R.; Loveday, D.; Hair, D. Applications of ethylene vinyl acetate copolymers (EVA) in drug delivery systems. *J. Control. Release* **2017**, *262*, 284–295. [CrossRef]
80. Arsac, A.; Carrot, C.; Guillet, J. Determination of primary relaxation temperatures and melting points of ethylene vinyl acetate copolymers. *J. Therm. Anal. Calorim.* **2000**, *61*, 681–685. [CrossRef]
81. Wang, X.; Venkatraman, S.S.; Boey, F.Y.; Loo, J.S.; Tan, L.P. Controlled release of sirolimus from a multilayered PLGA stent matrix. *Biomaterials* **2006**, *27*, 5588–5595. [CrossRef]
82. Zhang, Q.; Lin, W.; Yang, G.; Chen, Q. Studies on the phase structure of ethylene-vinyl acetate copolymers by solid-state 1H and 13C NMR. *J. Polym. Sci. Part B Polym. Phys.* **2002**, *40*, 2199–2207. [CrossRef]
83. Wang, L.; Fang, P.; Ye, C.; Feng, J. Solid-state NMR characterizations on phase structures and molecular dynamics of poly(ethylene-co-vinyl acetate). *J. Polym. Sci. Part B Polym. Phys.* **2006**, *44*, 2864–2879. [CrossRef]
84. Almeida, A.; Possemiers, S.; Boone, M.N.; De Beer, T.; Quinten, T.; Van Hoorebeke, L.; Remon, J.P.; Vervaet, C. Ethylene vinyl acetate as matrix for oral sustained release dosage forms produced via hot-melt extrusion. *Eur. J Pharm. Biopharm.* **2011**, *77*, 297–305. [CrossRef] [PubMed]
85. McKeen, L.W. *The Effect of Sterilization on Plastics and Elastomers*; William Andrew: Norwich, NY, USA, 2018.
86. Gaaz, T.S.; Sulong, A.B.; Akhtar, M.N.; Kadhum, A.A.; Mohamad, A.B.; Al-Amiery, A.A. Properties and Applications of Polyvinyl Alcohol, Halloysite Nanotubes and Their Nanocomposites. *Molecules* **2015**, *20*, 22833–22847. [CrossRef]
87. Song, S.I.; Kim, B.C. Characteristic rheological features of PVA solutions in water-containing solvents with different hydration states. *Polymer* **2004**, *45*, 2381–2386. [CrossRef]
88. Marin, E.; Rojas, J.; Ciro, Y. A review of polyvinyl alcohol derivatives: Promising materials for pharmaceutical and biomedical applications. *Afr. J. Pharm. Pharmacol.* **2014**, *8*, 674–684.
89. Rowe, R.C.; Sheskey, P.; Quinn, M. *Handbook of Pharmaceutical Excipients*; Libros Digitales-Pharmaceutical Press: London, UK, 2009.
90. Dediu, V.; Busila, M.; Tucureanu, V.; Bucur, F.I.; Iliescu, F.S.; Brincoveanu, O.; Iliescu, C. Synthesis of ZnO/Au nanocomposite for antibacterial applications. *Nanomaterials* **2022**, *12*, 3832. [CrossRef]
91. Fatema, U.K.; Rahman, M.M.; Islam, M.R.; Mollah, M.Y.A.; Susan, M.A.B.H. Silver/poly (vinyl alcohol) nanocomposite film prepared using water in oil microemulsion for antibacterial applications. *J. Colloid Interface Sci.* **2018**, *514*, 648–655. [CrossRef]
92. Meira, J.; Madeira, C.; Falcão-Reis, F.; Figueira, L. Sustained control from recurring non-infectious uveitic macular edema with 0.19 mg fluocinolone acetonide intravitreal implant—A case report. *Ophthalmol. Ther.* **2019**, *8*, 635–641. [CrossRef]
93. Wang, L.L.; Highley, C.B.; Yeh, Y.C.; Galarraga, J.H.; Uman, S.; Burdick, J.A. Three-dimensional extrusion bioprinting of single-and double-network hydrogels containing dynamic covalent crosslinks. *J. Biomed. Mater. Res. Part A* **2018**, *106*, 865–875. [CrossRef]
94. Liu, Z.; Zhang, J.; Liu, J.; Long, Y.; Fang, L.; Wang, Q.; Liu, T. Highly compressible and superior low temperature tolerant supercapacitors based on dual chemically crosslinked PVA hydrogel electrolytes. *J. Mater. Chem. A* **2020**, *8*, 6219–6228. [CrossRef]
95. Chao, Y.; Chen, Q.; Liu, Z. Smart injectable hydrogels for cancer immunotherapy. *Adv. Funct. Mater.* **2020**, *30*, 1902785. [CrossRef]
96. Lanman, T.; Martin, N.; Vinters, H. The pathology of encephalic arteriovenous malformations treated by prior embolotherapy. *Neuroradiology* **1988**, *30*, 1–10. [CrossRef]
97. Hulman, G.; Kirkham, J. Ivalon sponge presenting as an extrarectal mass. *J. Pathol.* **1990**, *160*, A171.
98. Fassi, A.; Naidoo, N. Irritation associated with tear-replacement ophthalmic drops-a pharmaceutical and subjective investigation. *S. Afr. Med. J.* **1989**, *75*, 233–235.

99. Singh, B.; Varshney, L.; Francis, S. Designing tragacanth gum based sterile hydrogel by radiation method for use in drug delivery and wound dressing applications. *Int. J. Biol. Macromol.* **2016**, *88*, 586–602. [CrossRef]
100. Nasef, S.M.; Khozemy, E.E.; Kamoun, E.A.; El-Gendi, H. Gamma radiation-induced crosslinked composite membranes based on polyvinyl alcohol/chitosan/$AgNO_3$/vitamin E for biomedical applications. *Int. J. Biol. Macromol.* **2019**, *137*, 878–885. [CrossRef] [PubMed]
101. Yuan, H.J.; Wei, Z.J.; Yu, X.Z. Study on Polyvinyl Alcohol Hydrogel Materials for Improving the Performance of Artificial Articular Cartilage. *Adv. Mater. Res.* **2013**, *703*, 29–32. [CrossRef]
102. DeMerlis, C.; Schoneker, D. Review of the oral toxicity of polyvinyl alcohol (PVA). *Food Chem. Toxicol.* **2003**, *41*, 319–326. [CrossRef]
103. Rodwell, D.; Kelly, C.; DeMerlis, C.; Schoneker, D.; Borzelleca, J. Effects of polyvinyl alcohol administered in the diet to rats on fertility, early embryonic development, growth and development. *Food Chem. Toxicol.* **2003**, *41*, 729–737. [CrossRef] [PubMed]
104. Kelly, C.; DeMerlis, C.; Schoneker, D.; Borzelleca, J. Subchronic toxicity study in rats and genotoxicity tests with polyvinyl alcohol. *Food Chem. Toxicol.* **2003**, *41*, 719–727. [CrossRef]
105. Sun, W.; Tian, J.; Chen, L.; He, S.; Wang, J. Improvement of biodegradability of PVA-containing wastewater by ionizing radiation pretreatment. *Environ. Sci. Pollut. Res.* **2012**, *19*, 3178–3184. [CrossRef] [PubMed]
106. Chou, W.-L.; Wang, C.-T.; Huang, K.-Y. Investigation of process parameters for the removal of polyvinyl alcohol from aqueous solution by iron electrocoagulation. *Desalination* **2010**, *251*, 12–19. [CrossRef]
107. Jabbari, E.; Khakpour, M. Morphology of and release behavior from porous polyurethane microspheres. *Biomaterials* **2000**, *21*, 2073–2079. [CrossRef]
108. Kääriä, K.; Hirvonen, A.; Norppa, H.; Piirilä, P.; Vainio, H.; Rosenberg, C. Exposure to 4,4′-methylenediphenyl diisocyanate (MDI) during moulding of rigid polyurethane foam: Determination of airborne MDI and urinary 4,4′-methylenedianiline (MDA). *Analyst* **2001**, *126*, 476–479. [CrossRef]
109. Wang, X.; Wang, Y.; Wei, K.; Zhao, N.; Zhang, S.; Chen, J. Drug distribution within poly (ε-caprolactone) microspheres and in vitro release. *J. Mater. Process. Technol.* **2009**, *209*, 348–354. [CrossRef]
110. Thompson, D.G.; Osborn, J.C.; Kober, E.M.; Schoonover, J.R. Effects of hydrolysis-induced molecular weight changes on the phase separation of a polyester polyurethane. *Polym. Degrad. Stab.* **2006**, *91*, 3360–3370. [CrossRef]
111. Kaur, M.; Gupta, K.M.; Poursaid, A.E.; Karra, P.; Mahalingam, A.; Aliyar, H.A.; Kiser, P.F. Engineering a degradable polyurethane intravaginal ring for sustained delivery of dapivirine. *Drug Deliv. Transl. Res.* **2011**, *1*, 223–237. [CrossRef] [PubMed]
112. Christenson, E.M.; Dadsetan, M.; Wiggins, M.; Anderson, J.M.; Hiltner, A. Poly (carbonate urethane) and poly (ether urethane) biodegradation: In vivo studies. *J. Biomed. Mater. Res. Part A Off. J. Soc. Biomater. Jpn. Soc. Biomater. Aust. Soc. Biomater. Korean Soc. Biomater.* **2004**, *69*, 407–416. [CrossRef]
113. Rychlý, J.; Lattuati-Derieux, A.; Lavédrine, B.; Matisová-Rychlá, L.; Malíková, M.; Csomorová, K.; Janigová, I. Assessing the progress of degradation in polyurethanes by chemiluminescence and thermal analysis. II. Flexible polyether-and polyester-type polyurethane foams. *Polym. Degrad. Stab.* **2011**, *96*, 462–469. [CrossRef]
114. Arifin, D.Y.; Lee, L.Y.; Wang, C.-H. Mathematical modeling and simulation of drug release from microspheres: Implications to drug delivery systems. *Adv. Drug Deliv. Rev.* **2006**, *58*, 1274–1325. [CrossRef] [PubMed]
115. Johnson, T.J.; Gupta, K.M.; Fabian, J.; Albright, T.H.; Kiser, P.F. Segmented polyurethane intravaginal rings for the sustained combined delivery of antiretroviral agents dapivirine and tenofovir. *Eur. J. Pharm. Sci.* **2010**, *39*, 203–212. [CrossRef] [PubMed]
116. Ritger, P.L.; Peppas, N.A. A simple equation for description of solute release II. Fickian and anomalous release from swellable devices. *J. Control. Release* **1987**, *5*, 37–42. [CrossRef]
117. Siepmann, J.; Siepmann, F. Mathematical modeling of drug delivery. *Int. J. Pharm.* **2008**, *364*, 328–343. [CrossRef] [PubMed]
118. Storey, R.F.; Hickey, T.P. Degradable polyurethane networks based on d, l-lactide, glycolide, ε-caprolactone, and trimethylene carbonate homopolyester and copolyester triols. *Polymer* **1994**, *35*, 830–838. [CrossRef]
119. Guelcher, S.; Srinivasan, A.; Hafeman, A.; Gallagher, K.; Doctor, J.; Khetan, S.; McBride, S.; Hollinger, J. Synthesis, in vitro biocompatibility and biodegradation, and mechanical properties of two-component polyurethane scaffolds: Effects of water and polyol composition. *Tissue Eng.* **2007**, *13*, 2321–2333. [CrossRef]
120. Labow, R.S.; Erfle, D.J.; Santerre, J.P. Elastase-induced hydrolysis of synthetic solid substrates: Poly (ester-urea-urethane) and poly (ether-urea-urethane). *Biomaterials* **1996**, *17*, 2381–2388. [CrossRef]
121. Ulery, B.D.; Nair, L.S.; Laurencin, C.T. Biomedical Applications of Biodegradable Polymers. *J. Polym. Sci. B Polym. Phys.* **2011**, *49*, 832–864. [CrossRef]
122. Feng, P.; Jia, J.; Liu, M.; Peng, S.; Zhao, Z.; Shuai, C. Degradation mechanisms and acceleration strategies of poly (lactic acid) scaffold for bone regeneration. *Mater. Des.* **2021**, *210*, 110066. [CrossRef]
123. Kuppermann, B.D.; Patel, S.S.; Boyer, D.S.; Augustin, A.J.; Freeman, W.R.; Kerr, K.J.; Guo, Q.; Schneider, S.; López, F.J. Phase 2 study of the safety and efficacy of brimonidine drug delivery system (brimo dds) generation 1 in patients with geographic atrophy secondary to age-related macular degeneration. *Retina* **2021**, *41*, 144–155. [CrossRef]
124. Pişkin, E. Biodegradable polymers as biomaterials. *J. Biomater. Sci. Polym. Ed.* **1995**, *6*, 775–795. [CrossRef]
125. Pisecky, L.; Luger, M.; Klasan, A.; Gotterbarm, T.; Klotz, M.C.; Hochgatterer, R. Bioabsorbable implants in forefoot surgery: A review of materials, possibilities and disadvantages. *EFORT Open Rev.* **2021**, *6*, 1132–1139. [CrossRef]
126. Wan, B.; Bao, Q.; Burgess, D. Long-acting PLGA microspheres: Advances in excipient and product analysis toward improved product understanding. *Adv. Drug Deliv. Rev.* **2023**, *198*, 114857. [CrossRef]

127. Blasi, P.; D'Souza, S.S.; Selmin, F.; DeLuca, P.P. Plasticizing effect of water on poly (lactide-co-glycolide). *J. Control. Release* **2005**, *108*, 1–9. [CrossRef] [PubMed]
128. Chandra, R.; Rustgi, R. Biodegradable polymers. *Prog. Polym. Sci.* **1998**, *23*, 1273–1335. [CrossRef]
129. Cayuela, J.; Bounor-Legaré, V.; Cassagnau, P.; Michel, A. Ring-opening polymerization of ε-caprolactone initiated with titanium n-propoxide or titanium phenoxide. *Macromolecules* **2006**, *39*, 1338–1346. [CrossRef]
130. Natta, F.J.v.; Hill, J.W.; Carothers, W.H. Studies of polymerization and ring formation. XXIII. 1 ε-Caprolactone and its polymers. *J. Am. Chem. Soc.* **1934**, *56*, 455–457. [CrossRef]
131. Woodruff, M.A.; Hutmacher, D.W. The return of a forgotten polymer—Polycaprolactone in the 21st century. *Prog. Polym. Sci.* **2010**, *35*, 1217–1256. [CrossRef]
132. Sinha, V.; Bansal, K.; Kaushik, R.; Kumria, R.; Trehan, A. Poly-ϵ-caprolactone microspheres and nanospheres: An overview. *Int. J. Pharm.* **2004**, *278*, 1–23. [CrossRef] [PubMed]
133. Wan, Y.; Lu, X.; Dalai, S.; Zhang, J. Thermophysical properties of polycaprolactone/chitosan blend membranes. *Thermochim. Acta* **2009**, *487*, 33–38. [CrossRef]
134. Perez, M.H.; Zinutti, C.; Lamprecht, A.; Ubrich, N.; Astier, A.; Hoffman, M.; Bodmeier, R.; Maincent, P. The preparation and evaluation of poly (ϵ-caprolactone) microparticles containing both a lipophilic and a hydrophilic drug. *J. Control. Release* **2000**, *65*, 429–438. [CrossRef]
135. Gorna, K.; Gogolewski, S. In vitro degradation of novel medical biodegradable aliphatic polyurethanes based on ϵ-caprolactone and Pluronics® with various hydrophilicities. *Polym. Degrad. Stab.* **2002**, *75*, 113–122. [CrossRef]
136. Ren, D.; Yi, H.; Wang, W.; Ma, X. The enzymatic degradation and swelling properties of chitosan matrices with different degrees of N-acetylation. *Carbohydr. Res.* **2005**, *340*, 2403–2410. [CrossRef]
137. Ravi Kumar, M.N.V. A review of chitin and chitosan applications. *React. Funct. Polym.* **2000**, *46*, 1–27. [CrossRef]
138. Mohammed, M.A.; Syeda, J.T.M.; Wasan, K.M.; Wasan, E.K. An Overview of Chitosan Nanoparticles and Its Application in Non-Parenteral Drug Delivery. *Pharmaceutics* **2017**, *9*, 53. [CrossRef]
139. Fonte, P.; Araújo, F.; Silva, C.; Pereira, C.; Reis, S.; Santos, H.A.; Sarmento, B. Polymer-based nanoparticles for oral insulin delivery: Revisited approaches. *Biotechnol. Adv.* **2015**, *33*, 1342–1354. [CrossRef]
140. Li, P.; Fujimoto, K.; Bourguingnon, L.; Yukl, S.; Deeks, S.; Wong, J.K. Exogenous and endogenous hyaluronic acid reduces HIV infection of CD4(+) T cells. *Immunol. Cell Biol.* **2014**, *92*, 770–780. [CrossRef] [PubMed]
141. Agrahari, V.; Meng, J.; Ezoulin, M.J.; Youm, I.; Dim, D.C.; Molteni, A.; Hung, W.T.; Christenson, L.K.; Youan, B.C. Stimuli-sensitive thiolated hyaluronic acid based nanofibers: Synthesis, preclinical safety and in vitro anti-HIV activity. *Nanomedicine* **2016**, *11*, 2935–2958. [CrossRef] [PubMed]
142. Jouyban, A.; Fakhree, M.A.A.; Shayanfar, A. Review of pharmaceutical applications of N-methyl-2-pyrrolidone. *J. Pharm. Pharm. Sci.* **2010**, *13*, 524–535. [CrossRef] [PubMed]
143. Indivior UK Limited. *Highlights of Prescribing Information—SublocadeTM*; Indivior UK Limited: Hull, UK, 2017.
144. Rivera-Briso, A.L.; Serrano-Aroca, Á. Poly(3-Hydroxybutyrate-co-3-Hydroxyvalerate): Enhancement Strategies for Advanced Applications. *Polymers* **2018**, *10*, 732. [CrossRef]
145. Avella, M.; Martuscelli, E.; Raimo, M. Review Properties of blends and composites based on poly (3-hydroxy) butyrate (PHB) and poly (3-hydroxybutyrate-hydroxyvalerate)(PHBV) copolymers. *J. Mater. Sci.* **2000**, *35*, 523–545. [CrossRef]
146. Mofokeng, J.; Luyt, A. Morphology and thermal degradation studies of melt-mixed poly (hydroxybutyrate-co-valerate)(PHBV)/poly (ε-caprolactone)(PCL) biodegradable polymer blend nanocomposites with TiO_2 as filler. *J. Mater. Sci.* **2015**, *50*, 3812–3824. [CrossRef]
147. Hedayati, A.; Yazdi, S.G.; Dehghankelishadi, P.; Javan, N.B.; Akbari, H.; Dorkoosh, F.A. Preparation, Optimization and Physicochemical Characterization of Aripiprazole Loaded Nano-porous in situ Forming Implant. *Pharm. Nanotechnol.* **2017**, *5*, 138–147. [CrossRef] [PubMed]
148. Acemoglu, M.; Nimmerfall, F.; Bantle, S.; Stoll, G.H. Poly(ethylene carbonate)s, part I: Syntheses and structural effects on biodegradation. *J. Control. Release* **1997**, *49*, 263–276. [CrossRef]
149. Sasanuma, Y.; Takahashi, Y. Structure–Property Relationships of Poly(ethylene carbonate) and Poly(propylene carbonate). *ACS Omega* **2017**, *2*, 4808–4819. [CrossRef] [PubMed]
150. Stoll, G.H.; Nimmerfall, F.; Acemoglu, M.; Bodmer, D.; Bantle, S.; Müller, I.; Mahl, A.; Kolopp, M.; Tullberg, K. Poly(ethylene carbonate)s, part II: Degradation mechanisms and parenteral delivery of bioactive agents. *J. Control. Release* **2001**, *76*, 209–225. [CrossRef] [PubMed]
151. Liu, Y.; Kemmer, A.; Keim, K.; Curdy, C.; Petersen, H.; Kissel, T. Poly(ethylene carbonate) as a surface-eroding biomaterial for in situ forming parenteral drug delivery systems: A feasibility study. *Eur. J. Pharm. Biopharm.* **2010**, *76*, 222–229. [CrossRef]
152. Panja, S.; Maji, S.; Maiti, T.K.; Chattopadhyay, S. A Smart Magnetically Active Nanovehicle for on-Demand Targeted Drug Delivery: Where van der Waals Force Balances the Magnetic Interaction. *ACS Appl. Mater. Interfaces* **2015**, *7*, 24229–24241. [CrossRef]
153. Panja, S.; Dey, G.; Bharti, R.; Mandal, P.; Mandal, M.; Chattopadhyay, S. Metal Ion Ornamented Ultrafast Light-Sensitive Nanogel for Potential In Vivo Cancer Therapy. *Chem. Mater.* **2016**, *28*, 8598–8610. [CrossRef]
154. Panja, S.; Dey, G.; Bharti, R.; Kumari, K.; Maiti, T.K.; Mandal, M.; Chattopadhyay, S. Tailor-Made Temperature-Sensitive Micelle for Targeted and On-Demand Release of Anticancer Drugs. *ACS Appl. Mater. Interfaces* **2016**, *8*, 12063–12074. [CrossRef]

155. Lan, T.; Guo, Q. Phenylboronic acid-decorated polymeric nanomaterials for advanced bio-application. *Nanotechnol. Rev.* **2019**, *8*, 548–561. [CrossRef]
156. Feng, R.; Zhu, L.; Teng, F.; Wang, M.; Chen, S.; Song, Z.; Li, H. Phenylboronic acid-modified polymaleic anhydride-F127 micelles for pH-activated targeting delivery of doxorubicin. *Colloids Surf. B Biointerfaces* **2022**, *216*, 112559. [CrossRef]
157. Mahalingam, A.; Jay, J.I.; Langheinrich, K.; Shukair, S.; McRaven, M.D.; Rohan, L.C.; Herold, B.C.; Hope, T.J.; Kiser, P.F. Inhibition of the transport of HIV in vitro using a pH-responsive synthetic mucin-like polymer system. *Biomaterials* **2011**, *32*, 8343–8355. [CrossRef]
158. Jay, J.I.; Shukair, S.; Langheinrich, K.; Hanson, M.C.; Cianci, G.C.; Johnson, T.J.; Clark, M.R.; Hope, T.J.; Kiser, P.F. Modulation of Viscoelasticity and HIV Transport as a Function of pH in a Reversibly Crosslinked Hydrogel. *Adv. Funct. Mater.* **2009**, *19*, 2969–2977. [CrossRef]
159. Folkman, J.; Long, D.M. The use of silicone rubber as a carrier for prolonged drug therapy. *J. Surg. Res.* **1964**, *4*, 139–142. [CrossRef] [PubMed]
160. Schlesinger, E.; Johengen, D.; Luecke, E.; Rothrock, G.; McGowan, I.; van der Straten, A.; Desai, T. A tunable, biodegradable, thin-film polymer device as a long-acting implant delivering tenofovir alafenamide fumarate for HIV pre-exposure prophylaxis. *Pharm. Res.* **2016**, *33*, 1649–1656. [CrossRef]
161. Li, L.; Johnson, L.M.; Krovi, S.A.; Demkovich, Z.R.; van der Straten, A. Performance and stability of tenofovir alafenamide formulations within subcutaneous biodegradable implants for HIV pre-exposure prophylaxis (PrEP). *Pharmaceutics* **2020**, *12*, 1057. [CrossRef]
162. Su, J.T.; Simpson, S.M.; Sung, S.; Tfaily, E.B.; Veazey, R.; Marzinke, M.; Qiu, J.; Watrous, D.; Widanapathirana, L.; Pearson, E. A subcutaneous implant of tenofovir alafenamide fumarate causes local inflammation and tissue necrosis in rabbits and macaques. *Antimicrob. Agents Chemother.* **2020**, *64*, 10-1128. [CrossRef]
163. Joiner, J.B.; Prasher, A.; Young, I.C.; Kim, J.; Shrivastava, R.; Maturavongsadit, P.; Benhabbour, S.R. Effects of Drug Physicochemical Properties on In-Situ Forming Implant Polymer Degradation and Drug Release Kinetics. *Pharmaceutics* **2022**, *14*, 1188. [CrossRef]
164. Romano, J.W.; Baum, M.M.; Demkovich, Z.R.; Diana, F.; Dobard, C.; Feldman, P.L.; Garcia-Lerma, J.G.; Grattoni, A.; Gunawardana, M.; Ho, D.K.; et al. Tenofovir Alafenamide for HIV Prevention: Review of the Proceedings from the Gates Foundation Long-Acting TAF Product Development Meeting. *AIDS Res. Hum. Retroviruses* **2021**, *37*, 409–420. [CrossRef] [PubMed]
165. Barrett, S.E.; Teller, R.S.; Forster, S.P.; Li, L.; Mackey, M.A.; Skomski, D.; Yang, Z.; Fillgrove, K.L.; Doto, G.J.; Wood, S.L.; et al. Extended-Duration MK-8591-Eluting Implant as a Candidate for HIV Treatment and Prevention. *Antimicrob. Agents Chemother.* **2018**, *62*, 10-1128. [CrossRef] [PubMed]
166. Malcolm, R.K.; Fetherston, S.M.; McCoy, C.F.; Boyd, P.; Major, I. Vaginal rings for delivery of HIV microbicides. *Int. J. Womens Health* **2012**, *4*, 595–605. [CrossRef]
167. Ullah Nayan, M.; Sillman, B.; Hasan, M.; Deodhar, S.; Das, S.; Sultana, A.; Thai Hoang Le, N.; Soriano, V.; Edagwa, B.; Gendelman, H.E. Advances in long-acting slow effective release antiretroviral therapies for treatment and prevention of HIV infection. *Adv. Drug Deliv. Rev.* **2023**, *200*, 115009. [CrossRef]
168. Karunakaran, D.; Simpson, S.M.; Su, J.T.; Bryndza-Tfaily, E.; Hope, T.J.; Veazey, R.; Dobek, G.; Qiu, J.; Watrous, D.; Sung, S.; et al. Design and Testing of a Cabotegravir Implant for HIV Prevention. *J. Control. Release* **2021**, *330*, 658–668. [CrossRef]
169. Nevoralová, M.; Koutný, M.; Ujčić, A.; Starý, Z.; Šerá, J.; Vlková, H.; Šlouf, M.; Fortelný, I.; Kruliš, Z. Structure Characterization and Biodegradation Rate of Poly(ε-caprolactone)/Starch Blends. *Front. Mater.* **2020**, *7*, 141. [CrossRef]
170. Markowitz, M.; Sarafianos, S.G. EFdA (4′-ethynyl-2-fluoro-2′-deoxyadenosine, MK-8591): A novel HIV-1 reverse transcriptase translocation inhibitor. *Curr. Opin. HIV AIDS* **2018**, *13*, 294. [CrossRef]
171. Wang, X.; Burgess, D.J. Drug release from in situ forming implants and advances in release testing. *Adv. Drug Deliv. Rev.* **2021**, *178*, 113912. [CrossRef]
172. Phaechamud, T.; Senarat, S.; Puyathorn, N.; Praphanwittaya, P. Solvent exchange and drug release characteristics of doxycycline hyclate-loaded bleached shellac in situ-forming gel and-microparticle. *Int. J. Biol. Macromol.* **2019**, *135*, 1261–1272. [CrossRef]
173. Amini-Fazl, M.S. Biodegradation study of PLGA as an injectable in situ depot-forming implant for controlled release of paclitaxel. *Polym. Bull.* **2021**, *79*, 2763–2776. [CrossRef]
174. Kempe, S.; Mäder, K. In situ forming implants—An attractive formulation principle for parenteral depot formulations. *J. Control. Release* **2012**, *161*, 668–679. [CrossRef]
175. Kanwar, N.; Sinha, V.R. In situ forming depot as sustained-release drug delivery systems. *Crit. Rev. Ther. Drug Carr. Syst.* **2019**, *36*, 93–136. [CrossRef]
176. Yu, L.; Ding, J. Injectable hydrogels as unique biomedical materials. *Chem. Soc. Rev.* **2008**, *37*, 1473–1481. [CrossRef]
177. Ibrahim, T.M.; Ayoub, M.M.; El-Bassossy, H.M.; El-Nahas, H.M.; Gomaa, E. Investigation of Alogliptin-Loaded In Situ Gel Implants by 23 Factorial Design with Glycemic Assessment in Rats. *Pharmaceutics* **2022**, *14*, 1867. [CrossRef] [PubMed]
178. Wex, J.; Sidhu, M.; Odeyemi, I.; Abou-Setta, A.M.; Retsa, P.; Tombal, B. Leuprolide acetate 1-, 3-and 6-monthly depot formulations in androgen deprivation therapy for prostate cancer in nine European countries: Evidence review and economic evaluation. *Clin. Outcomes Res.* **2013**, *5*, 257–269. [CrossRef]
179. Kim, M.; Johnson, C.E.; Schmalstig, A.A.; Annis, A.; Wessel, S.E.; Van Horn, B.; Schauer, A.; Exner, A.A.; Stout, J.E.; Wahl, A. A long-acting formulation of rifabutin is effective for prevention and treatment of Mycobacterium tuberculosis. *Nat. Commun.* **2022**, *13*, 4455. [CrossRef]

180. Benhabbour, S.R.; Kovarova, M.; Jones, C.; Copeland, D.J.; Shrivastava, R.; Swanson, M.D.; Sykes, C.; Ho, P.T.; Cottrell, M.L.; Sridharan, A. Ultra-long-acting tunable biodegradable and removable controlled release implants for drug delivery. *Nat. Commun.* **2019**, *10*, 4324. [CrossRef]
181. Cox, M.C.; Scripture, C.D.; Figg, W.D. Leuprolide acetate given by a subcutaneous extended-release injection: Less of a pain? *Expert Rev. Anticancer Ther.* **2005**, *5*, 605–611. [CrossRef] [PubMed]
182. Agarwal, P.; Rupenthal, I.D. Injectable implants for the sustained release of protein and peptide drugs. *Drug Discov Today* **2013**, *18*, 337–349. [CrossRef]
183. Packhaeuser, C.; Schnieders, J.; Oster, C.; Kissel, T. In situ forming parenteral drug delivery systems: An overview. *Eur. J. Pharm. Biopharm.* **2004**, *58*, 445–455. [CrossRef] [PubMed]
184. Griffin, J.B.; Ridgeway, K.; Montgomery, E.; Torjesen, K.; Clark, R.; Peterson, J.; Baggaley, R.; van der Straten, A. Vaginal ring acceptability and related preferences among women in low-and middle-income countries: A systematic review and narrative synthesis. *PLoS ONE* **2019**, *14*, e0224898. [CrossRef] [PubMed]
185. MACHT, D.I. On the absorption of drugs and poisons through the vagina. *J. Pharmacol. Exp. Ther.* **1918**, *10*, 509–522.
186. Malcolm, R.K.; Boyd, P.J.; McCoy, C.F.; Murphy, D.J. Microbicide vaginal rings: Technological challenges and clinical development. *Adv. Drug Deliv. Rev.* **2016**, *103*, 33–56. [CrossRef] [PubMed]
187. McConville, C.; Andrews, G.P.; Laverty, T.P.; Woolfson, A.D.; Malcolm, R.K. Rheological evaluation of the isothermal cure characteristics of medical grade silicone elastomers. *J. Appl. Polym. Sci.* **2010**, *116*, 2320–2327. [CrossRef]
188. Monteiro, I.; Guazzelli, C.F.; Bahamondes, L. Advances in contraceptive vaginal rings: What does the future hold? *Expert Opin. Pharmacother.* **2018**, *19*, 1685–1691. [CrossRef] [PubMed]
189. Speroff, L.; Group, U.S.V.I. Efficacy and tolerability of a novel estradiol vaginal ring for relief of menopausal symptoms. *Obstet. Gynecol.* **2003**, *102*, 823–834.
190. Woolfson, A.; Malcolm, R.; Gallagher, R. Design of a silicone reservoir intravaginal ring for the delivery of oxybutynin. *J. Control. Release* **2003**, *91*, 465–476. [CrossRef]
191. Leonhard, M.; Moser, D.; Reumueller, A.; Mancusi, G.; Bigenzahn, W.; Schneider-Stickler, B. Comparison of biofilm formation on new phonax and provox 2 voice prostheses—A pilot study. *Head Neck* **2010**, *32*, 886–895. [CrossRef]
192. MacCallum, N.; Howell, C.; Kim, P.; Sun, D.; Friedlander, R.; Ranisau, J.; Ahanotu, O.; Lin, J.J.; Vena, A.; Hatton, B. Liquid-infused silicone as a biofouling-free medical material. *ACS Biomater. Sci. Eng.* **2015**, *1*, 43–51. [CrossRef]
193. Fundeanu, I.; Klee, D.; Schouten, A.J.; Busscher, H.J.; Van der Mei, H.C. Solvent-free functionalization of silicone rubber and efficacy of PAAm brushes grafted from an amino-PPX layer against bacterial adhesion. *Acta Biomater.* **2010**, *6*, 4271–4276. [CrossRef]
194. Rodrigues, L.; Van der Mei, H.; Teixeira, J.A.; Oliveira, R. Biosurfactant from Lactococcus lactis 53 inhibits microbial adhesion on silicone rubber. *Appl. Microbiol. Biotechnol.* **2004**, *66*, 306–311. [CrossRef]
195. Everaert, E.; Belt-Gritter, B.V.D.; Van der Mei, H.; Busscher, H.; Verkerke, G.; Dijk, F.; Mahieu, H.; Reitsma, A. In vitro and in vivo microbial adhesion and growth on argon plasma-treated silicone rubber voice prostheses. *J. Mater. Sci. Mater. Med.* **1998**, *9*, 147–157. [CrossRef]
196. Everaert, E.P.; Mahieu, H.F.; van de Belt-Gritter, B.; Peeters, A.J.C.; Verkerke, G.J.; van der Mei, H.C.; Busscher, H.J. Biofilm formation in vivo on perfluoro-alkylsiloxane–modified voice prostheses. *Arch. Otolaryngol. Head Neck Surg.* **1999**, *125*, 1329–1332. [CrossRef]
197. Traore, Y.L.; Chen, Y.; Bernier, A.-M.; Ho, E.A. Impact of hydroxychloroquine-loaded polyurethane intravaginal rings on lactobacilli. *Antimicrob. Agents Chemother.* **2015**, *59*, 7680–7686. [CrossRef]
198. Roddy, R.E.; Zekeng, L.; Ryan, K.A.; Tamoufe, U.; Weir, S.S.; Wong, E.L. A controlled trial of nonoxynol 9 film to reduce male-to-female transmission of sexually transmitted diseases. *N. Engl. J. Med.* **1998**, *339*, 504–510. [CrossRef]
199. Malcolm, R.K.; Woolfson, A.D.; Toner, C.F.; Morrow, R.J.; McCullagh, S.D. Long-term, controlled release of the HIV microbicide TMC120 from silicone elastomer vaginal rings. *J. Antimicrob. Chemother.* **2005**, *56*, 954–956. [CrossRef]
200. Nel, A.; Smythe, S.; Young, K.; Malcolm, K.; McCoy, C.; Rosenberg, Z.; Romano, J. Safety and Pharmacokinetics of Dapivirine Delivery from Matrix and Reservoir Intravaginal Rings to HIV-Negative Women. *JAIDS J. Acquir. Immune Defic. Syndr.* **2009**, *51*, 416–423. [CrossRef]
201. Nel, A.; Bekker, L.-G.; Bukusi, E.; Hellström, E.; Kotze, P.; Louw, C.; Martinson, F.; Masenga, G.; Montgomery, E.; Ndaba, N. Safety, acceptability and adherence of dapivirine vaginal ring in a microbicide clinical trial conducted in multiple countries in Sub-Saharan Africa. *PLoS ONE* **2016**, *11*, e0147743. [CrossRef] [PubMed]
202. Baeten, J.M.; Palanee-Phillips, T.; Brown, E.R.; Schwartz, K.; Soto-Torres, L.E.; Govender, V.; Mgodi, N.M.; Matovu Kiweewa, F.; Nair, G.; Mhlanga, F. Use of a vaginal ring containing dapivirine for HIV-1 prevention in women. *N. Engl. J. Med.* **2016**, *375*, 2121–2132. [CrossRef]
203. Johnson, T.J.; Clark, M.R.; Albright, T.H.; Nebeker, J.S.; Tuitupou, A.L.; Clark, J.T.; Fabian, J.; McCabe, R.T.; Chandra, N.; Doncel, G.F.; et al. A 90-day tenofovir reservoir intravaginal ring for mucosal HIV prophylaxis. *Antimicrob. Agents Chemother.* **2012**, *56*, 6272–6283. [CrossRef] [PubMed]
204. Mesquita, P.M.; Rastogi, R.; Segarra, T.J.; Teller, R.S.; Torres, N.M.; Huber, A.M.; Kiser, P.F.; Herold, B.C. Intravaginal ring delivery of tenofovir disoproxil fumarate for prevention of HIV and herpes simplex virus infection. *J. Antimicrob. Chemother.* **2012**, *67*, 1730–1738. [CrossRef]

205. Smith, J.M.; Rastogi, R.; Teller, R.S.; Srinivasan, P.; Mesquita, P.M.; Nagaraja, U.; McNicholl, J.M.; Hendry, R.M.; Dinh, C.T.; Martin, A.; et al. Intravaginal ring eluting tenofovir disoproxil fumarate completely protects macaques from multiple vaginal simian-HIV challenges. *Proc. Natl. Acad. Sci. USA* **2013**, *110*, 16145–16150. [CrossRef]
206. Su, J.T.; Teller, R.S.; Srinivasan, P.; Zhang, J.; Martin, A.; Sung, S.; Smith, J.M.; Kiser, P.F. A Dose Ranging Pharmacokinetic Evaluation of IQP-0528 Released from Intravaginal Rings in Non-Human Primates. *Pharm. Res.* **2017**, *34*, 2163–2171. [CrossRef]
207. Quinn, H.L.; Kearney, M.-C.; Courtenay, A.J.; McCrudden, M.T.; Donnelly, R.F. The role of microneedles for drug and vaccine delivery. *Expert Opin. Drug Deliv.* **2014**, *11*, 1769–1780. [CrossRef]
208. Permana, A.D.; McCrudden, M.T.; Donnelly, R.F. Enhanced intradermal delivery of nanosuspensions of antifilariasis drugs using dissolving microneedles: A proof of concept study. *Pharmaceutics* **2019**, *11*, 346. [CrossRef]
209. Harvey, A.J.; Kaestner, S.A.; Sutter, D.E.; Harvey, N.G.; Mikszta, J.A.; Pettis, R.J. Microneedle-based intradermal delivery enables rapid lymphatic uptake and distribution of protein drugs. *Pharm. Res.* **2011**, *28*, 107–116. [CrossRef]
210. Moga, K.A.; Bickford, L.R.; Geil, R.D.; Dunn, S.S.; Pandya, A.A.; Wang, Y.; Fain, J.H.; Archuleta, C.F.; O'Neill, A.T.; DeSimone, J.M. Rapidly–dissolvable microneedle patches via a highly scalable and reproducible soft lithography approach. *Adv. Mater.* **2013**, *25*, 5060–5066. [CrossRef]
211. Römgens, A.; Bader, D.; Bouwstra, J.; Baaijens, F.; Oomens, C. Monitoring the penetration process of single microneedles with varying tip diameters. *J. Mech. Behav. Biomed. Mater.* **2014**, *40*, 397–405. [CrossRef]
212. Gill, H.; Prausnitz, M. Effect of microneedle design on pain in human subjects. *Clin. J. Pain* **2008**, *24*, 585–594. [CrossRef]
213. Moffatt, K.; Quinn, C.; McCague, P.J.; Donnelly, R.F. Exploration into the opinions of patients with HIV, healthcare professionals and the lay public of the use of microneedles in clinical practice: Highlighting the translational potential for their role in HIV infection. *Drug Deliv. Transl. Res.* **2021**, *11*, 1199–1217. [CrossRef]
214. Chen, M.; Quan, G.; Sun, Y.; Yang, D.; Pan, X.; Wu, C. Nanoparticles-encapsulated polymeric microneedles for transdermal drug delivery. *J. Control. Release* **2020**, *325*, 163–175. [CrossRef]
215. Xie, L.; Zeng, H.; Sun, J.; Qian, W. Engineering microneedles for therapy and diagnosis: A survey. *Micromachines* **2020**, *11*, 271. [CrossRef]
216. Lee, J.W.; Park, J.-H.; Prausnitz, M.R. Dissolving microneedles for transdermal drug delivery. *Biomaterials* **2008**, *29*, 2113–2124. [CrossRef]
217. Ito, Y.; Hirono, M.; Fukushima, K.; Sugioka, N.; Takada, K. Two-layered dissolving microneedles formulated with intermediate-acting insulin. *Int. J. Pharm.* **2012**, *436*, 387–393. [CrossRef]
218. Chu, L.Y.; Choi, S.-O.; Prausnitz, M.R. Fabrication of dissolving polymer microneedles for controlled drug encapsulation and delivery: Bubble and pedestal microneedle designs. *J. Pharm. Sci.* **2010**, *99*, 4228–4238. [CrossRef]
219. Chen, W.; Wang, C.; Yan, L.; Huang, L.; Zhu, X.; Chen, B.; Sant, H.J.; Niu, X.; Zhu, G.; Yu, K. Improved polyvinylpyrrolidone microneedle arrays with non-stoichiometric cyclodextrin. *J. Mater. Chem. B* **2014**, *2*, 1699–1705. [CrossRef]
220. Vora, L.K.; Vavia, P.R.; Larrañeta, E.; Bell, S.E.; Donnelly, R.F. Novel nanosuspension-based dissolving microneedle arrays for transdermal delivery of a hydrophobic drug. *J. Interdiscip. Nanomed.* **2018**, *3*, 89–101. [CrossRef]
221. Park, J.-H.; Allen, M.G.; Prausnitz, M.R. Polymer microneedles for controlled-release drug delivery. *Pharm. Res.* **2006**, *23*, 1008–1019. [CrossRef]
222. Abdelghany, S.; Tekko, I.A.; Vora, L.; Larrañeta, E.; Permana, A.D.; Donnelly, R.F. Nanosuspension-based dissolving microneedle arrays for intradermal delivery of curcumin. *Pharmaceutics* **2019**, *11*, 308. [CrossRef]
223. Li, W.; Terry, R.N.; Tang, J.; Feng, M.R.; Schwendeman, S.P.; Prausnitz, M.R. Rapidly separable microneedle patch for the sustained release of a contraceptive. *Nat. Biomed. Eng.* **2019**, *3*, 220–229. [CrossRef]
224. Zhao, X.; Zhang, S.; Yang, G.; Zhou, Z.; Gao, Y. Exploring Trehalose on the release of levonorgestrel from implantable PLGA microneedles. *Polymers* **2020**, *12*, 59. [CrossRef] [PubMed]
225. He, M.; Yang, G.; Zhao, X.; Zhang, S.; Gao, Y. Intradermal implantable PLGA microneedles for etonogestrel sustained release. *J. Pharm. Sci.* **2020**, *109*, 1958–1966. [CrossRef]
226. Than, A.; Liang, K.; Xu, S.; Sun, L.; Duan, H.; Xi, F.; Xu, C.; Chen, P. Transdermal delivery of anti-obesity compounds to subcutaneous adipose tissue with polymeric microneedle patches. *Small Methods* **2017**, *1*, 1700269. [CrossRef]
227. Chen, M.-C.; Wang, K.-W.; Chen, D.-H.; Ling, M.-H.; Liu, C.-Y. Remotely triggered release of small molecules from LaB6@ SiO_2-loaded polycaprolactone microneedles. *Acta Biomater.* **2015**, *13*, 344–353. [CrossRef]
228. Mc Crudden, M.T.C.; Larrañeta, E.; Clark, A.; Jarrahian, C.; Rein-Weston, A.; Creelman, B.; Moyo, Y.; Lachau-Durand, S.; Niemeijer, N.; Williams, P.; et al. Design, Formulation, and Evaluation of Novel Dissolving Microarray Patches Containing Rilpivirine for Intravaginal Delivery. *Adv. Healthc. Mater.* **2019**, *8*, 1801510. [CrossRef]
229. Abbate, M.T.A.; Ramöller, I.K.; Sabri, A.H.; Paredes, A.J.; Hutton, A.J.; McKenna, P.E.; Peng, K.; Hollett, J.A.; McCarthy, H.O.; Donnelly, R.F. Formulation of antiretroviral nanocrystals and development into a microneedle delivery system for potential treatment of HIV-associated neurocognitive disorder (HAND). *Int. J. Pharm.* **2023**, *640*, 123005. [CrossRef]
230. Zhang, C.; Vora, L.K.; Tekko, I.A.; Volpe-Zanutto, F.; Peng, K.; Paredes, A.J.; McCarthy, H.O.; Donnelly, R.F. Development of dissolving microneedles for intradermal delivery of the long-acting antiretroviral drug bictegravir. *Int. J. Pharm.* **2023**, *642*, 123108. [CrossRef]

231. Donnelly, R.F.; McCrudden, M.T.; Zaid Alkilani, A.; Larrañeta, E.; McAlister, E.; Courtenay, A.J.; Kearney, M.-C.; Singh, T.R.R.; McCarthy, H.O.; Kett, V.L. Hydrogel-forming microneedles prepared from "super swelling" polymers combined with lyophilised wafers for transdermal drug delivery. *PLoS ONE* **2014**, *9*, e111547. [CrossRef]
232. Wang, M.; Hu, L.; Xu, C. Recent advances in the design of polymeric microneedles for transdermal drug delivery and biosensing. *Lab Chip* **2017**, *17*, 1373–1387. [CrossRef]
233. Chai, Q.; Jiao, Y.; Yu, X. Hydrogels for biomedical applications: Their characteristics and the mechanisms behind them. *Gels* **2017**, *3*, 6. [CrossRef]
234. Lin, C.-C.; Metters, A.T. Hydrogels in controlled release formulations: Network design and mathematical modeling. *Adv. Drug Deliv. Rev.* **2006**, *58*, 1379–1408. [CrossRef]
235. Donnelly, R.F.; Singh, T.R.R.; Garland, M.J.; Migalska, K.; Majithiya, R.; McCrudden, C.M.; Kole, P.L.; Mahmood, T.M.T.; McCarthy, H.O.; Woolfson, A.D. Hydrogel-forming microneedle arrays for enhanced transdermal drug delivery. *Adv. Funct. Mater.* **2012**, *22*, 4879–4890. [CrossRef]
236. Blasi, P. Poly (lactic acid)/poly (lactic-co-glycolic acid)-based microparticles: An overview. *J. Pharm. Investig.* **2019**, *49*, 337–346. [CrossRef]
237. Abdelkader, H.; Fathalla, Z.; Seyfoddin, A.; Farahani, M.; Thrimawithana, T.; Allahham, A.; Alani, A.W.G.; Al-Kinani, A.A.; Alany, R.G. Polymeric long-acting drug delivery systems (LADDS) for treatment of chronic diseases: Inserts, patches, wafers, and implants. *Adv. Drug Deliv. Rev.* **2021**, *177*, 113957. [CrossRef] [PubMed]
238. Ghosh, S.K. Functional coatings and microencapsulation: A general perspective. In *Functional Coatings: By Polymer Microencapsulation*; Wiley-VCH: Weinheim, Germany, 2006; pp. 1–28. [CrossRef]
239. Yeo, Y.; Baek, N.; Park, K. Microencapsulation methods for delivery of protein drugs. *Biotechnol. Bioprocess Eng.* **2001**, *6*, 213–230. [CrossRef]
240. Tomaro-Duchesneau, C.; Saha, S.; Malhotra, M.; Kahouli, I.; Prakash, S. Microencapsulation for the therapeutic delivery of drugs, live mammalian and bacterial cells, and other biopharmaceutics: Current status and future directions. *J. Pharm.* **2013**, *2013*, 103527. [CrossRef] [PubMed]
241. Jyothi, N.V.N.; Prasanna, P.M.; Sakarkar, S.N.; Prabha, K.S.; Ramaiah, P.S.; Srawan, G. Microencapsulation techniques, factors influencing encapsulation efficiency. *J. Microencapsul.* **2010**, *27*, 187–197. [CrossRef]
242. Bansode, S.; Banarjee, S.; Gaikwad, D.; Jadhav, S.; Thorat, R. Microencapsulation: A review. *Int. J. Pharm. Sci. Rev. Res.* **2010**, *1*, 38–43.
243. Dubey, R. Microencapsulation technology and applications. *Def. Sci. J.* **2009**, *59*, 82.
244. Sharifi, F.; Otte, A.; Yoon, G.; Park, K. Continuous in-line homogenization process for scale-up production of naltrexone-loaded PLGA microparticles. *J. Control. Release* **2020**, *325*, 347–358. [CrossRef]
245. Chereddy, K.K.; Vandermeulen, G.; Préat, V. PLGA based drug delivery systems: Promising carriers for wound healing activity. *Wound Repair Regen.* **2016**, *24*, 223–236. [CrossRef] [PubMed]
246. Ramazani, F.; Chen, W.; van Nostrum, C.F.; Storm, G.; Kiessling, F.; Lammers, T.; Hennink, W.E.; Kok, R.J. Strategies for encapsulation of small hydrophilic and amphiphilic drugs in PLGA microspheres: State-of-the-art and challenges. *Int. J. Pharm.* **2016**, *499*, 358–367. [CrossRef] [PubMed]
247. Lee, J.; Kwon, H.J.; Ji, H.; Cho, S.H.; Cho, E.-H.; Han, H.D.; Shin, B.C. Marbofloxacin-encapsulated microparticles provide sustained drug release for treatment of veterinary diseases. *Mater. Sci. Eng. C* **2016**, *60*, 511–517. [CrossRef] [PubMed]
248. Shahjin, F.; Patel, M.; Machhi, J.; Cohen, J.D.; Nayan, M.U.; Yeapuri, P.; Zhang, C.; Waight, E.; Hasan, M.; Abdelmoaty, M.M.; et al. Multipolymer microsphere delivery of SARS-CoV-2 antigens. *Acta Biomater.* **2023**, *158*, 493–509. [CrossRef] [PubMed]
249. van de Weert, M.; Hennink, W.E.; Jiskoot, W. Protein instability in poly (lactic-co-glycolic acid) microparticles. *Pharm. Res.* **2000**, *17*, 1159–1167. [CrossRef] [PubMed]
250. Cabezas, L.I.; Gracia, I.; de Lucas, A.; Rodríguez, J.F. Novel model for the description of the controlled release of 5-fluorouracil from PLGA and PLA foamed scaffolds impregnated in supercritical CO_2. *Ind. Eng. Chem. Res.* **2014**, *53*, 15374–15382. [CrossRef]
251. Khuroo, T.; Dharani, S.; Mohamed, E.M.; Immadi, S.; Wu, Z.; Khan, M.A.; Lu, D.; Nehete, P.; Rahman, Z. Ultra-long acting prodrug of dolutegravir and delivery system—Physicochemical, pharmacokinetic and formulation characterizations. *Int. J. Pharm.* **2021**, *607*, 120889. [CrossRef]
252. Collett, B.M. Scanning electron microscopy: A review and report of research in wood science. *Wood Fiber Sci.* **1970**, *2*, 113–133.
253. Doty, A.C.; Weinstein, D.G.; Hirota, K.; Olsen, K.F.; Ackermann, R.; Wang, Y.; Choi, S.; Schwendeman, S.P. Mechanisms of in vivo release of triamcinolone acetonide from PLGA microspheres. *J. Control. Release* **2017**, *256*, 19–25. [CrossRef]
254. Shah, R.B.; Schwendeman, S.P. A biomimetic approach to active self-microencapsulation of proteins in PLGA. *J. Control. Release* **2014**, *196*, 60–70. [CrossRef]
255. Srikar, G.; Rani, A.P. Study on influence of polymer and surfactant on in vitro performance of biodegradable aqueous-core nanocapsules of tenofovirdisoproxil fumarate by response surface methodology. *Braz. J. Pharm. Sci.* **2019**, *55*, e18736. [CrossRef]
256. Kazazi-Hyseni, F.; Landin, M.; Lathuile, A.; Veldhuis, G.J.; Rahimian, S.; Hennink, W.E.; Kok, R.J.; van Nostrum, C.F. Computer modeling assisted design of monodisperse PLGA microspheres with controlled porosity affords zero order release of an encapsulated macromolecule for 3 months. *Pharm. Res.* **2014**, *31*, 2844–2856. [CrossRef]

257. Samadi, N.; Abbadessa, A.; Di Stefano, A.; Van Nostrum, C.; Vermonden, T.; Rahimian, S.; Teunissen, E.; Van Steenbergen, M.; Amidi, M.; Hennink, W. The effect of lauryl capping group on protein release and degradation of poly (d, l-lactic-co-glycolic acid) particles. *J. Control. Release* **2013**, *172*, 436–443. [CrossRef]
258. Mandal, S.; Kang, G.; Prathipati, P.K.; Zhou, Y.; Fan, W.; Li, Q.; Destache, C.J. Nanoencapsulation introduces long-acting phenomenon to tenofovir alafenamide and emtricitabine drug combination: A comparative pre-exposure prophylaxis efficacy study against HIV-1 vaginal transmission. *J. Control. Release* **2019**, *294*, 216–225. [CrossRef] [PubMed]
259. Mandal, S.; Prathipati, P.K.; Kang, G.; Zhao, Y.; Yuan, Z.; Fan, W.; Li, Q.; Destache, C.J. Tenofovir alafenamide and elvitegravir loaded nanoparticles for long-acting prevention of HIV-1 vaginal transmission. *AIDS* **2017**, *31*, 469. [CrossRef] [PubMed]
260. Mandal, S.; Prathipati, P.K.; Sunagawa, S.W.; Destache, C.J. A concept evaluation study of a new combination bictegravir plus tenofovir alafenamide nanoformulation with prolonged sustained-drug-release potency for HIV-1 preexposure prophylaxis. *Antimicrob. Agents Chemother.* **2021**, *65*, 940. [CrossRef] [PubMed]
261. Mandal, S.; Kang, G.; Prathipati, P.K.; Fan, W.; Li, Q.; Destache, C.J. Long-acting parenteral combination antiretroviral loaded nano-drug delivery system to treat chronic HIV-1 infection: A humanized mouse model study. *Antivir. Res.* **2018**, *156*, 85–91. [CrossRef]
262. Prathipati, P.K.; Mandal, S.; Pon, G.; Vivekanandan, R.; Destache, C.J. Pharmacokinetic and Tissue Distribution Profile of Long Acting Tenofovir Alafenamide and Elvitegravir Loaded Nanoparticles in Humanized Mice Model. *Pharm. Res.* **2017**, *34*, 2749–2755. [CrossRef]
263. DeYoung, M.B.; MacConell, L.; Sarin, V.; Trautmann, M.; Herbert, P. Encapsulation of exenatide in poly-(D, L-lactide-co-glycolide) microspheres produced an investigational long-acting once-weekly formulation for type 2 diabetes. *Diabetes Technol. Ther.* **2011**, *13*, 1145–1154. [CrossRef]
264. Jindal, A.B.; Bhide, A.R.; Salave, S.; Rana, D.; Benival, D. Long-acting parenteral drug delivery systems for the treatment of chronic diseases. *Adv. Drug Deliv. Rev.* **2023**, *198*, 114862. [CrossRef]
265. Trezza, C.; Ford, S.L.; Spreen, W.; Pan, R.; Piscitelli, S. Formulation and pharmacology of long-acting cabotegravir. *Curr. Opin. HIV AIDS* **2015**, *10*, 239. [CrossRef] [PubMed]
266. Overton, E.T.; Richmond, G.; Rizzardini, G.; Jaeger, H.; Orrell, C.; Nagimova, F.; Bredeek, F.; García Deltoro, M.; Swindells, S.; Andrade-Villanueva, J.F.; et al. Long-acting cabotegravir and rilpivirine dosed every 2 months in adults with HIV-1 infection (ATLAS-2M), 48-week results: A randomised, multicentre, open-label, phase 3b, non-inferiority study. *Lancet* **2021**, *396*, 1994–2005. [CrossRef] [PubMed]
267. Delany-Moretlwe, S.; Hughes, J.P.; Bock, P.; Ouma, S.G.; Hunidzarira, P.; Kalonji, D.; Kayange, N.; Makhema, J.; Mandima, P.; Mathew, C. Cabotegravir for the prevention of HIV-1 in women: Results from HPTN 084, a phase 3, randomised clinical trial. *Lancet* **2022**, *399*, 1779–1789. [CrossRef] [PubMed]
268. Landovitz, R.J.; Donnell, D.; Clement, M.E.; Hanscom, B.; Cottle, L.; Coelho, L.; Cabello, R.; Chariyalertsak, S.; Dunne, E.F.; Frank, I. Cabotegravir for HIV prevention in cisgender men and transgender women. *N. Engl. J. Med.* **2021**, *385*, 595–608. [CrossRef]
269. Deodhar, S.; Sillman, B.; Bade, A.N.; Avedissian, S.N.; Podany, A.T.; McMillan, J.M.; Gautam, N.; Hanson, B.; Dyavar Shetty, B.L.; Szlachetka, A. Transformation of dolutegravir into an ultra-long-acting parenteral prodrug formulation. *Nat. Commun.* **2022**, *13*, 3226. [CrossRef]
270. Sillman, B.; Bade, A.N.; Dash, P.K.; Bhargavan, B.; Kocher, T.; Mathews, S.; Su, H.; Kanmogne, G.D.; Poluektova, L.Y.; Gorantla, S. Creation of a long-acting nanoformulated dolutegravir. *Nat. Commun.* **2018**, *9*, 443. [CrossRef]
271. Zhou, T.; Su, H.; Dash, P.; Lin, Z.; Shetty, B.L.D.; Kocher, T.; Szlachetka, A.; Lamberty, B.; Fox, H.S.; Poluektova, L. Creation of a nanoformulated cabotegravir prodrug with improved antiretroviral profiles. *Biomaterials* **2018**, *151*, 53–65. [CrossRef]
272. Cobb, D.A.; Smith, N.; Deodhar, S.; Bade, A.N.; Gautam, N.; Shetty, B.L.D.; McMillan, J.; Alnouti, Y.; Cohen, S.M.; Gendelman, H.E. Transformation of tenofovir into stable ProTide nanocrystals with long-acting pharmacokinetic profiles. *Nat. Commun.* **2021**, *12*, 5458. [CrossRef]
273. Guo, D.; Zhou, T.; Araínga, M.; Palandri, D.; Gautam, N.; Bronich, T.; Alnouti, Y.; McMillan, J.; Edagwa, B.; Gendelman, H.E. Creation of a long-acting nanoformulated 2′, 3′-dideoxy-3′-thiacytidine. *J. Acquir. Immune Defic. Syndr. (1999)* **2017**, *74*, e75. [CrossRef]
274. Ibrahim, I.M.; Bade, A.N.; Lin, Z.; Soni, D.; Wojtkiewicz, M.; Dyavar Shetty, B.L.; Gautam, N.; McMillan, J.M.; Alnouti, Y.; Edagwa, B.J. Synthesis and characterization of a long-acting emtricitabine prodrug nanoformulation. *Int. J. Nanomed.* **2019**, *14*, 6231–6247. [CrossRef]
275. Ogunnaike, M.; Das, S.; Raut, S.S.; Sultana, A.; Nayan, M.U.; Ganesan, M.; Edagwa, B.J.; Osna, N.A.; Poluektova, L.Y. Chronic Hepatitis B Infection: New Approaches towards Cure. *Biomolecules* **2023**, *13*, 1208. [CrossRef] [PubMed]
276. Das, S.; Wang, W.; Ganesan, M.; Fonseca-Lanza, F.; Cobb, D.A.; Bybee, G.; Sun, Y.; Guo, L.; Hanson, B.; Cohen, S.M. An ultralong-acting tenofovir ProTide nanoformulation achieves monthslong HBV suppression. *Sci. Adv.* **2022**, *8*, eade9582. [CrossRef] [PubMed]
277. Spreen, W.; Ford, S.L.; Chen, S.; Wilfret, D.; Margolis, D.; Gould, E.; Piscitelli, S. GSK1265744 Pharmacokinetics in Plasma and Tissue after Single-Dose Long-Acting Injectable Administration in Healthy Subjects. *JAIDS J. Acquir. Immune Defic. Syndr.* **2014**, *67*, 481–486. [CrossRef]

278. Ford, S.; Chiu, J.; Lovern, M.; Spreen, W.; Kim, J. Population PK approach to predict cabotegravir (CAB, GSK1265744) long-acting injectable doses for phase 2b. In Proceedings of the 54th Interscience Conference on Antimicrobial Agents and Chemotherapy, Washington DC, USA, 30 October–2 November 2004; pp. 5–9.
279. Soriano, V.; Barreiro, P.; Benitez, L.; Peña, J.M.; de Mendoza, C. New antivirals for the treatment of chronic hepatitis B. *Expert Opin. Investig. Drugs* **2017**, *26*, 843–851. [CrossRef]
280. Scarsi, K.K. Chasing the cabotegravir tail: Implications for prevention. *Lancet HIV* **2020**, *7*, e451–e453. [CrossRef]

MDPI AG
Grosspeteranlage 5
4052 Basel
Switzerland
Tel.: +41 61 683 77 34

Pharmaceutics Editorial Office
E-mail: pharmaceutics@mdpi.com
www.mdpi.com/journal/pharmaceutics

www.ingramcontent.com/pod-product-compliance
Lightning Source LLC
Chambersburg PA
CBHW061610210926
51850CB00009B/2275
* 9 7 8 3 7 2 5 8 7 9 7 3 1 *